Lecture Notes in Mathematics

Edited by A. Dold and B. Eckmann

1199

Analytic Theory of Continued Fractions II

Proceedings of a Seminar-Workshop
held in Pitlochry and Aviemore, Scotland
June 13–29, 1985

Edited by W. J. Thron

Springer-Verlag
Berlin Heidelberg New York London Paris Tokyo

Editor

Wolfgang J. Thron
Department of Mathematics, University of Colorado, Boulder
Campus Box 426, Boulder, Colorado 80309, USA

Mathematics Subject Classification (1980): 30B70, 33A40, 65D99

ISBN 3-540-16768-4 Springer-Verlag Berlin Heidelberg New York
ISBN 0-387-16768-4 Springer-Verlag New York Heidelberg Berlin

Printed in Germany

Printing and binding: Beltz Offsetdruck, Hemsbach/Bergstr.
2146/3140-543210

PREFACE

The success of the workshop held in Loen, Norway in the summer of 1981 (see Springer Lecture Notes in Mathematics No. 932) encouraged us to arrange a second workshop in Pitlochry and Aviemore, Scotland in the summer of 1985. Most of the organizational work was done by Haakon Waadeland. Local arrangements were in the hands of John McCabe.

There were both continuity and progress in the topics treated at the two conferences. In these proceedings most contributions fall into two subareas. They are: convergence theory of continued fractions and continued fraction methods in the solution of strong moment problems. Under the first topic limit periodic continued fractions with $a_n \to -1/4$ or ∞ receive most attention. Modified convergence also plays an important role. In the proofs element regions and value (limit) regions are frequently used. Many of the element regions are Cartesian ovals. Truncation error estimates are obtained whenever feasible. In the second subarea Stieltjes, Hamburger and trigonometric moment problems are studied and various types of continued fractions, useful in solving these problems, are investigated. These continued fractions correspond to power series both at 0 and at ∞. The two-point Padé tables, known as M-tables are analyzed for some of these continued fractions. Szegö polynomials coming up in connection with trigonometric moment problems are orthogonal on the unit circle. In addition there are contributions dealing with the location of the zeros of polynomials and multi-point Padé tables and applications to special functions. Applicability of results is emphasized in almost all articles.

There is one survey article in this volume (pp. 127 - 158). All other papers contain original research. All contributions were refereed and we appreciate the efforts of those who helped with this task.

Grateful acknowledgement is made for the financial support of the Seminar-Workshop from a number of sources. Support for various individuals was received from their respective universities and, in some instances, from the Norges Allmen Vitenskaplige Forskningsråd, the London Mathematical Society, the U.S. National Science Foundation and the Fridtjof Nansen Foundation. The latter organization also contributed to the workshop as a whole. Finally, we would like to thank Professor B. Eckmann for accepting this volume for publication in the LECTURE NOTES IN MATHEMATICS.

CONTENTS

LIST OF CONTRIBUTORS AND PARTICIPANTS

CHRISTOPHER BALTUS, Department of Mathematics, University of Northern Colorado, Greeley, Colorado 80639, USA.

SANDRA CLEMENT COOPER, Department of Mathematics, Colorado State University, Fort Collins, Colorado 80523, USA.

ROY M. ISTAD, Department of Mathematics and Statistics, University of Trondheim (AVH), N-7055 Dragvoll, Norway.

LISA JACOBSEN, Department of Mathematics and Statistics, University of Trondheim (AVH), N-7055 Dragvoll, Norway.

WILLIAM B. JONES, Department of Mathematics, University of Colorado, Boulder, Colorado 80309, USA.

N.J. KALTON, Department of Mathematics, University of Missouri, Columbia, Missouri 65211, USA.

L.J. LANGE, Department of Mathematics, University of Missouri, Columbia, Missouri 65211, USA.

ARNE MAGNUS, Department of Mathematics, Colorado State University, Fort Collins, Colorado 80523, USA.

JOHN McCABE, Department of Applied Mathematics, University of St. Andrews, North Haugh, St. Andrews, Fife, Scotland, KY169SS, U.K.

OLAV NJÅSTAD, Department of Mathematics, University of Trondheim (NTH), N-7034 Trondheim, Norway.

A. SRI RANGA, Department of Mathematics, Universidade de São Paulo, Instituto de Ciências Matematicas de São Carlos, Av. Dr. Carlos Botelho, 1465, 13560-São Carlos, SP, Brazil.

ELLEN SØRSDAL, Department of Mathematics and Statistics, University of Trondheim (AVH), N-7055 Dragvoll, Norway.

W.J. THRON, Department of Mathematics, University of Colorado, Boulder, Colorado 80309, USA.

HAAKON WAADELAND, Department of Mathematics and Statistics, University of Trondheim (AVH), N-7055 Dragvol, Norway.

A FAMILY OF BEST VALUE REGIONS FOR MODIFIED CONTINUED FRACTIONS

Christopher Baltus
Department of Mathematics and
and Applied Statistics
University of Northern Colorado
Greeley, CO 80639 U.S.A.

William B. Jones*
Department of Mathematics
Campus Box 426
University of Colorado
Boulder, CO 80309-0426 U.S.A.

1. Introduction. In the analytic theory of continued fractions, value regions have played an important role both for convergence theory and truncation error analysis. A sequence $E = \{E_n\}$ of non-empty subsets of $\mathbb{C}$ is called a sequence of element regions for a continued fraction $K(a_n/1)$ if $a_n \in E_n$, $n \geq 1$. A sequence $V = \{V_n\}$ of subsets of $\hat{\mathbb{C}}$ is called a sequence of value regions corresponding to E if

$$E_n \subseteq V_{n-1} \quad \text{and} \quad \frac{E_n}{1+V_n} \subseteq V_{n-1}, \quad n = 1,2,3,\ldots .$$

If f_n denotes the nth approximant of $K(a_n/1)$ and

$$S_n(w) := \frac{a_1}{1} + \frac{a_2}{1} + \cdots + \frac{a_{n-1}}{1} + \frac{a_n}{1+w},$$

and if we assume that $0 \in V_n$, $n \geq 0$, then it can be readily seen that

$$f_n = S_n(0) \in S_n(V_n) \subseteq S_{n-1}(V_{n-1}), \quad n = 1,2,3,\ldots,$$

where $S_0(w) := w$. It follows that $\{S_n(V_n)\}$ is a nested sequence of non-empty sets such that, for $n = 1,2,3,\ldots$ and $m = 0,1,2,\ldots$,

$$f_{n+m} \quad S_n(V_n) \quad \text{and} \quad |f_{n+m}-f_n| \leq \text{diam } S_n(V_n) .$$

In many cases, suitable conditions have been found for the element regions E to insure that $\lim_{n\to\infty} \text{diam } S_n(V_n) = 0$, thus proving the convergence of the continued fraction to the value $f = \lim f_n$ [1, Chap. 4]. Moreover, sharp estimates for $\text{diam } S_n(V_n)$ provide useful bounds for the truncation error $|f-f_n|$ of the nth approximant f_n [1, Chap. 8]. Value regions have also been used to investigate the numerical stability of the backward recurrence algorithm for evaluating continued fraction approximants [1, Chap. 10].

For a modified continued fraction $K(a_n,1;\ w_n)$ (with reference continued fraction $K(a_n/1)$), one considers the nth approximant $g_n := S_n(w_n)$, where the modifying factor w_n may or may not be zero. Since

---

*W.B.J. was supported in part by the U.S. National Science Foundation under Grant DMS-8401717.

$$f = \lim_{n\to\infty} f_n = S_n(f^{(n)}) ,$$

where $f^{(n)}$ denotes the nth tail of $K(a_n/1)$,

$$f^{(n)} := \frac{a_{n+1}}{1} + \frac{a_{n+2}}{1} + \frac{a_{n+3}}{1} + \cdots ,$$

we see that, if w_n is closer to $f^{(n)}$ than 0 is, then $g_n = S_n(w_n)$ may be a better approximation of $f = S_n(f^{(n)})$ than $f_n = S_n(0)$ is. It has been shown (see, for example, [3]), that a judicious choice of the w_n can accelerate the convergence of the continued fraction; that is, $\{S_n(0)\}$ and $\{S_n(w_n)\}$ satisfy

$$\lim_{n\to\infty} \frac{f - S_n(w_n)}{f - S_n(0)} = 0 .$$

Value regions also play an important role in the study of modified continued fractions. The purpose of this paper is to investigate value regions for modified continued fractions, particularly best value regions. In Section 2 we give a characterization of best value regions (Theorem 2.1) and a result (Theorem 2.2) that is helpful in proving that value regions are best. Section 3 is used to investigate a family of circular value regions corresponding to sequences of element regions that are bounded by Cartesian ovals. It is shown (Theorem 3.1) that these value regions are best. Some interesting properties of Cartesian ovals are described in Lemmas 3.3, 3.4, 3.5, 3.7, and 3.8. Examples of ovals are illustrated by Figure 2. It is expected that the results of this paper will be directly applicable to the problem of finding best truncation error bounds for modified continued fractions. We now summarize some basic definitions and concepts that are subsequently used.

By a <u>modified continued fraction</u> we mean an ordered pair

(1.1) $\quad <<\{a_n\}_1^\infty, \{b_n\}_0^\infty; \{w_n\}_0^\infty>, \{g_n\}_0^\infty>$

where the <u>elements</u> a_n and b_n are complex numbers with $a_n \neq 0$, the nth <u>modifying factor</u> w_n lies in $\hat{\mathbb{C}} := \mathbb{C} \cup [\infty]$ and the nth <u>approximant</u> g_n is defined by

(1.2) $\quad g_n := S_n(w_n), \quad n = 0,1,2,\ldots,$

where $\{S_n\}$ is the sequence of linear fractional transformations (l.f.t.) defined by

(1.3a) $$s_0(w) := b_0 + w, \quad s_n(w) := \frac{a_n}{b_n + w}, \quad n = 1,2,3,\ldots,$$

(1.3b) $$S_0(w) := s_0(w), \quad S_n(w) := S_{n-1}(s_n(w)), \quad n = 1,2,3,\ldots .$$

For convenience we denote the modified continued fraction (1.1) by one of the symbols

(1.4) $$b_0 + K(a_n, b_n; w_n) \quad \text{or} \quad b_0 + \mathop{K}_{n=1}^{\infty} (a_n, b_n; w_n) .$$

The nth approximant g_n may be expressed by the symbol

(1.5) $$g_n = b_0 + \frac{a_1}{b_1} + \frac{a_2}{b_2} + \cdots + \frac{a_{n-1}}{b_{n-1}} + \frac{a_n}{b_n + w_n} .$$

A modified continued fraction (1.1) is said to <u>converge</u> to a <u>value</u> g if its sequence of approximants $\{g_n\}$ converges in the larger sense to g in $\hat{\mathbb{C}}$.

Every modified continued fraction (1.1) has an underlying <u>reference continued fraction</u>

(1.6) $$b_0 + K\left(\frac{a_n}{b_n}\right) = b_0 + \frac{a_1}{b_1} + \frac{a_2}{b_2} + \frac{a_3}{b_3} + \cdots .$$

The nth <u>numerator</u> A_n and <u>denominator</u> B_n of (1.6) are defined by the <u>difference equations</u> [1, (2.1.6)]

(1.7a) $$A_{-1} := 1, \quad A_0 := b_0, \quad B_{-1} := 0, \quad B_0 := 1,$$

(1.7b) $$A_n = b_n A_{n-1} + a_n A_{n-2}, \quad n = 1,2,3,\ldots,$$

(1.7c) $$B_n = b_n B_{n-1} + a_n B_{n-2}, \quad n = 1,2,3,\ldots .$$

In terms of A_n and B_n we have the well known expressions

(1.8) $$S_n(w) = \frac{A_n + wA_{n-1}}{B_n + wB_{n-1}}, \quad n = 1,2,3,\ldots$$

and the <u>determinant formulas</u> [1, (2.1.9)]

(1.9) $$A_n B_{n-1} - A_{n-1} B_n = (-1)^{n-1} \prod_{j=1}^{n} a_j, \quad n = 1,2,3,\ldots .$$

For $m = 1,2,3,\ldots,$ the mth <u>tail</u> of a modified continued fraction (1.1) is defined to be the modified continued fraction

$$\mathop{K}_{n=m+1}^{\infty}(a_n,b_n;\ w_n) = \langle\langle\{a_n\}_{m+1}^{\infty},\ \{b_n\}_{m+1}^{\infty}\rangle,\ \{w_n\}_{m+1}^{\infty}\rangle,\ \{g_n^{(m)}\}_{m+1}^{\infty}\rangle\ .$$

The 0th tail of (1.1) is defined to be (1.1) itself. The nth approximant $g_n^{(m)}$ of the mth tail is given by

$$(1.10a)\qquad g_0^{(0)} := S_0^{(0)}(w_0),\quad g_n^{(m)} = S_n^{(m)}(w_{n+m}),\quad m,n = 1,2,3,\ldots,$$

where

$$(1.10b)\qquad S_n^{(0)}(w) := S_n(w),\quad n = 0,1,2,\ldots$$

$$(1.10c)\qquad S_n^{(m)}(w) := \frac{a_{m+1}}{b_{m+1}} + \frac{a_{m+2}}{b_{m+2}} + \cdots + \frac{a_{m+n-1}}{b_{m+n-1}} + \frac{a_{m+n}}{b_{m+n}+w},$$

$$m,n = 1,2,3,\ldots\ .$$

It can be seen that

$$(1.11)\qquad S_{m+n}(w) = S_m(S_n^{(m)}(w)),\quad m,n = 1,2,3,\ldots$$

and hence

$$(1.12)\qquad g_{m+n} := S_{m+n}(w_{m+n}) = S_m(S_n^{(m)}(w_{m+n})) = S_m(g_n^{(m)}),$$

$$m,n = 1,2,3,\ldots\ .$$

If an mth tail converges to a limit $g^{(m)} := \lim_{n\to\infty} g_n^{(m)}$, then all tails converge and we have

$$(1.13)\qquad g := \lim_{n\to\infty} g_n = S_m(g^{(m)}),\quad m = 1,2,3,\ldots\ .$$

Our attention in this paper is restricted to modified continued fractions (1.1) where $b_0 = 0$ and $b_n = 1$, $n = 1,2,3,\ldots$; that is, modified continued fractions of the form $K(a_n,1;\ w_n)$.

A sequence $E = \{E_n\}_1^{\infty}$ of subsets of $\mathbb{C}$ such that

$$(1.14)\qquad E_n - [0] \neq \emptyset,\quad n = 1,2,3,\ldots$$

is called a <u>sequence of element regions</u> for a modified continued fraction $K(a_n,1;\ w_n)$ if

$$(1.15)\qquad a_n \in E_n,\quad n = 1,2,3,\ldots\ .$$

A sequence $V = \{V_n\}_0^\infty$ of subsets of $\hat{\mathbb{C}}$ is called a <u>sequence of value regions corresponding to a sequence of element regions</u> $E = \{E_n\}$ <u>at</u> $W = \{w_n\}$ if

$$(1.16)\qquad \frac{E_n}{1+w_n} \subseteq V_{n-1} \quad \text{and} \quad \frac{E_n}{1+V_n} \subseteq V_{n-1}, \quad n = 1,2,3,\ldots .$$

We denote by $\mathcal{V}(E,W)$ the family of all sequences of value regions corresponding to E at W. The following statements are readily verified:

$$(1.17)\qquad \left[w_{n-1} \in V_{n-1} \text{ and } \frac{E_n}{1+V_n} \subseteq V_{n-1}, \; n \geq 1\right] \Rightarrow \{V_n\} \in \mathcal{V}(E,W),$$

$$(1.18)\qquad \{V_n\} \in \mathcal{V}(E,W) \Rightarrow \{c(V_n)\} \in \mathcal{V}(E,W),$$

$$(1.19)\qquad \left[\{V_n^{(\alpha)}\} \in \mathcal{V}(E,W), \; \alpha \in A\right] \Rightarrow \{\bigcap_{\alpha\, A} V_n^{(\alpha)}\} \in \mathcal{V}(E,w) .$$

Here $c(V_n)$ denotes the closure of V_n. If

$$(1.20a)\qquad \{U_n(E,W)\} \in \mathcal{V}(E,W)$$

and

$$(1.20b)\qquad \{V_n\} \in \mathcal{V}(E,W) \Rightarrow U_n(E,W) \subseteq V_n, \quad n = 0,1,2,\ldots,$$

then $\{U_n(E,W)\}$ is called the <u>best sequence of value regions corresponding to</u> E <u>at</u> W. We shall denote the best sequence of value regions corresponding to E at W by

$$U(E,W) := \{U_n(E,W)\}$$

In Section 2 (Theorem 2.1) it is shown that $U(E,W)$ exists and is unique. Theorem 2.2 provides a set of sufficient conditions on $\{V_n\}$ to insure that

$$(1.21)\qquad V_n = cU_n(E,W), \quad n = 0,1,2,\ldots .$$

This result is used in Section 3 to establish (1.21) for certain sequences of disks V_n and Cartesian ovals E_n (Theorem 3.1).

By diam S and $c(S)$ we mean the diameter and closure, respectively, of the set S.

2. <u>Best Value Regions</u>.

Throughout this section we let $E = \{E_n\}$ denote a given sequence of element regions and $W = \{w_n\}$ a sequence of modifying factors for modified continued fractions $K(a_n,1;\, w_n)$.

<u>Theorem</u> 2.1. <u>The best sequence of value regions</u> $\{U_n(E,W)\}$ <u>corresponding to</u> E <u>at</u> W <u>exists and is given, for</u> $n = 0,1,2,\ldots,$ <u>by</u>

$$(2.1)\qquad U_n(E,W) = \left[\frac{a_{n+1}}{1} + \frac{a_{n+2}}{1} + \cdots + \frac{a_{m-1}}{1} + \frac{a_m}{1+w_m} : 0 \neq a_k \in E_k,\right.$$

$$\left. n+1 \leq k \leq m, \quad m \geq n+1\right].$$

<u>Proof</u>. Let U_n^* denote the right side of (2.1). It follows that

$$(2.2)\qquad \frac{E_n}{1+w_n} \subseteq U_{n-1}^*, \quad n = 1,2,3,\ldots .$$

Let $n \geq 1$ be given and let a_n and u_n be arbitrary (fixed) elements in E_n and U_n^*, respectively. Then there exists an $m \geq n+1$ and $a_k \in E_k$ for $k = n+1,n+2,\ldots,m$ such that

$$(2.3)\qquad u_n = \frac{a_{n+1}}{1} + \cdots + \frac{a_{m-1}}{1} + \frac{a_m}{1+w_m} .$$

Hence

$$\frac{a_n}{1+u_n} = \frac{a_n}{1} + \frac{a_{n+1}}{1} + \cdots + \frac{a_{m-1}}{1} + \frac{a_m}{1+w_m} \in U_{n-1}^* .$$

It follows that

$$(2.4)\qquad \frac{E_n}{1+U_n^*} \subseteq U_{n-1}^* , \quad n = 1,2,3,\ldots .$$

From (2.2) and (2.4) we see that

$$(2.5)\qquad \{U_n^*\} \in \mathcal{V}(E,W) .$$

Now let $\{V_n\}$ be an arbitrary member of $\mathcal{V}(E,W)$. Let $n \geq 0$ be given, and let u_n denote an arbitrary (fixed) element of U_n^*. Then there exist $m \geq n+1$ and $a_k \in E_k$, $k = n+1,n+2,\ldots,m$ such that (2.3) holds. We define $v_m, v_{m-1}, v_{m-2},\ldots, v_n$ by

$$(2.6)\qquad v_m := w_m, \quad v_k := \frac{a_{k+1}}{1+v_{k+1}} , \quad k = m-1,\ m-2,\ \ldots,\ n.$$

It follows that

$$(2.7)\qquad v_k \in V_k, \quad k = m-1,\ m-2,\ \ldots,\ n.$$

Hence by (2.3), (2.6) and (2.7)

$$u_n = \frac{a_{n+1}}{1+v_{n+1}} = v_n \in V_n .$$

Thus

(2.8) $\quad U_n^* \subseteq V_n, \quad n = 0,1,2,\ldots .$

Combining (2.5) and (2.8) yields the assertion that $U_n^* = U_n(E,W)$, $n \geq 0$. □

<u>Theorem</u> 2.2. <u>Let</u> $V = \{V_n\}$ <u>be a sequence of closed value regions corresponding to</u> E <u>at</u> W, <u>such that for some non-negative integer</u> k

(2.9) $$\frac{E_n}{1+V_n} = V_{n-1}, \quad n = k+1, k+2, k+3, \ldots$$

<u>and</u>

(2.10a) $$\lim_{n\to\infty} \operatorname{diam} S_n^{(m)}(V_{n+m}) = 0, \quad m = k, k+1, k+2, \ldots,$$

<u>for every sequence</u> $\{a_n\}$ <u>such that</u>

(2.10b) $$0 \neq a_n \in E_n, \quad n = k+1, k+2, k+3, \ldots$$

<u>where</u>

(2.10c) $$S_n^{(m)}(w) := \frac{a_{m+1}}{1} + \frac{a_{m+2}}{1} + \cdots + \frac{a_{m+n-1}}{1} + \frac{a_{m+n}}{1+w} .$$

<u>Then</u>

(2.11) $$V_n = c(U_n(E,W)), \quad n = k, k+1, k+2, \ldots .$$

<u>Proof</u>. By the definition of best value regions, $U_m(E,W) \subseteq V_m$, $m = 0,1,2,\ldots$. Since each V_m is a closed set, it follows that

(2.12) $$c(U_m(E,W)) \subseteq V_m, \quad m = 0,1,2,\ldots .$$

It remains to show that

(2.13) $$V_m \subseteq c(U_m(E,W)), \quad m = k, k+1, k+2, \ldots .$$

Let $m \geq k$ be given and let v_m denote an arbitrary (fixed) element of V_m. Then by (2.9) there exist sequences $\{a_j\}_{m+1}^{\infty}$ and $\{v_j\}_{m+1}^{\infty}$ such that

(2.14a) $$0 \neq a_j \in E_j, \quad v_j \in V_j, \quad j = m+1, m+2, \ldots,$$

and

(2.14b) $$v_m = \frac{a_{m+1}}{1+v_{m+1}} = \frac{a_{m+1}}{1} + \frac{a_{m+2}}{1} + \cdots + \frac{a_{m+n-1}}{1} + \frac{a_{m+n}}{1+v_{m+n}},$$

$$n = 2,3,4,\ldots .$$

For $n \geq 2$,

$$g_n^{(m)} := S_n^{(m)}(w_{m+n}) \in S_{n-1}^{(m)}(V_{m+n-1}) \subseteq S_1^{(m)}(V_{m+1}) \subseteq V_m .$$

It follows from this and (2.10) that $\overset{\infty}{\underset{j=m+1}{K}} (a_j, 1; w_j)$ converges to the finite value

$$g^{(m)} = \lim_{n\to\infty} g_n^{(m)} \in V_m .$$

By (2.1) $g^{(m)} \in c(U_m(E,W))$. By (2.14b)

$$v_m = S_n^{(m)}(v_{m+n}) \in S_n^{(m)}(V_{m+n}), \quad n = 2,3,4,\ldots .$$

Thus (2.10) implies that

$$v_m = \lim_{n\to\infty} S_n^{(m)}(v_{m+n}) = g^{(m)} \in c(U_m(E,W)) .$$

Hence (2.13) holds. □

3. Value Regions Corresponding to Cartesian Ovals. Throughout this section we let k denote a given (fixed) non-negative integer.

Theorem 3.1. Let $a \in \mathbb{C} - (-\infty, -1/4]$ be a given complex number. Let x_1 and $x_2 = -(x_1+1)$ denote the fixed points of the l.f.t. $T(w) := a/(1+w)$ such that $|x_1| < |x_2|$. Let $\{\rho_n\}$ be a sequence of positive numbers satisfying:

(3.1a) $$\rho_n < |x_2|, \quad n = k, k+1, k+2, \ldots,$$

(3.1b) $$\rho_n \rho_{n-1} < \rho_{n-1}|x_2| - \rho_n|x_1|, \quad n = k+1, k+2, k+3, \ldots,$$

and (A) if $a \in \mathbb{C} - (-\infty, 0]$ then

(3.1c) $$\rho_{n-1}|x_2| - \rho_n|x_1| < \sqrt{|a|}\cos(\alpha/2), \; \alpha := \arg a, \; n = k+1, k+2, \ldots,$$

(3.1d) $$\rho_n \leq \tfrac{1}{2}\cos(\alpha/2) + \mathrm{Re}(x_1 e^{-i\alpha/2}) = \mathrm{Re}(\sqrt{a + \tfrac{1}{4}}\, e^{-i\alpha/2}),$$

$$n = k, k+1, k+2, \ldots,$$

and (B) if $-\frac{1}{4} < a \leq 0$ then

(3.1') $$(\rho_{n-1} + |x_1|)(|x_2| - \rho_n) \leq \tfrac{1}{4}, \quad n = k+1, k+2, k+3, \ldots,$$

(3.1d') $\rho_n \le \sqrt{a + \frac{1}{4}} = x_1 + \frac{1}{2}$, $n = k, k+1, k+2, \ldots$.

<u>Let</u> $E = \{E_n\}$, $V = \{V_n\}$, $W = \{w_n\}$ <u>be defined by</u>

(3.2a) $V_n := [v: |v-x_1| \le \rho_n]$, $n = k, k+1, k+2, \ldots$,

(3.2b) $E_n := [w: |w\bar{x}_2 + x_1(|x_2|^2-\rho_n^2)| + \rho_n|w| \le \rho_{n-1}(|x_2|^2-\rho_n^2)]$,

(3.3c) $w_n := x_1$, $n = k, k+1, k+2, \ldots$.

Then

(3.3) $V_n = c(U_n(E,W))$, $n = k, k+1, k+2, \ldots$.

The numbers x_1, x_2 are given by

$$x_1 = -\frac{1}{2} + |a+\frac{1}{4}|^{1/2} e^{i\frac{\text{Arg}(a+\frac{1}{4})}{2}}, \quad x_2 = -\frac{1}{2} - |a+\frac{1}{4}|^{1/2} e^{i\frac{\text{Arg}(a+\frac{1}{4})}{2}},$$

where $-\pi < \text{Arg}(a + \frac{1}{4}) < \pi$. We note that (3.1a) implies $-1 \notin V_n$, $n \ge k$, and (3.1b) implies $a = -x_1x_2 = x_1(x_1+1) \in \text{Int } E_n$, $n \ge k+1$. Our proof of Theorem 3.1 makes use of several lemmas, some of which, for future use, are proved in greater generality than is needed for Theorem 3.1. We shall consider sequences $\{\Gamma_n\}$ and $\{\rho_n\}$ that satisfy

(3.4a) $0 < \rho_n < |1 + \Gamma_n|$, $\Gamma_n \in ¢$, $n = k, k+1, k+2, \ldots$

and

(3.4b) $\rho_n\rho_{n-1} < \rho_{n-1}|1 + \Gamma_n| - \rho_n|\Gamma_{n-1}|$, $n = k+1, k+1, k+3, \ldots$,

and define $V = \{V_n\}$, $E = \{E_n\}$, and $W = \{w_n\}$ by

(3.5a) $V_n := [v: |v-\Gamma_n| \le \rho_n]$, $n = k, k+1, k+2, \ldots$,

(3.5b) $E_n := [w: |w(1+\bar{\Gamma}_n)-\Gamma_{n-1}(|1+\Gamma_n|^2-\rho_n^2)|+\rho_n|w| \le \rho_{n-1}(|1+\Gamma_n|^2)]-\rho_n^2)]$,

$$n = k+1, k+2, k+3, \ldots,$$

and

(3.5c) $w_n = \Gamma_n$, $n = k, k+1, k+2, \ldots$.

To apply our results to the proof of Theorem 3.1, we shall set $\Gamma_n = x_1$ so that $x_2 = -(1 + \Gamma_n)$ and $a = \Gamma_{n-1}(1 + \Gamma_n)$. We note that (3.5b) can also be written in the form

(3.5d) $E_n = [w: |w-c_n| + p_n|w| \le q_n]$,

where

(3.5e) $c_n := \Gamma_{n-1}(1+\Gamma_n)\left(1 - \dfrac{\rho_n^2}{|1+\Gamma_n|^2}\right), \quad p_n := \dfrac{\rho_n}{|1+\Gamma_n|},$

$$q_n := \frac{\rho_{n-1}}{|1+\Gamma_n|}\left(|1+\Gamma_n|^2 - \rho_n^2\right).$$

It can be shown that (3.4b) implies c_n Int (E_n).

Lemma 3.2. _Let_ $V = \{V_n\}$, $E = \{E_n\}$ _and_ $W = \{w_n\}$ _be defined as in_ (3.4) _and_ (3.5). _Then_: (A)

(3.6) $\Gamma_{n-1}(1+\Gamma_n) \in \text{Int}\,(E_n), \quad n = k+1, k+2, k+3, \ldots .$

(B)

(3.7) $\dfrac{E_n}{1+w_n} \subseteq V_{n-1}$ _and_ $\dfrac{E_n}{1+V_n} \subseteq V_{n-1}, \quad n = k+1, k+2, k+3, \ldots .$

(C) _For each_ [illegible] $\geq k+1$, E_n _is a closed_, _bounded_, _convex subset of of_ $\mathbb{C}$; _if_ $\Gamma_{n-1} \neq 0$, _then_ E_n _is symmetric with respect to the ray_ arg w = arg [illegible]$(1+\Gamma_n)$; _if_ $\Gamma_{n-1} = 0$, _then by_ (3.5b), $E_n = [w: |w| \leq \rho_{n-1}(|1+\Gamma_n|$ [illegible]$)]$.

Proof. [illegible] proof is basically Lane's method [1, Theorem 4.3]. (A) follows by substituting $w = \Gamma_{n-1}(1+\Gamma_n)$ into the inequality of (3.5b) and applying (3.4b).

(B): First we note that

$$\frac{1}{1+V_n} = \left[u: \left|u - \frac{1+\overline{\Gamma}_n}{|1+\Gamma_n|^2-\rho_n^2}\right| \leq \frac{\rho_n}{|1+\Gamma_n|^2-\rho_n^2}\right]$$

and hence

(3.8a) $D_n(w) := \dfrac{w}{1+V_n} = [v: |v - \gamma_n(w)| \leq r_n(w)]$

where

(3.8b) $\gamma_n(w) := \dfrac{(1+\overline{\Gamma}_n)w}{|1+\Gamma_n|^2-\rho_n^2}$ and $r_n(w) := \dfrac{\rho_n|w|}{|1+\Gamma_n|^2-\rho_n^2}.$

One can then readily see that $w/(1+V_n) \subseteq V_{n-1}$ if and only if

(3.9) $|\gamma_n(w) - \Gamma_{n-1}| + r_n(w) \leq \rho_{n-1}.$

Since (3.9) is equivalent to the inequality of (3.5b), we have proved the second inclusion relation in (3.7). The first inclusion relation in (3.7) follows from this, since $w_n = \Gamma_n \in V_n$.

(C): That E_n is closed and bounded is an immediate consequence of (3.5b). Convexity of E_n follows from (3.5); for if $w_1, w_2 \in E_n$, then by (3.5), $w_1 t + (1-t)w_2 \in E_n$, $0 \le t \le 1$. The assertion about symmetry can be deduced from (3.5d,e). (See also Lemma 3.7.) □

Lemma 3.3. *Let* V, E *and* W *be defined as in Lemma* 3.2 *and let* $\gamma_n(w)$ *be given by* (3.8b), $n > k+1$. *As* w *traverses the boundary* ∂E_n *of* E_n *one time counter-clockwise,* $\gamma_n(w)$ *moves around* Γ_{n-1} *one time.*

Proof. Suppose that $\Gamma_{n-1} = 0$. Then $0 = \Gamma_{n-1}(1+\Gamma_n) \in \text{Int}\ (E_n)$ by (3.6). Thus as w traverses ∂E_n once counter-clockwise, $\arg w$ increases by 2π. Hence by (3.8b), $\arg \gamma_n(w)$ increases by 2π, so that $\gamma_n(w)$ moves one time around $0 = \Gamma_{n-1}$.

Suppose that $\Gamma_{n-1} \neq 0$. Let $c_n \neq 0$ be given by (3.5e). Then by (3.8), the center $\gamma_n(c_n)$ of the disk $c_n/(1+V_n)$ is Γ_{n-1}. From (3.5e) and (3.8b) we have

$$\gamma_n(w) = \frac{\Gamma_{n-1}}{c_n} w \quad \text{and} \quad r_n(w) = p_n \left| \frac{\Gamma_{n-1} w}{c_n} \right| .$$

Hence

$$\arg\,(\gamma_n(w) - \Gamma_{n-1}) = \arg[\Gamma_{n-1}(\frac{w-c_n}{c_n})] = \arg(w-c_n) + \arg \Gamma_{n-1} - \arg c_n .$$

Since $c_n \in \text{Int}\ E_n$, $\arg\,(w-c_n)$ increases by 2π as w traverses ∂E_n once counter-clockwise. □

Lemma 3.4. *Let* V, E *and* W *be as in Lemma* 3.2. *Then for* $n \ge k+1$, *the disk* $D_n(w) = w/(1+V_n)$ *is tangent to* ∂V_{n-1} *if and only if* $w \in \partial E_n$.

Proof. It can be seen from (3.8) that $D_n(w)$ is tangent to ∂V_{n-1} if and only if

$$(3.10) \qquad |\gamma_n(w) - \Gamma_{n-1}| + r_n(w) = \rho_{n-1} .$$

It follows from (3.5b) and (3.8b) that (3.10) holds if and only if $w \in \partial E_n$. □

It is readily seen that $\partial D_n(w)$ and ∂V_{n-1} can be tangent at only a single point. For suppose $\partial D_n(w) = \partial V_{n-1}$; since the centers coincide, we would have $\gamma_n(w) = \Gamma_{n-1}$, which means that $w = c_n$. A simple calculation shows that $\partial D_n(c_n) = \partial V_{n-1}$ if and only if $\rho_n|\Gamma_{n-1}| = \rho_{n-1}|1+\Gamma_n|$. Since this contradicts (3.4b), we conclude that $\partial D_n(w)$ and ∂V_{n-1} can be tangent in at most one point, if $w \in E_n$.

For $w \in \partial E_n$ we let $T_n(w)$ denote the unique point at which the two circles $\partial D_n(w)$ and ∂V_{n-1} are tangent.

Lemma 3.5. Let V, E and W be as in Lemma 3.2. Let n satisfy $n > k+1$. Let $\alpha, \beta \in \partial E_n$ be such that the line segment $[T_n(\alpha), T_n(\beta)]$ is a diameter of ∂V_{n-1}. For every $v \in [\gamma_n(\alpha), \gamma_n(\beta)]$ there exists a point $w \in [\alpha, \beta] \subseteq E_n$ such that $v = \gamma_n(w)$. (See Figure 1.)

Proof. Let $v \in [\gamma_n(\alpha), \gamma_n(\beta)]$ be given. Then for some t with $0 \leq t \leq 1$ we have $v = t\gamma_n(\alpha) + (1-t)\gamma_n(\beta)$. Let $w := t\alpha + (1-t)\beta$. By convexity of E_n we have $w \in [\alpha, \beta] \subseteq E_n$. Moreover

$$\gamma_n(w) = \frac{(1+\overline{\Gamma}_n)(t\alpha+(1-t)\beta)}{|1+\Gamma_n|^2 - \rho_n^2} = t\gamma_n(\alpha) + (1-t)\gamma_n(\beta) = v . \square$$

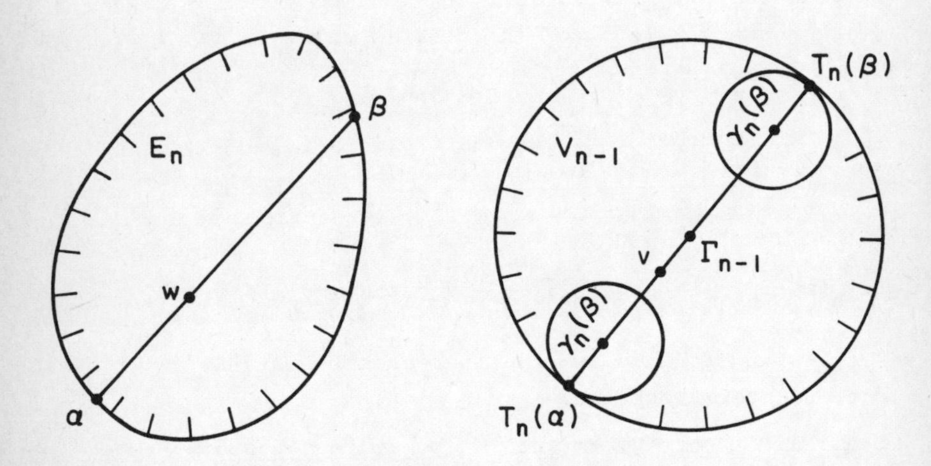

Figure 1

Lemma 3.6. Let V, E and W be as in Lemma 3.2. Then

$$\frac{E_n}{1+V_n} = V_{n-1}, \quad n = k+1, k+2, k+3, \ldots . \tag{3.11}$$

Proof. Let $n \geq k+1$ be given. By Lemma 3.2, $E_n/(1+V_n) \subseteq V_{n-1}$. Thus it suffices to show that

$$V_{n-1} \subseteq \frac{E_n}{1+V_n} . \tag{3.12}$$

Let $v \in V_{n-1}$ be given. By Lemma 3.3, 3.4 and 3.5, there exist $\alpha, \beta \in \partial E_n$ such that $v \in [T_n(\alpha), T_n(\beta)]$. If $v \in D_n(\alpha) \cup D_n(\beta)$, then $v \in E_n/(1+V_n)$. If $v \notin D_n(\alpha) \cup D_n(\beta)$, then at least we have $v \in [\gamma_n(\alpha), \gamma_n(\beta)]$. Hence by Lemma 3.5, there exists $w \in [\alpha, \beta] \subseteq E_n$ such that

$$v = \gamma_n(w) \in D_n(w) = \frac{w}{1+V_n} \subseteq \frac{E_n}{1+V_n} .$$

We have shown that (3.12) holds. □

Our next lemma expresses E_n in polar coordinates. We consider now the restricted case with $\Gamma_n = x_1$, $x_2 = -(1+\Gamma_n)$, $a = -x_1 x_2$ in Theorem 3.1.

Lemma 3.7. *Let* $E = \{E_n\}$ *be defined as in Theorem* 3.1 *with* $a \in \mathbb{C} - (-\infty,-1/4]$. *Let* $a = |a|e^{i\alpha}$, $\alpha := \arg a$ *if* $a \neq 0$. *Then for* $n = k+1, k=2, k+3, \ldots$: (A)

(3.13a) $$E_n = [\rho e^{i\phi}: t_n(\phi) - \sqrt{[t_n(\phi)]^2-K_n} \leq \rho \leq t_n(\phi) + \sqrt{[t_n(\phi)]^2-K_n} ,\quad \rho \geq 0] ,$$

where

(3.13b) $$t_n(\phi) := \begin{cases} |x_1 x_2|\cos(\phi-\alpha) - \rho_n\rho_{n-1}, & \text{if } a \neq 0 \\ -\rho_n\rho_{n-1}, & \text{if } a = 0 , \end{cases}$$

(3.13c) $$K_n := (|x_1|^2 - \rho_{n-1}^2)(|x_2|^2 - \rho_n^2) .$$

(B) *If* $0 < \rho_{n-1} < |x_1|$, *then* $K_n > 0$ *and*

(3.14a) $$0 < \frac{\rho_n\rho_{n-1}+\sqrt{K_n}}{|a|} < 1;$$

if, in addition, $\rho e^{i\phi} \in E_n$, *then*

(3.14b) $$0 \leq |\phi - \alpha| \leq \operatorname{Cos}^{-1} \left(\frac{\rho_n\rho_{n-1}+\sqrt{K_n}}{|a|}\right).$$

(C) *If* $0 < \rho_{n-1} = |x_1|$, *then* $K_n = 0$ *and*

(3.15a) $$E_n = [\rho e^{i\phi}: 0 \leq \rho \leq 2|x_1|(|x_2|\cos(\phi-\alpha) - \rho_n)] ,$$

where

(3.15b) $$0 \leq |\phi-\alpha| \leq \operatorname{Cos}^{-1} (\rho_n/|x_2|) .$$

(D) *If* $\rho_{n-1} > |x_1|$, *then* $K_n < 0$ *and*

(3.16a) $$E_n = [\rho e^{i\phi}: 0 \leq \rho \leq t_n(\phi) + \sqrt{[t_n(\phi)]^2-K_n}] ,$$

where

(3.16b) $$0 \leq |\phi - \alpha| \leq \pi.$$

Proof. (A): From (3.5) we see that $w \in \partial E_n$ if and only if

(3.17a) $$|w - c_n| = q_n - p_n|w|$$

where

(3.17b) $$c_n = -x_1x_2\left(\frac{|x_2|^2-\rho_n^2}{|x_2|^2}\right), \quad p_n = \frac{\rho_n}{|x_2|}, \quad q_n = \rho_{n-1}\left(\frac{|x_2|^2-\rho_n^2}{|x_2|}\right) .$$

By the law of cosines

(3.18) $$|w - c_n|^2 = |w|^2 + |c_n|^2 - 2|c_n w|\cos(\phi-\alpha) ,$$

where $\phi := \arg w$, $\alpha := \arg(-x_1x_2) = \arg c_n$. Setting $|w| = \rho$, squaring both sides of (3.17a) and applying (3.18), we see that $w \in \partial E_n$ if and only if

(3.19a) $$q_n - p_n\rho \geq 0$$

and

(3.19b) $$\rho^2 + |c_n|^2 - 2\rho|c_n|\cos(\phi-\alpha) = (q_n-p_n\rho)^2 .$$

We shall show that (3.1b) and (3.19b) imply $q_n - p_n\rho > 0$; hence $w \in \partial E_n$ if and only if (3.19b) holds. In fact, (3.19b) implies

(3.20) $$(\rho-|c_n|^2) \leq \rho^2 + |c_n|^2 - 2\rho|c_n|\cos(\phi-\alpha) = (q_n-p_n\rho)^2 < \left(\rho - \frac{q_n}{p_n}\right)^2$$

since $0 < p_n < 1$ by (3.1a). By (3.1b) we have $|x_1|/\rho_{n-1} < |x_2|/\rho_n$ and hence

(3.21) $$|c_n| = \frac{q_n|x_1|}{\rho_{n-1}} < \frac{q_n|x_2|}{\rho_n} = \frac{q_n}{p_n} .$$

It follows that $\rho < q_n/p_n$; for if $\rho \geq q_n/p_n$, then by (3.20)

$$\rho - \frac{q_n}{p_n} = |\rho - \frac{q_n}{p_n}| > |\rho - |c_n|| \geq \rho - |c_n|$$

which contradicts (3.21). We have shown that $w \in \partial E_n$ if and only if (3.19b) holds. Solving for ρ in (3.19b) yields

$$\rho = t_n(\phi) \pm \sqrt{[t_n(\phi)]^2-K_n} .$$

Equation (3.13a) follows from this and the fact that E_n is bounded and convex. This proves (A).

(B): Suppose $0 < \rho_{n-1} < |x_1|$. Then $K_n > 0$. Moreover, if $w \in E_n$, then $t_n(\phi) \geq \sqrt{K_n} > 0$, for otherwise the inequality in (3.13a) cannot be satisfied. The condition $t_n(\phi) \geq \sqrt{K_n}$ is equivalent to (3.14b). The first inequality in (3.14a) is immediate. The second is equivalent to

$$\sqrt{(|x_1^2| - \rho_{n-1}^2)(|x_2|^2 - \rho_n^2)} < |x_1 x_2| - \rho_n \rho_{n-1} \text{ .}$$

Since the right side of this is positive, we obtain an equivalent inequality by squaring both sides. Rearranging the terms yields the equivalent inequality

$$(|x_1|\rho_n - |x_2|\rho_{n-1})^2 > 0$$

the truth of which follows from (3.1b). This establishes (B).

(C): Suppose $\rho_{n-1} = |x_1| > 0$. Then clearly $K_n = 0$. Moreover, $t_n(\phi) \geq 0$ if $w \in E_n$; for otherwise the inequality in (3.13a) becomes $-2|t_n(\phi)| \leq \rho \leq 0$ so that E_n contains only a single point 0. However Lemma 3.2 asserts that E_n contains $a = -x_1 x_2$ as an interior point. The formulation (3.15a) thus follows from (3.13a), and (3.15b) follows from (3.14b).

(D): Suppose $\rho_{n-1} > |x_1|$. Then $K_n < 0$. Therefore

$$[t_n(\phi)]^2 - K_n > [t_n(\phi)]^2 \geq 0$$

and hence

$$t_n(\phi) - \sqrt{[t_n(\phi)]^2 - K_n} < 0 \text{ .}$$

The expression (3.16a) follows and there is no restriction on ϕ. □

The next lemma provides us with sufficient conditions to insure that the ovals E_n of Theorem 3.1 are contained in parabolic regions associated with a convergence theorem due to Thron [2].

<u>Lemma</u> 3.8. <u>Let</u> $E = \{E_n\}$ <u>be defined as in Theorem</u> 3.1 <u>and</u> (3.1). <u>For each</u> $a \in \mathbb{C} - (-\infty, -1/4]$ <u>we let</u> P_a <u>denote the parabolic region</u>

$$\text{(3.22a)} \quad P_a := \begin{cases} P(\alpha), & \text{if } a \in \mathbb{C} - (-\infty, 0] \text{ , } \quad \alpha := \arg a \\ P(0), & \text{if } -1/4 < a \leq 0, \end{cases}$$

<u>where</u>

$$\text{(3.22b)} \quad P(\alpha) := [w: |w| - \text{Re}(we^{-i\alpha}) \leq \tfrac{1}{2}\cos^2(\alpha/2)], \quad -\pi < \alpha < \pi \text{ .}$$

Then

(3.23) $E_n \subseteq P_a$, $n = k+1, k+2, k+3, \ldots$.

Proof: (A) Suppose $a \in ℂ - (-\infty,0]$, $P_a = P(\alpha)$. Then the boundary $\partial P(\alpha)$ of $P_a := P(\alpha)$ is a parabola with focus at the origin, axis $\arg w = \alpha$, and vertex $v := -\rho e^{i\alpha}$ where $\rho = (1/4)\cos^2(\alpha/2)$; $\partial P(\alpha)$ passes through $-1/4$. It can be seen from (3.22b) that $\rho e^{i\theta} \in \partial P(\alpha)$ if and only if

$$\rho = \frac{(1/2)\cos^2(\alpha/2)}{1 - \cos(\theta-\alpha)} .$$

From this and Lemma 3.7 it follows that a point $\rho e^{i\theta}$ lies on $\partial P(\alpha) \cap \partial E_n$ if and only if

$$(3.24) \quad \frac{\cos^2\beta}{2(1-\cos\lambda)} = |a|\cos\lambda - \rho_n\rho_{n-1} \pm \sqrt{(|a|\cos\lambda - \rho_n\rho_{n-1})^2 - K_n} ,$$

where $\beta := \alpha/2$, $\lambda := \theta - \alpha$. In (3.24), we multiply both sides by $2(1-\cos\lambda)$, isolate the radical, square both sides, subtract $4(1-\cos\lambda)^2(|a|\cos\lambda - \rho_n\rho_{n-1})^2$ from both sides, and divide both sides by 4 to obtain

$$(K_n + |a|\cos^2\beta)\cos^2\lambda - (2K_n + \rho_n\rho_{n-1}\cos^2\beta + |a|\cos^2\beta)\cos\lambda$$
$$+ (K_n + \rho_n\rho_{n-1}\cos^2\beta + (1/4)\cos^4\beta) = 0 .$$

This is a quadratic equation in $\cos\lambda$. We let its discriminant be denoted by $\mathcal{D}_n$, and obtain after simplification

$$\mathcal{D}_n = \cos^4\beta[(\rho_n|x_1| - \rho_{n-1}|x_2|)^2 - |a|\cos^2\beta], \quad a = -x_1x_2 .$$

Since $\cos\beta = \cos(\alpha/2) > 0$, we have $\mathcal{D}_n < 0$ if and only if

$$\rho_{n-1}|x_2| - \rho_n|x_1| < \sqrt{|a|}\cos(\alpha/2)$$

which is condition (3.1c) of Theorem 3.1. It follows that $\partial P(\alpha) \cap \partial E_n = \emptyset$. Hence (3.23) holds, since it is easily verified that

$$a \in P(\alpha) \cap E_n, \quad n = k+1, k+2, k+3, \ldots .$$

(B) Suppose $-1/4 < a \le 0$, $P_a = P(0)$. From (3.5d) it can be seen that, for $n \ge k+1$,

$$E_n = \bigcup_{0 \le s \le q_n} L_n(s) =: E_n^*$$

where

$$L_n(s) := [w: |w-c_n| \leq s] \cap [w: |w| \leq \frac{q_n-s}{p_n}].$$

If fact, $E_n^* \subseteq E_n$ follows immediately from (3.5d). If $w \in E_n$ we set $s := |w-c_n|$ so that by (3.5d) $s \leq q_n$ and $p_n|w| \leq q_n-s$; hence $w \in E_n^*$ and thus $E_n \subseteq E_n^*$. We wish to show that $E_n \subseteq P(0)$. We note that $\partial P(0)$ is a parabola with focus at 0 and vertex at -1/4, and we recall from [1, p. 109] that if x is any point on the line segment joining the vertex and focus (i.e., if $-1/4 < x \leq 0$), then the point on $\partial P(0)$ nearest to x is the focus (i.e., $d(\partial P(0),x) = x + 1/4$). It follows that -1/4 is the point on $\partial P(0)$ nearest to c_n, since $-1/4 < c_n \leq 0$. Thus if $L_n(s)$ (for some $0 \leq s \leq q_n$) contains a point x_0 not in $P(0)$, then $-1/4 \in \text{Int } L_n(s) \subseteq \text{Int } E_n$. Hence if $-1/4 \notin \text{Int } E_n$, then $E_n \subseteq P(0)$. From (3.13) we see that $-1/4 \notin \text{Int } E_n$ if and only if

$$t_n(\pi) + \sqrt{[t_n(\pi)]^2-K_n} = (\rho_{n-1}+|x_1|)(|x_2|-\rho_n) \leq 1/4$$

which is (3.1c'). □

<u>Lemma</u> 3.9. <u>Let</u> $V = \{V_n\}$ <u>be defined as in Theorem</u> 3.1. <u>For each</u> $a \notin ¢ - (-\infty,-1/4]$ <u>let</u> H <u>denote the half-plane</u>

$$\text{(3.25a)} \quad H_a := \begin{cases} H(\alpha), & \text{if } a \in ¢ - (-\infty,0], \quad \alpha := \arg a, \\ H(0), & \text{if } -\frac{1}{4} < a \leq 0, \end{cases}$$

where

$$\text{(3.25b)} \quad H(\alpha) := [v: \text{Re}(ve^{-i(\alpha/2)}) \geq -\frac{1}{2}\cos(\alpha/2)], \quad -\pi < \alpha < \pi .$$

Then

$$\text{(3.26)} \quad V_n \subseteq H_a, \quad n = k, k+1, k+2, \ldots .$$

<u>Proof</u>. (A): Suppose $a \in C - (-\infty,0]$, $H_a = H(\alpha)$. Then (3.26) holds if and only if

$$\text{(3.27)} \quad e^{-i(\alpha/2)} V_n \subseteq e^{-i(\alpha/2)} H(\alpha)$$

where

$$e^{-i(\alpha/2)} V_n = [u: |u - x_1 e^{-i(\alpha/2)}| \leq \rho_n],$$

$$e^{-i(\alpha/2)} H(\alpha) = [u: \text{Re}(u) \geq -\frac{1}{2}\cos(\alpha/2)] .$$

It can be readily shown that (3.27) holds if and only if (3.1d) holds.

(B): Suppose $-1/4 < a \leq 0$, $H_a = H(0)$. Then it is easily seen that (3.26) holds if and only if (3.1d') holds. □

Proof of Theorem 3.1. In addition to the above lemmas we shall also make use of the following important result of Thron [2, Theorem B]: *For* $-\pi < \alpha < \pi$, *let* $P(\alpha)$ *and* $H(\alpha)$ *be defined by* (3.22b) *and* (3.25b). *Let* $\{a_n\}$ *be a sequence of complex numbers satisfying*

$$\text{(3.28)} \qquad a_n \in P(\alpha) \cap [w: |w| \leq M]$$

for some $M > 0$. *For* $m = 0,1,2,\ldots$, *let*

$$S_n^{(m)}(w) := \frac{a_{m+1}}{1} + \frac{a_{m+2}}{1} + \cdots + \frac{a_{m+n-1}}{1} + \frac{a_{m+n}}{1+w}, \quad n = 1,2,3,\ldots .$$

Then for $m \geq 0$, $\{S_n^{(m)}(H(\alpha))\}_{n=1}^{\infty}$ *is a nested sequence of closed circular disks and*

$$\lim_{n\to\infty} \operatorname{diam} S_n^{(m)}(V(\alpha)) = 0, \quad m = 0,1,2,\ldots .$$

It follows from (3.2b) that there exists a number $M > 0$ such that $E_n \subseteq [w: |w| \leq M]$ for $n \geq k+1$. Hence by Lemma 3.8 $E_n \subseteq P_a \cap [w: |w| \leq M]$, $n = k+1, k+2, \ldots$, so that $0 \neq a_n \in E_n$, $n \geq k+1$, implies that (3.28) holds. By Lemma 3.9

$$S_n^{(m)}(V_{n+m}) \quad S_n^{(m)}(H_a), \quad n = 1,2,3,\ldots, \quad m = 0,1,2,\ldots .$$

Thus by Thron's theorem stated above, condition (2.10a) of Theorem 2.2 holds. Condition (2.9) of Theorem 2.2 holds by Lemma 3.6 (with $\Gamma_n = x_1$, $w_n = x_1$). By Lemma 3.2, $V = \{V_n\}$ is a sequence of closed value regions corresponding to E at W. Hence the assertion (3.3) of Theorem 3.1 follows from Theorem 2.2. □

We conclude by considering some examples of Cartesian ovals whose graphs are given in Figure 2. Figure 2 contains four ovals with $x_1 = 1$, $x_2 = -2$, $a = -x_1 x_2 = 2$, and the parabola $\partial P(0)$. In case (a), $\rho_{n-1} = 1.4$ and $\rho_n = 1.5$ so that $\mathcal{D}_n < 0$ (see formula following (3.24)) and $K_n < 0$; we obtain $E_n \subset P(0)$ as shown. We note, however, that (3.1b) does not hold in this case. In case (b), $\rho_{n-1} = 1.5$ and $\rho_n = 1.1$ so that $\mathcal{D}_n > 0$ and $K_n < 0$; thus the oval ∂E_n intersects the parabola $\partial P(0)$ in two points. In case (c) $\rho_{n-1} = 1.5$ and and $\rho_n = 1.0$ so that $\mathcal{D}_n = 0$ and $K_n < 0$; hence $\partial E_n \cap \partial P(0)$ consists of two points. In case (d) $\rho_{n-1} = .9$ and $\rho_n = .3$ so that $\mathcal{D}_n > 0$ and $K_n > 0$; here we have $\partial E_n \cap \partial P(0)$ consists of 4 points.

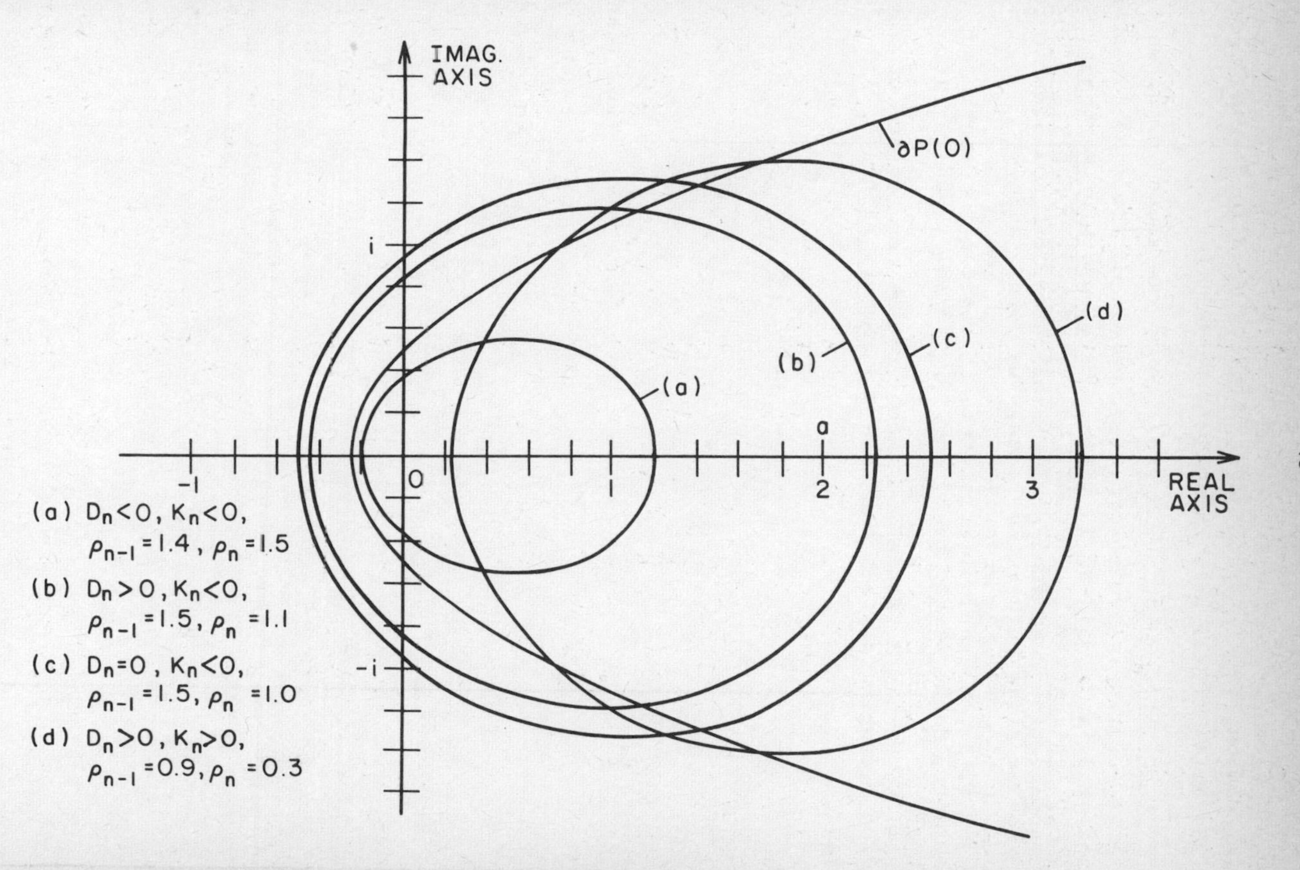

IMAG. AXIS
∂P(O)
(a)
(b)
(c)
(d)
a
i
-i
-1
0
1
2
3
REAL AXIS
(a) $D_n<0$, $K_n<0$, $\rho_{n-1}=1.4$, $\rho_n=1.5$
(b) $D_n>0$, $K_n<0$, $\rho_{n-1}=1.5$, $\rho_n=1.1$
(c) $D_n=0$, $K_n<0$, $\rho_{n-1}=1.5$, $\rho_n=1.0$
(d) $D_n>0$, $K_n>0$, $\rho_{n-1}=0.9$, $\rho_n=0.3$

Figure 2

Acknowledgements. The authors wish to thank Robert B. Jones for his assistance in preparing the graphs in Figure 2.

References.

1. Jones, William B. and W. J. Thron, Encyclopedia of Mathematics and Its Applications, vol. 11, Continued Fractions: Analytic Theory and Applications, Addison-Wesley Publishing Co., now distributed by The Cambridge University Press, New York, (1980).

2. Thron, W. J., On parabolic convergence regions for continued fractions, Math. Zeitschr. Bd. 69, S. 173-182 (1958).

3. Thron, W. J. and H. Waadeland, Accelerating convergence of limit periodic continued fractions $K(a_n/1)$, Numer. Math. 34, 155-170 (1980).

ON M-TABLES ASSOCIATED WITH STRONG MOMENT PROBLEMS

S. Clement Cooper*
Department of Mathematics
Colorado State University
Fort Collins, CO 80523

1. Introduction

This article is concerned with bi-infinite sequences $C = \{c_n\}_{n=-\infty}^{\infty}$ of complex numbers. With such a sequence one can associate M-tables, continued fractions and moment problems. We will be examining certain interrelationships between these three topics. The moment problems under consideration will be the strong Stieltjes moment problem (SSMP) and the strong Hamburger moment problem (SHMP).

A continued fraction which plays a dominant role in this analysis is the T-fraction. A general T-fraction [4] is a continued fraction of the form

$$\frac{F_1 z}{1 + G_1 z} + \frac{F_2 z}{1 + G_2 z} + \frac{F_3 z}{1 + G_3 z} + \dots, \; F_n \neq 0, \; n = 1,2, \dots . \quad [1.1]$$

The T-fraction can also be written in the form

$$\frac{z}{e_1 + d_1 z} + \frac{z}{e_2 + d_2 z} + \dots, \text{ where } e_n \neq 0 \text{ for } n = 1,2,\dots . \quad [1.2]$$

The constants e_n, d_n, F_n, G_n, $n = 1, 2, \dots$ are related by

$$e_1 = \frac{1}{F_1}, \quad e_{2n-1} = \frac{\prod_{k=1}^{n-1} F_{2k}}{\prod_{k=1}^{n} F_{2k-1}}, \; n = 2, 3, 4, \dots, \quad [1.3a]$$

$$e_{2n} = \frac{\prod_{k=1}^{n} F_{2k-1}}{\prod_{k=1}^{n} F_{2k}}, \; n = 1, 2, 3, \dots, \quad [1.3b]$$

$$d_n = G_n e_n, \; n = 1, 2, 3, \dots . \quad [1.3c]$$

Let $\{R_n(z)\}$ be a sequence of rational functions holomorphic at $z = 0$. Then we say the sequence $\{R_n(z)\} = \{P_n(z)/Q_n(z)\}$ <u>corresponds weakly</u> at $z = 0$ to the formal Laurent series (fLs) $L(z) = \sum_{k=0}^{\infty} \alpha_k z^k$ if the

*Research supported in part by the U. S. National Science Foundation under grant no. DMS-8401717, in part by the Nansen Foundation of Norway, and in part by Colorado State University.

fLs $Q_n(z)L(z) - P_n(z)$ is such that

$$Q_n(z)L(z) - P_n(z) = g_{m_n} z^{m_n} + \dots$$

where $g_{m_n} \neq 0$ and $m_n \to \infty$ as $n \to \infty$. The constant m_n is called the degree of weak correspondence. Analogously, we say the sequence $\{R_n(z)\} = \{P_n(z)/Q_n(z)\}$ <u>corresponds weakly</u> at $z = \infty$ to the fLs $L^*(z) = \sum_{k=0}^{\infty} \alpha_{-k} z^{-k}$ if the fLs $Q_n(z)L^*(z) - P_n(z)$ is such that

$$Q_n(z)L^*(z) - P_n(z) = g_{-m_n} z^{-m_n} + g_{-(m_n+1)} z^{-(m_n+1)} + \dots$$

where $g_{-m_n} \neq 0$ and $m_n + \deg Q_n(z) \to \infty$ as $n \to \infty$. The constant $m_n + \deg Q(z)$ is called the degree of weak correspondence or the weak order of correspondence. A continued fraction $K\left[\frac{a_n(z)}{b_n(z)}\right]$ is said to correspond weakly to a fLs if the sequence of approximants corresponds weakly to the fLs. Conditions for weak correspondence between T-fractions and certain pairs of fLs can be given in terms of Hankel determinants. Let $C = \{c_n\}_{n=-\infty}^{\infty}$ be a bi-infinite sequence of complex numbers, then Hankel determinants for C are defined by

$$\begin{cases} H_0^{(k)}(C) := 1 \text{ for } k = 0, 1, \dots \text{ and} \\ H_n^{(k)}(C) := \begin{vmatrix} c_k & c_{k+1} & \cdots & c_{n+k-1} \\ c_{k+1} & c_{k+2} & & c_{n+k} \\ \vdots & & & \\ c_{n+k-1} & c_{n+k} & \cdots & c_{2n+k-2} \end{vmatrix} & \begin{matrix} n = 1, 2, \dots, \\ k = 0, \pm 1, \pm 2, \dots . \end{matrix} \end{cases} \qquad [1.4]$$

We have the following theorem relating T-fractions to pairs of fLs.

<u>THEOREM 1</u>. <u>Let</u> $C = \{c_n\}_{n=-\infty}^{\infty}$ <u>be a bi-infinite sequence of complex numbers. From</u> C, <u>generate the pair of fLs</u>,

$$L_0(C) = \sum_{m=1}^{\infty} - c_{-m} z^m \quad \text{and} \quad L_\infty(C) = \sum_{m=0}^{\infty} c_m z^{-m}. \qquad [1.5]$$

<u>Then there exists a general T-fraction</u> [1.1] <u>with</u> $F_n \neq 0$ <u>and</u> $G_n \neq 0$, $n = 1, 2, 3, \dots$ <u>which corresponds weakly to</u> $L_0(C)$ <u>at</u> $z = 0$ <u>and to</u> $L_\infty(C)$ <u>at</u> $z = \infty$ <u>if and only if</u>

$$H_n^{(-n+1)}(C) \neq 0 \quad \text{and} \quad H_n^{(-n)}(C) \neq 0, \; n = 0, 1, 2, \dots . \qquad [1.6]$$

<u>The weak order of correspondence is</u> $n + 1$ <u>at</u> $z = 0$ <u>and</u> n <u>at</u> $z = \infty$. <u>Furthermore</u>,

$$F_1 = -H_1^{(1)}(C),\ F_n = -\frac{H_{n-2}^{(-n+3)}(C)H_n^{(-n)}(C)}{H_{n-1}^{(-n+2)}(C)H_{n-1}^{(-n+1)}(C)},\ n=2,3,\ldots, \qquad [1.7a]$$

$$G_1 = -\frac{H_1^{(-1)}(C)}{H_1^{(0)}(C)},\ G_n = -\frac{H_{n-1}^{(-n+2)}(C)H_n^{(-n)}(C)}{H_n^{(-n+1)}(C)H_{n-1}^{(-n+1)}(C)},\ n=2,3,\ldots\ . \qquad [1.7b]$$

_If_ $A_n(z)$ _and_ $B_n(z)$ _are the_ n^{th} _numerator and_ n^{th} _denominator, respectively, of the T-fraction then_

$$A_1(z) = 1,\ A_n(z) = a_{n,1}z + \ldots + a_{n,n}z^n,\ n=2,3,\ldots \qquad [1.8a]$$

where $a_{n,1} = e_2e_3 \ldots e_n,\ a_{n,n} = d_2d_3 \ldots d_n$, _and_ [1.8b]

$$B_n(z) = b_{n,0} + b_{n,1}z + \ldots + b_{n,n}z^n,\ n=1,2,\ldots \qquad [1.8c]$$

where $b_{n,0} = e_1e_2 \ldots e_n,\ b_{n,n} = d_1d_2 \ldots d_n$. [1.8d]

(_Note that_ e_n _and_ d_n _are given by [1.3]_).

Proof: See the statement and proof of Theorem 2.1 [6].

Theorem 1 is the basis for the connection between certain T-fractions associated with each moment problem and the M-table for a particular pair of fLs generated from a bi-infinite sequence of numbers.

In [2] the M-table was defined for a pair of fLs,

$$f^{(M)}(z) = \sum_{n=0}^{\infty} a_n z^n \quad \text{and} \quad g^{(M)}(z) = \sum_{n=1}^{\infty} -a_{-n}z^{-n}. \qquad [1.9]$$

In Section 1 we generalize this definition to accomodate any pair of fLs generated from a bi-infinite sequence C of complex numbers where

$$f(z)=c_0^{(0)}+\sum_{n=1}^{\infty} c_n z^n,\ g(z)=-c_0^{(\infty)}-\sum_{n=1}^{\infty} c_{-n}z^{-n} \text{ and } c_0=c_0^{(0)}+c_0^{(\infty)}. \qquad [1.10]$$

Using results from [2] we prove that the entries in this "expanded" M-table exist and are unique. The table also inherits the characteristic that equal entries appear as square blocks in the table.

In Section 2 we start with a bi-infinite sequence of real numbers $C = \{c_n\}_{n=-\infty}^{\infty}$. With this sequence we formulate the strong Stieltjes moment problem (SSMP). It is pointed out that under certain conditions a particular _positive_ T-fraction associated with C in the sense of [1.7] plays an important role in the question of uniqueness of the solution of the SSMP. From C we also generate a pair of fLs. It will be shown that under certain conditions the approximants of the positive T-fraction appear as entries in the M-table for the pair generated. Also, the M-table in question is shown to be normal.

Again we start with a given bi-infinite sequence $C = \{c_n\}_{n=-\infty}^{\infty}$ of real numbers in Section 3. This time we formulate the strong Hamburger moment problem (SHMP). Here we point out that an APT-fraction (a special T-fraction defined in Section 3) plays a role similar to that of the positive T-fraction in the previous problem. Under certain conditions, there is also a connection between the M-table for a given pair of fLs and the APT-fraction involved in the SHMP. We get partial information about the block structure of the M-table for this pair of fLs.

2. A Further Generalization of the M-Table

Let $C = \{c_n\}_{n=-\infty}^{\infty}$ be a bi-infinite sequence of complex constants. With C, we can associate infinitely many pairs of fLs of the form

$$f(z)=c_0^{(0)}+\sum_{n=1}^{\infty} c_n z^n,\ g(z)=-c_0^{(\infty)}-\sum_{n=1}^{\infty} c_{-n} z^{-n},\ c_0=c_0^{(0)}+c_0^{(\infty)}. \qquad [2.1]$$

To date, the most general form of the M-table is found in [2]. An earlier version can be found in [7]. The M-table found in [2] is a table of rational approximations to fLs of the form

$$f^{(M)}(z) = \sum_{n=0}^{\infty} c_n z^n \quad \text{and} \quad g^{(M)}(z) = \sum_{n=1}^{\infty} -c_{-n} z^{-n}. \qquad [2.2]$$

This case is subsumed in the more general approach we propose above. We will define the entries in the M-table. Using a simple relation between the pair of fLs [2.1] and the pair [2.2], we then exploit results known about the M-table for [2.2] to obtain results about the M-table for [2.1]. In particular, we are interested in existence and uniqueness of the entries as well as the block structure of equal entries.

Definition 2. For every integer k and nonnegative integer n the (n, k) entry

$$M_{n,k}(z) = A_{n,k}(z)/B_{n,k}(z)$$

in the M-table for the fLs in [2.1] is determined by

$$\begin{cases} B_{n,k}(z)f(z) - A_{n,k}(z) = O(z^{n+k}) \\ B_{n,k}(z)g(z) - A_{n,k}(z) = O_-(z^{k-1}) \end{cases} \qquad [2.3]$$

where

$$B_{n,k}(z) = \sum_{i=0}^{n} \beta_i z^i \not\equiv 0, \text{ and} \qquad [2.4]$$

$$A_{n,k}(z) = \begin{cases} \sum_{i=0}^{n} \alpha_i z^i, & -n \le k, \\ z^{k+n} \sum_{i=0}^{-k} \alpha_i z^i, & k \le -n \le 0,\ k < 0, \end{cases} \qquad [2.5]$$

and with $M_{0,0} = -c_0^{(\infty)}$.

The expression $O(z^{n+k})$ denotes a fLs in increasing powers of z starting with a term in z^{n+k} while $O_-(z^{k-1})$ denotes a fLs in decreasing powers of z beginning with a term in z^{k-1}.

The definition of the M-table for the fLs [2.2] as given in [2] reads as follows.

<u>Definition 3</u>. For every integer k and every nonnegative integer n the (n, k) entry

$M_{n,k}(z) = A_{n,k}(z)/B_{n,k}(z)$

in the M-table for the pair of fLs [2.2] is determined by

$$\begin{cases} B_{n,k}(z)f^{(M)}(z) - A_{n,k}(z) = O(z^{n+k}) \\ B_{n,k}(z)g^{(M)}(z) - A_{n,k}(z) = O_-(z^{k-1}) \end{cases} \qquad [2.6]$$

where

$$B_{n,k}(z) = \sum_{i=0}^{n} \beta_i z^i \not\equiv 0 \qquad [2.7]$$

and

$$A_{n,k}(z) = \begin{cases} \sum_{i=0}^{k-1} \alpha_i z^i, & 0 \le n \le k,\ k > 0 \\ \sum_{i=0}^{n-1} \alpha_i z^i, & -n < k < n, \\ z^{n+k} \sum_{i=0}^{-k-1} \alpha_i z^i, & k \le -n \le 0,\ k < 0 \end{cases} \qquad [2.8]$$

and with $A_{0,0} \equiv 0$.

Comparing the definitions we see that the only differences are in the $A_{n,k}(z)$'s and $M_{0,0}(z)$. Upon reflection, it is clear that if both $c_0^{(0)}$ and $c_0^{(\infty)}$ are nonzero, then in order for [2.3] to hold the $\deg(A_{n,k}(z)) = \deg(B_{n,k}(z))$. (If $A_{n,k}(z)$ contains negative powers of z as is possible when $k \le -n \le 0$, $k < 0$, we use the convention that

$\deg(A_{n,k}(z))$ = largest power of z, not the largest power in absolute value.) This accounts for the differences in degrees seen in the $A_{n,k}(z)$'s. The change from $A_{0,0}(z) \equiv 0$ to $M_{0,0}(z) = -c_0^{(\infty)}$ is demanded by [2.3].

In order to establish the existence of the entries in the "expanded" M-table we first note that given two fLs obtained from C of the form [2.1], they are related to a pair of fLs of the form [2.2] by

$$f(z) = f^{(M)}(z) - c_0^{(\infty)} \text{ and } g(z) = g^{(M)}(z) - c_0^{(\infty)}.$$

This observation leads us to

<u>THEOREM 4</u>. <u>The entries in the M-table for [2.1] as given by Definition 2 exist. Furthermore, the</u> (n, k) <u>entry</u>, $M_{n,k}(z)$, <u>for [2.1] is related to the</u> (n, k) <u>entry</u>, $M_{n,k}^{(M)}(z)$, <u>for [2.2] by</u>

$$M_{n,k}(z) = M_{n,k}^{(M)}(z) - c_0^{(\infty)}.$$

<u>Proof</u>: The original definition of the M-table involves three regions and $M_{0,0}^{(M)}(z)$. Thus, it is natural to divide the argument accordingly. The existence of $M_{n,k}^{(M)}$ was established in [4] for all permissible values of n and k. Clearly, $M_{0,0}(z) = M_{0,0}^{(M)}(z) - c_0^{(\infty)}$.

Now consider $M_{n,k}(z)$ for $0 \le n \le k$, $k > 0$. The claim is that $M_{n,k}(z) = M_{n,k}^{(M)}(z) - c_0^{(\infty)}$. From Definition 3 we see that

$$M_{n,k}^{(M)}(z) = \frac{A_{n,k}^{(M)}(z)}{B_{n,k}^{(M)}(z)} = \frac{\sum_{i=0}^{k-1} \alpha_i^{(M)} z^i}{\sum_{i=0}^{n} \beta_i^{(M)} z^i}, \quad B_{n,k}^{(M)}(z) \neq 0$$

where

$$B_{n,k}^{(M)}(z) f^{(M)}(z) - A_{n,k}^{(M)}(z) = O(z^{n+k}) \text{ and}$$
$$B_{n,k}^{(M)}(z) g^{(M)}(z) - A_{n,k}^{(M)}(z) = O_-(z^{k-1}).$$

Thus, $M_{n,k}^{(M)}(z) - c_0^{(\infty)} = \dfrac{A_{n,k}^{(M)}(z) - c_0^{(\infty)} B_{n,k}^{(M)}(z)}{B_{n,k}^{(M)}(z)}$. Since the $\deg(B_{n,k}^{(M)}(z)) \le n$, the $\deg(A_{n,k}^{(M)}(z) - c_0^{(\infty)} B_{n,k}^{(M)}(z)) \le n$. Let $A_{n,k}(z) = A_{n,k}^{(M)}(z) - c_0^{(\infty)} B_{n,k}^{(M)}(z)$ and $B_{n,k}(z) = B_{n,k}^{(M)}(z)$. The degrees are correct and we also have

$$B_{n,k}(z)f(z) - A_{n,k}(z)$$
$$= B_{n,k}^{(M)}(z)f(z) - (A_{n,k}^{(M)}(z) - c_0^{(\infty)}B_{n,k}^{(M)}(z))$$
$$= B_{n,k}^{(M)}(z)(c_0^{(\infty)} + f(z)) - A_{n,k}^{(M)}(z)$$
$$= B_{n,k}^{(M)}(z)f^{(M)}(z) - A_{n,k}^{(M)}(z) = O(z^{n+k})$$

and $B_{n,k}(z)g(z) - A_{n,k}(z)$

$$= B_{n,k}^{(M)}(z)(c_0^{(\infty)} + g(z)) - A_{n,k}^{(M)}(z)$$
$$= B_{n,k}^{(M)}(z)g^{(M)}(z) - A_{n,k}^{(M)}(z) = O\_(z^{k-1}).$$

Thus, the correspondence requirements are also satisfied.

The arguments for the remaining two regions, $-n < k < n$ and $k \leq -n \leq 0$, $k < 0$ are completely analogous so we omit them. ■

With the question of existence of the entries settled it is now natural to ask about the uniqueness.

<u>THEOREM 5</u>. <u>The entries in the M-table for [2.1] as given by Definition 2 are unique</u>.

<u>Proof</u>: Assume that both $M_{n,k}(z) = A_{n,k}(z)/B_{n,k}(z)$ and $M'_{n,k} = A'_{n,k}/B'_{n,k}$ are entries in the M-table for [2.1]. Then

$$B'_{n,k}(z)(B_{n,k}(z)f(z) - A_{n,k}(z)) = O(z^{n+k}) \text{ and}$$
$$B_{n,k}(z)(B'_{n,k}(z)f(z) - A'_{n,k}(z)) = O(z^{n+k}).$$

Subtraction yields

$$B_{n,k}(z)A'_{n,k}(z) - B'_{n,k}(z)A_{n,k}(z) = O(z^{n+k}).$$

Similarly,

$$B_{n,k}(z)A'_{n,k}(z) - B'_{n,k}(z)A_{n,k}(z) = O\_(z^{n+k-1}).$$

Therefore,

$$B_{n,k}(z)A'_{n,k}(z) - B'_{n,k}(z)A_{n,k}(z) \equiv 0$$

which implies that $M_{n,k}(z) = M'_{n,k}(z)$ since $B_{n,k}(z)B'_{n,k}(z) \neq 0$. ■

The M-table as given by Definition 3 was shown (in [2]) to exhibit the square block structure of equal entries which is familiar to those acquainted with the original Padé table [1]. This structure is inherited by the "expanded" M-table.

<u>THEOREM 6</u>. <u>If a rational function</u> R <u>appears more than once in the M-table, then the set</u>

$$E_R = \{(n, k) | M_{n,k}(z) \equiv R\}$$

<u>forms a square block in the table. The square block may be infinite.</u>

Proof: This is a direct consequence of the relationship between $M_{n,k}(z)$ and $M^{(M)}_{n,k}(z)$ (see Theorem 4) and Theorem 4 in [2].

We say an entry $M_{n,k}(z)$ in the M-table for [2.1] is normal if the square block containing $M_{n,k}(z)$ is of dimension 1. An M-table is normal if every entry is normal. The following lemma and its corollary will be useful in the next two sections.

Lemma 7. *Let $M_{n,k}(z)$ be the (n, k) entry in the M-table for [2.1]. If*

$$H_n^{(-n+k)}(C) \neq 0 \text{ and } H_n^{(-n+k+1)}(C) \neq 0 \qquad [2.9]$$

then $M_{n,k}(z)$ is either a normal entry or is on the left edge of a square block.

A direct consequence of the lemma is

Corollary 8. *Let $C = \{c_n\}_{n=-\infty}^{\infty}$ be a bi-infinite sequence of complex numbers. If $H_n^{(k)}(C) \neq 0$ for $n = 0, 1, \ldots, k = 0, \pm 1, \pm 2, \ldots$ then the M-table for [2.1] is normal.*

Proof of Lemma 7: Let $M_{n,k}(z)$ be the (n, k) entry in the M-table for [2.1]. By [2.3] we have

$$B_{n,k}(z)f(z) - A_{n,k}(z) = O(z^{n+k}) \quad \text{and}$$

$$B_{n,k}(z)g(z) - A_{n,k}(z) = O_-(z^{k-1}).$$

Subtraction yields

$$B_{n,k}(z)(f(z) - g(z)) = O(z^{n+k}) - O_-(z^{k-1})$$

from which we obtain the system

$$\begin{cases} \beta_n c_{-n+k} + \beta_{n-1} c_{-n+k+1} + \cdots + \beta_1 c_{k-1} = -\beta_0 c_k \\ \beta_n c_{-n+k+1} + \beta_{n-1} c_{-n+k+2} + \cdots + \beta_1 c_k = -\beta_0 c_{k+1} \\ \qquad \vdots \\ \beta_n c_{k-1} + \beta_{n-1} c_k + \cdots + \beta_1 c_{n+k-2} = -\beta_0 c_{n+k-1} \end{cases} \qquad [2.10]$$

which has a non-trivial solution. Since $\Delta_k = H_n^{(-n+k)}(c) \neq 0$ we can use Cramer's rule to obtain

$$\beta_j = (-1)^j \beta_0 \hat{H}_j / \Delta_k \quad \text{where} \qquad [2.11]$$

$$\hat{H}_j = \begin{vmatrix} c_{-n+k} & c_{-n+k+1} & \cdots & \hat{c}_{-j+k} & \cdots & c_k \\ c_{-n+k+1} & c_{-n+k+2} & \cdots & c_{-j+k+1} & \cdots & c_{k+1} \\ \vdots & \vdots & & & & \\ c_{k-1} & c_k & \cdots & c_{-j+k+n-1} & \cdots & c_{n+k-1} \end{vmatrix}$$

(the indicated column is omitted). A nontrivial solution exists and to get it we must choose $\beta_0 \neq 0$. Thus,

$$\beta_n = \frac{(-1)^n \beta_0 \hat{H}_n}{\Delta_k} = \frac{(-1)^n \beta_0 H_n^{(-n+k+1)}(C)}{\Delta_k}$$

is also nonzero under the hypothesis that $H_n^{(-n+k+1)}(C) \neq 0$. Therefore, $B_{n,k}(z) = \beta_0 + \beta_1 z + \ldots + \beta_n z^n$ where $\beta_0 \beta_n \neq 0$. Notice that $B_{n,k}(z)$ is unique up to a constant factor.

In order to reach a contradiction, suppose $M_{n,k}(z)$ is an entry in a square block but is not on the left edge. Then $M_{n-1,k}(z) = M_{n,k}(z)$. From Definition 2 we know that the $\deg(B_{n-1,k}(z)) \leq n - 1$. However, the coefficients of $B_{n-1,k}(z)$ must satisfy [2.10] which in turn implies that $B_{n-1,k}(z) = cB_{n,k}(z)$ where c is some constant. This contradicts the fact that the $\deg(B_{n-1,k}(z)) \leq n - 1$ and the $\deg(B_{n,k}(z)) = n$. The result follows. ∎

We conclude this section with a brief summary. The M-table is now defined for any pair of fLs of the form [2.1] associated with a given bi-infinite sequence of complex constants. The entries as given by Definition 2 exist and are unique. Furthermore, equal entries appearing in the table form a square block (of finite or infinite dimension). Finally, we considered a lemma and a corollary to the lemma which will be useful in later sections.

3. The M-table and the Strong Stieltjes Moment Problem

Let $C = \{c_n\}_{n=-\infty}^{\infty}$ be a bi-infinite sequence of real numbers. The strong Stieltjes moment problem (SSMP) can be stated as follows.

Does there exist a real-valued, bounded, monotonically increasing function $\psi(t)$ with infinitely many points of increase on $[0, \infty)$ such that for every integer n,

$$c_n = \int_0^\infty (-t)^n d\psi(t)?$$

Thron, Jones and Waadeland showed that a solution existed if and only if the bi-infinite sequence C satisfies the following condition,

$$H_{n+1}^{(-n)}(C)>0,\ n=0,1,2,\ldots,\quad H_{2n}^{(-2n)}(C)>0,\quad H_{2n-1}^{(-2n+1)}(C)<0,\ n=1,2,\ldots, \qquad [3.1]$$

(Theorem 6.1, [6]).

Let

$$L_0(C) = \sum_{m=1}^{\infty} -c_{-m} z^m \quad \text{and} \quad L_\infty(C) = \sum_{m=0}^{\infty} c_m z^{-m} \qquad [3.2]$$

be two fLs generated by C. It has also been shown (Theorem 3.1, [6]) that there exists a positive T-fraction corresponding weakly to $L_0(C)$ at $z = 0$ and $L_\infty(C)$ at $z = \infty$ if and only if [3.1] holds. A <u>positive T-fraction</u> is a T-fraction

$$\frac{F_1 z}{1 + G_1 z} + \frac{F_2 z}{1 + G_2 z} + \frac{F_3 z}{1 + G_3 z} + \cdots$$

where $F_n > 0$, $G_n > 0$, $n = 1, 2, \ldots$. In addition, the SSMP has exactly one solution (essentially equal solutions are considered equal) if the positive T-fraction which corresponds weakly to $L_0(C)$ at $z = 0$ and $L_\infty(C)$ at $z = \infty$ is convergent on compact subsets of $R = [z: |\arg z| < \pi]$ (Theorem 6.2, [6]).

Several different concepts are related here. We start with a bi-infinite sequence C of real numbers from which we can form a pair of fLs [3.2]. In Section 2, we saw that from such a pair of fLs one could form an M-table. (Note that according to Definition 2 we generated the pair of fLs from $C' = \{-c_{-n}\}_{n=-\infty}^{\infty}$, not from $C = \{c_n\}_{n=-\infty}^{\infty}$.) Under certain conditions on the Hankel determinants associated with C, namely [3.1], the SSMP has a solution <u>and</u> there is a positive T-fraction which corresponds weakly to $L_0(C)$ at $z = 0$ and $L_\infty(C)$ at $z = \infty$. It will be shown that in this case the approximants of the positive T-fraction occur as entries in the M-table for $L_0(C)$ and $L_\infty(C)$ and that this M-table is <u>normal</u>.

<u>THEOREM 9</u>. <u>The nth approximant, $A_n(z)/B_n(z)$, $n = 1, 2, \ldots$ of the positive T-fraction associated with the bi-infinite sequence of real numbers C (in the sense of [1.7]) satisfying [3.1] appears in the M-table for [3.2]</u>.

<u>Proof</u>: Let $A_n(z)/B_n(z)$ be the nth approximant of the positive T-fraction associated with C in the sense of [1.7]. Then $\{A_n(z)/B_n(z)\}$ corresponds weakly to $L_0(C)$ at $z = 0$ and $L_\infty(C)$ at $z = \infty$. Since $F_n > 0$, $n = 1, 2, \ldots$, $e_{2n-1} = \prod_{k=1}^{n-1} F_{2k} / \prod_{k=1}^{n} F_{2k-1}$, $n = 2, 3, \ldots$ and $e_{2n} = \prod_{k=1}^{n} F_{2k-1} / \prod_{k=1}^{n} F_{2k}$, $n = 1, 2, \ldots$ we see from [1.8] that the $\deg(A_n(z)) = n$. Also, since $G_n > 0$ and $d_n = G_n e_n$, $n = 1, 2, 3, \ldots$ we see from [1.8] that the $\deg(B_n(z)) = n$. From Theorem 1 we also have

$$B_n(z)L_0(C) - A_n(z) = O(z^{n+1}) \text{ and}$$

$$B_n(z)L_\infty(C) - A_n(z) = O_-(z^0).$$

Thus, $A_n(z)/B_n(z)$ satisfies Definition 2 for the (n, 1) entry,

$n = 1, 2, \ldots$, in the M-table for $L_0(C)$ and $L_\infty(C)$. By uniqueness of the entries we have $M_{n,1}(z) = A_n(z)/B_n(z)$ for $n = 0, 1, 2, \ldots$. ■

Thus, we have shown that the entries in the first row of the M-table are approximants of the positive T-fraction. It is not hard to see that they must be distinct. In Theorem 10 we assert that all of the entries in the M-table associated with this positive T-fraction are distinct.

<u>THEOREM 10</u>. <u>Let</u> $C = \{c_n\}_{n=-\infty}^{\infty}$ <u>generate the following pair of fLs</u>,

$$L_0(C) = \sum_{m=1}^{\infty} -c_{-m}z^m \text{ and } L_\infty(C) = \sum_{m=0}^{\infty} c_m z^{-m}. \qquad [3.4]$$

<u>If the following condition on</u> C,

$$H_{n+1}^{(-n)}(C)>0,\ n=0,1,2,\ldots,\ H_{2n}^{(-2n)}(C)>0,\ H_{2n-1}^{(-2n+1)}(C)<0,\ n=1,2,\ldots \qquad [3.5]$$

<u>are met, then the M-table for [3.4] is normal</u>.

<u>Proof</u>: By Corollary 8, the M-table for $L_0(C)$ and $L_\infty(C)$ is normal if $H_n^{(s)}(C) \neq 0$ for $n = 0, 1, 2, \ldots$, $s = 0, \pm1, \pm2, \ldots$. From the definition of $H_n^{(s)}(C)$ we have $H_0^{(s)} = 1 > 0$ for $s = 0, \pm1, \pm2, \ldots$. We will show that $H_n^{(s)}(C) \neq 0$ for $n = 1, 2, \ldots$ by showing that $H_n^{(-(n+k))}(C) \neq 0$, $k = 0, \pm1, \pm2, \ldots$, $n = 1, 2, \ldots$ first by induction on k and then on $-k$. From [3.5] we have

(1) $H_{n+1}^{(-((n+1)-1))}(C) > 0$, $n = 0, 1, 2, \ldots$,

(2) $H_{2n}^{(-2n)}(C) > 0$, $n = 1, 2, \ldots$, and

(3) $H_{2n-1}^{(-(2n-1))}(C) < 0$, $n = 1, 2, \ldots$.

We claim that

$$H_{2m}^{(-(2m+2k))}(C) > 0,\ H_{2m}^{(-(2m+(2k+1)))}(C) > 0,\ H_{2m+1}^{(-((2m+1)+2k))}(C) < 0$$
$$\text{and } H_{2m+1}^{(-((2m+1)+(2k+1)))}(C) > 0$$

for $k = 0, 1, 2, \ldots$, $m = 0, 1, 2, \ldots$. We begin by pointing out that $H_{2m}^{(-2m)}(C) > 0$ by (2) and $H_{2m+1}^{(-(2m+1))}(C) < 0$ by (3), for $m = 1, 2, \ldots$. Using Jacobi's identity we see that

$$\left[H_{2m}^{(-2m)}(C)\right]^2 = H_{2m}^{(-(2m+1))}(C)H_{2m}^{(-(2m-1))}(C) - H_{2m+1}^{(-(2m+1))}(C)H_{2m-1}^{(-(2m-1))}(C),$$

for $m = 1, 2, \ldots$. Since $\left[H_{2m}^{(-2m)}(C)\right]^2 > 0$ by (2), $H_{2m}^{(-(2m-1))}(C) > 0$ by (1) and $H_{2m+1}^{(-(2m+1))}(C)\cdot H_{2m-1}^{(-(2m-1))}(C) > 0$ by (3), we can conclude that

$H_{2m}^{(-(2m+1))}(C) > 0$, $m = 1, 2, \ldots$. Similarly, we see that $H_{2m+1}^{(-((2m+1)+1))}(C) > 0$, $m = 0, 1, 2, \ldots$. Thus, $H_{2m}^{(-2m)}(C) > 0$, $H_{2m}^{(-(2m+1))}(C) > 0$, $H_{2m+1}^{(-(2m+1))}(C) < 0$ and $H_{2m+1}^{(-((2m+1)+1))}(C) > 0$ for $m = 0, 1, 2, \ldots$.

Let $\ell = 2s$, $s > 0$. Now assume that

(a) $H_{2m}^{(-(2m+2k))}(C) > 0$,

(b) $H_{2m}^{(-(2m+(2k+1)))}(C) > 0$,

(c) $H_{2m+1}^{(-((2m+1)+2k))}(C) < 0$ and

(d) $H_{2m+1}^{(-((2m+1)+(2k+1)))}(C) > 0$

for $k < s$, $m = 0, 1, 2, \ldots$. Four applications of Jacobi's identity together with the induction hypothesis suffice to show that $H_{2m}^{(-(2m+2s))}(C) > 0$, $H_{2m}^{(-(2m+(2s+1)))}(C) > 0$, $H_{2m+1}^{(-((2m+1)+2s))}(C) < 0$ and $H_{2m+1}^{(-((2m+1)+(2s+1)))}(C) > 0$ for $m = 0, 1, 2, \ldots$. Therefore, by induction, $H_n^{(-(n+k))}(C) \neq 0$ for $n = 0, 1, 2, \ldots, k = 0, 1, 2, \ldots$.

A similar argument gives you $H_n^{(-(n-k))}(C) \neq 0$ for $n = 0, 1, 2, \ldots$, and $k = 0, 1, 2, \ldots$.

Thus, we have established that $H_n^{(s)}(C) \neq 0$ for $n = 0, 1, 2, \ldots$, $s = 0, \pm 1, \pm 2, \ldots$ and hence the M-table for [3.4] under the condition [3.5] for C is normal. ∎

In conclusion, we have shown that when the SSMP has a solution the M-table for $L_0(C) = \sum_{m=1}^{\infty} - c_{-m}z^m$ and $L_\infty(C) = \sum_{m=0}^{\infty} c_m z^{-m}$ contains the approximants of the positive T-fraction corresponding weakly to $L_0(C)$ at $z = 0$ and $L_\infty(C)$ at $z = \infty$. Furthermore, it is a <u>normal</u> table.

4. The M-table and the Strong Hamburger Moment Problem

Once again we find moments, continued fractions, fLs and their M-tables intimately related. As before, consider a bi-infinite sequence of real numbers $C = \{c_n\}_{n=-\infty}^{\infty}$. The strong Hamburger moment problem can be stated as follows.

Given $C = \{c_n\}_{n=-\infty}^{\infty}$, does there exist a real-valued, bounded, monotonically increasing function $\psi(t)$ on $(-\infty, \infty)$ with infinitely many points of increase such that for each n,

$$c_n = \int_{-\infty}^{\infty} (-t)^n d\psi(t) \quad (n = 0, \pm 1, \pm 2, \ldots)?$$

In [5], Jones and Thron showed that if C satisfies the condition

$H^{(-2n)}_{2n+1}(C) > 0$ and $H^{(-2n)}_{2n}(C) > 0$ for $n = 0, 1, \ldots$ [4.1a]

then the SHMP has a solution. If, in addition to satisfying [4.1a], C satisfies

$H^{(-2n+1)}_{2n}(C) \neq 0$ and $H^{(-(2n+1))}_{2n+1}(C) \neq 0$ for $n = 0,1,2,\ldots$ [4.1b]

then there is a continued fraction which plays an important role in the question of uniqueness of the solution [3]. This continued fraction is an APT-fraction which is a general T-fraction satisfying

$F_n \in \mathbb{R},\ 0 \neq G_n \in \mathbb{R},\ F_{2n-1}F_{2n} > 0,\ F_{2n-1}/G_{2n-1} > 0$ [4.2]

for $n = 1, 2, 3, \ldots$. The question of uniqueness is partially settled by the following result. If C satisfies [4.1] then the SHMP has a unique solution if and only if the APT-fraction associated with C in the sense of [1.7] converges completely (Theorem 5.3, [3]).

By Theorem 1, if [4.1] is satisfied then the APT-fraction corresponds weakly to a pair of fLs

$$L_0(C) = \sum_{m=1}^{\infty} - c_{-m}z^m \quad \text{and} \quad L_\infty(C) = \sum_{m=0}^{\infty} c_m z^{-m}. \qquad [4.3]$$

We claim that the approximants of the APT-fraction appear as entries in the M-table for [4.3]. Also, some information about the block structure is obtained.

<u>THEOREM 11</u>. <u>Let</u> $C = \{c_n\}_{n=-\infty}^{\infty}$ <u>be a bi-infinite sequence of real numbers. If</u> C <u>satisfies [3.1], namely</u>

$H^{(-2n+1)}_{2n}(C)\neq 0,\ H^{(-(2n+1))}_{2n+1}(C)\neq 0,\ H^{(-2n)}_{2n+1}(C)>0,$ <u>and</u> $H^{(-2n)}_{2n}(C)>0$

<u>for</u> $n = 0, 1, 2, \ldots$, <u>then there exists an APT-fraction associated with</u> C <u>(in the sense of [1.7]) such that the approximants appear as entries in the first row of the M-table for the pair [4.3]</u>.

<u>Proof</u>: Since $H^{(-n+1)}_n(C) \neq 0$ and $H^{(-n)}_n(C) \neq 0$ there exists a general T-fraction

$$\frac{F_1 z}{1 + G_1 z} + \frac{F_2 z}{1 + G_2 z} + \cdots$$

such that the nth approximant is a rational function $A_n(z)/B_n(z)$ where $A_n(z)$ and $B_n(z)$ are both polynomials of degree less than or equal to n. In fact, since $F_n\neq 0$, $G_n\neq 0$ for $n=1,2,\ldots$ the $\deg(A_n(z)) = n = \deg(B_n(z))$. Furthermore, by Theorem 1

$B_n(z)L_0(C) - A_n(z) = O(z^{n+1})$ and

$B_n(z)L_\infty(C) - A_n(z) = O\_(z^0)$.

Therefore, by Definition 2 and Theorem 5, $A_n(z)/B_n(z)$ is the $(n, 1)$ entry in the M-table for [4.3]. ■

Unfortunately, we cannot be quite as conclusive about the block structure in the M-table for [4.3] where C satisfies [4.1] as we were when C satisfied [3.1]. However, we can state the following.

THEOREM 12. Let $C = \{c_n\}_{n=-\infty}^{\infty}$ be a bi-infinite sequence of real numbers. If C satisfies [4.1], then the M-table for [4.3]

(i) has normal entries in rows 0 and 1, and

(ii) admits at most 2×2 squares in rows -1 and 2. In row 2 any 2×2 blocks which occur must start in an "odd position" and extend downwards in the table (i.e., $M_{2n+1,2} = M_{2n+2,2} = M_{2n+1,3} = M_{2n+2,3}$ for some $n = 0, 1, 2, \ldots$). Also, any 2×2 blocks occurring in row -1 must start in an "odd position" and extend upwards into the $(-2)^{nd}$ row (i.e., $M_{2n+1,-2} = M_{2n+2,-2} = M_{2n+1,-3} = M_{2n+2,-3}$).

The information appears to be inconclusive for all other rows.

Proof: We have already shown that the approximants of the APT-fraction associated with C (in the sense of [1.7]) are entries in the first row. Explicitly,

$$A_n(z)/B_n(z) = M_{n,1}(z) \quad \text{for } n = 1, 2, \ldots .$$

In the proof of Theorem 11 we showed that the $\deg(B_n(z)) = n$ and from [1.8] and [4.2] we see that $\beta_0\beta_n \neq 0$ (where β_0 and β_n are the constant and leading coefficients, respectively, of $B_n(z)$). From [4.1] we see that $H_n^{(-n+1)}(C) \neq 0$ for $n = 1, 2, \ldots$. Since $B_n(z) = B_{n,1}(z)$, we have

$$\beta_n = (-1)^n \frac{\beta_0 H_n^{(-n+2)}(C)}{H_n^{(-n+1)}(C)}$$

and hence $H_n^{(-n+2)}(C) \neq 0$ for $n = 1, 2, \ldots$. Therefore, by Lemma 7, each entry in the first row must be normal.

We were given

(1) $H_{2n}^{(-2n+1)}(C) \neq 0$,

(2) $H_{2n+1}^{(-(2n+1))}(C) \neq 0$,

(3) $H_{2n+1}^{(-2n)}(C) > 0$,

(4) $H_{2n}^{(-2n)}(C) > 0$

for $n = 0, 1, 2, \ldots$. Using Jacobi's identity we obtain

$$\left[H_{2n}^{(-(2n-1))}(C)\right]^2 = H_{2n}^{(-2n)}(C)H_{2n}^{(-(2n-2))}(C) - H_{2n+1}^{(-2n)}(C)H_{2n-1}^{(-(2n-2))}(C)$$

and since $\left[H_{2n}^{(-(2n-1))}(C)\right]^2 > 0$ by (1), $H_{2n}^{(-2n)}(C) > 0$ by (4), $H_{2n+1}^{(-2n)}(C) > 0$ by (3) and $H_{2n-1}^{(-(2n-1))}(C) > 0$ by (3) we see that

(5) $H_{2n}^{(-(2n-2))}(C) > 0$ for $n = 1, 2, \ldots$.

All we can conclude about $H_{2n+1}^{(-((2n+1)-2))}(C) = H_{2n+1}^{(-(2n-1))}(C)$ is that it is nonzero since the entries in the first row are normal. Thus, we have

(6) $H_{2n+1}^{(-(2n-1))}(C) \neq 0$.

In the 0th row we have $\Delta_0 = H_n^{(-n)}(C)$ and $\Delta_0\beta_n = (-1)^n\beta_0 H_n^{(-(n-1))}$. We know that $\Delta_0 \neq 0$ by (2) and (4). In the even case, $H_{2n}^{(-2n+1)}(C) \neq 0$ by (1) so applying Lemma 7 we see that $M_{2n,0}(z)$ is either normal or on the left edge of a square block for $n = 1, 2, \ldots$. Similarly, $H_{2n+1}^{(-2n)}(C) \neq 0$ by (3) and hence $M_{2n+1,0}(z)$ is either normal or on the left edge of a square block for $n = 0, 1, 2, \ldots$. These two observations allow us to conclude that all of the entries in the 0th row are normal.

In the second row we have $\Delta_2 = H_n^{(-n+2)}(C)$ and $\Delta_2\beta_n = (-1)^n\beta_0 H_n^{(-n+3)}(C)$. By (5) and (6) we see that $\Delta_2 \neq 0$ in both the even and odd cases. Using Jacobi's identity we obtain

$$\left[H_{2n+1}^{(-(2n-1))}(C)\right]^2 = H_{2n+1}^{(-2n)}(C)H_{2n+1}^{(-(2n-2))}(C) - H_{2n+2}^{(-2n)}(C)H_{2n}^{(-(2n-2)}(C)$$

and since $\left[H_{2n+1}^{(-(2n-1))}(C)\right]^2 > 0$ by (6), $H_{2n+1}^{(-2n)}(C) > 0$ by (3), $H_{2n+2}^{(-2n)}(C) > 0$ by (5) and $H_{2n}^{(-(2n-2))}(C) > 0$ by (5) we see that $H_{2n+1}^{(-(2n-2))}(C) > 0$. Thus, $M_{2n+1,2}(z)$, $n = 0, 1, 2, \ldots$, is either a normal entry or on the left edge of a square block. The information about $H_{2n}^{(-(2n-2))}(C)$ is inconclusive so we do not get any information on the even entries, $M_{2n,2}(z)$, $n = 1, 2, \ldots$. However, with the information from the odd entries and the fact that the entries in the first row are normal we can conclude that there are at most 2×2 blocks extending downward from the second row and any 2×2 block must contain $M_{2n+1,2}(z)$ and $M_{2n+2,2}(z)$ for some $n = 0, 1, 2, \ldots$.

The argument for the $(-1)^{st}$ row is completely analogous to that above so we omit it. ∎

In conclusion, if a bi-infinite sequence of real numbers $C = \{c_n\}_{n=-\infty}^{\infty}$ satisfies [4.1], then the SHMP has a solution. There is also an APT-fraction corresponding to C in the sense of [1.7] which gives information about the uniqueness of the solution if the fraction converges completely. Regardless of whether or not the APT-fraction converges, it corresponds to a pair of fLs [4.3]. The approximants of the fraction appear as entries in the first row of the M-table for [4.3]. The 0th and 1st rows of the M-table for [4.3] are normal and the 2nd and $(-1)^{st}$ rows contain at most 2 × 2 blocks extending downward or upward, respectively.

Acknowledgements. The author would like to thank W. J. Thron for suggesting the problem and Arne Magnus for the many helpful discussions in conjunction with this material.

REFERENCES

1. Baker, Jr., George A. and Peter Graves-Morris, Padé Approximants, Part I, Encyclopedia of Mathematics and Its Applications, No. 13, Addison-Wesley Publ. Co., Reading, MA (1981).

2. Cooper, S. Clement, Arne Magnus and J. H. McCabe, "On the non-normal two-point Padé table", submitted.

3. Jones, William B., Olav Njåstad and W. J. Thron, "Continued Fractions and Strong Hamburger Moment Problems", Proc. London Math. Soc. (3), 47 (1983), 363-384.

4. Jones, William B. and W. J. Thron, "Survey of Continued Fraction Methods of Solving Moment Problems and Related Topics", Analytic Theory of Continued Fractions, Proc., Loen, Norway, 1981, Lecture Notes in Mathematics 932 (Springer, Berlin, 1982), 4-37.

5. Jones, William B. and W. J. Thron, "Orthogonal Laurent Polynomials and the Strong Hamburger Moment Problem", J. Math. Anal. Appl. (2), 98 (1984), 528-554.

6. Jones, William B., W. J. Thron and H. Waadeland, "A strong Stieltjes moment problem", Trans. Amer. Math. Soc. (1980), 503-528.

7. McCabe, J. H., "A Formal Extension of the Padé Table to include two-point Padé quotients", J. Inst. Math. Appls. 15 (1975), 362-372.

8. Sri Ranga, A., "$\hat{J}$-fractions and Strong Moment Problems", these proceedings.

A STRATEGY FOR NUMERICAL COMPUTATION OF LIMITS REGIONS

Roy M. Istad and Haakon Waadeland
The University of Trondheim
Department of Mathematics and Statistics
N-7055 Dragvoll
Norway

1. The problem

In computing values of convergent continued fractions the situation is often the following: We want to compute the values of several continued fractions of the form

$$(1.1) \qquad \overset{\infty}{\underset{n=1}{K}} \left(\frac{a_n}{1}\right) = \frac{a_1}{1} + \frac{a_2}{1} + \cdots + \frac{a_n}{1} + \cdots \quad , \quad a_n \in \mathbb{C},$$

all of which are members of a certain family of continued fractions, given by the condition that all a_n are contained in a certain given set E, which is called an element region. We shall here assume all the continued fractions in question to convergence, i.e. that E is a convergence region, and in addition that all possible values are finite. We would like to know, in advance, something about how fast the continued fraction converges, in order to decide which n-values will suffice for the nth approximant

$$(1.2) \qquad f_n = \overset{\infty}{\underset{\nu=1}{K}} \left(\frac{a_\nu}{1}\right) = \frac{a_1}{1} + \frac{a_2}{1} + \cdots + \frac{a_n}{1}$$

to be of a desired accuracy as an approximation to the value f of the continued fraction.

Very helpful information related to this question is to know a set W, such that all possible values of (1.1) with all $a_n \in E$, are in W. How this is used to estimate the a priori truncation error (actually the modified a priori truncation error) is described in e.g. [4], where also some historical remarks and an extensive list of references are given.

The "naive" approach would be to choose a finite, large subset $E(N)$ of E with N elements, and for some fixed large n-value compute all f_n, where $a_n \in E(N)$. For $N = 10$ and $n = 100$ this would give 10^{100} values (points). Each value is computed in 100 steps by the backward recurrence algorithm. Even an enormous (and unrealistic) amount of computation of this kind, would normally at best lead to good guesses for the set of all values.

Another approach is to use the derivatives of a continued fraction with respect to the elements. This will enable us to find a linear expression which approximates the value of the continued fraction. This can be used to find an approximation to the set of values. This method, including an error estimate, is described and illustrated in some specific examples in [6]. It is primarily useful for small element regions.

A simpler approach is to choose W first and find an E such that W is the set of values of all continued fractions (1.1) with $a_n \in E$ for all n, see for example [2], [3, Thm. 4.3].

In the paper [4] the authors explored the possibility of finding a sparse set of continued fractions with elements in E which in a way may lead to a complete determination of the set of values. In the same paper examples are given in which such a sparse set of continued fractions determines a set, dense in the set of values, or determines the boundary of the set of values. In the particular cases studied in [4], periodic continued fractions, with period 2 and with elements from the boundary of E, play a crucial role.

The topic of the present article originates from one of the examples in [4]. There the element regions E were disks with center on the positive real axis and with a radius "not too large". The sparse set was the set of 2-periodic continued fractions with elements from the circle (boundary of E). Numerical computation indicated that for a given radius the closed convex hull of the values of these 2-periodic continued fractions is a good approximation to the set of values of all continued fractions $K(\frac{a_n}{1})$ with $a_n \in E$. The approximation seemed to be better for larger values of the center of E. As a matter of fact it turned out to be "asymptotically exact". The precise meaning of this expression will be explained later.

The present paper extends these observations in three ways:

a) The center is no longer required to be on the real axis.

b) A firm mathematical basis for the strategy is established.

c) The strategy from [4] is replaced by a simpler and cheaper one.

Instead of considering the class of all 2-periodic continued fractions with elements from the boundary in c) it suffices to use those continued fractions whose elements are opposite points of the same diagonal. This was first observed and proved in [1].

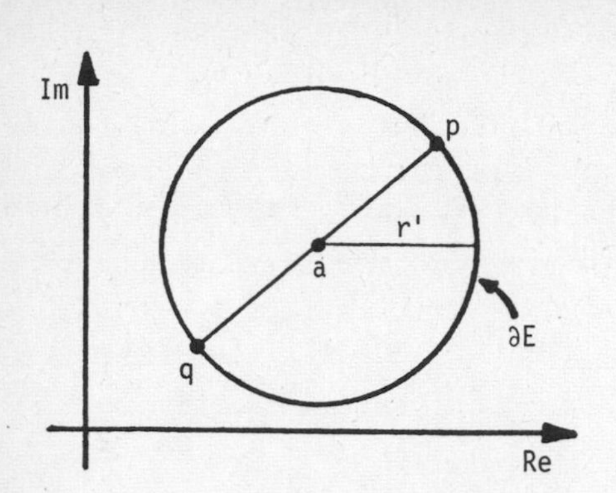

2. What we know

The element regions shall, in the following, be disks:

(2.1) $$|z-a| \leqq r',$$

and the strategy will be to use the continued fractions

(2.2) $$\frac{p}{1} + \frac{q}{1} + \frac{p}{1} + \frac{q}{1} + \cdots,$$

where

(2.3) $$p = a + r'e^{i\theta}, \quad q = a - r'e^{i\theta}.$$

We will assume that the disk (2.1) is contained in some parabolic region defined by

(2.4) $$P_\alpha : |z| - \mathrm{Re}(ze^{-i2\alpha}) \leqq \frac{1}{2}\cos^2\alpha \quad ; \alpha \in \langle -\frac{\pi}{2}, \frac{\pi}{2} \rangle,$$

in which case (2.1) is known to be a uniform convergence region [3, Thm. 4.40], [5].

We furthermore define, for complex c with $|c| < |1+c|$

(2.5) $$E_* : \left|z - c(1+c)\left(1 - \frac{\rho^2}{|1+c|^2}\right)\right| \leqq \rho(|1+c| - |c|)\left(1 - \frac{\rho}{|1+c|}\right)$$

(2.6) $$E^* : \left|z - c(1+c)\left(1 - \frac{\rho^2}{|1+c|^2}\right)\right| \leqq \rho(|1+c| - |c|)\left(1 + \frac{\rho}{|1+c|}\right)$$

(2.7) $$L : |w-c| \leqq \rho \quad ; \quad \rho < |c|$$

The following result from [2] will be used here, reformulated for our purpose.

Theorem 1. Let c be a complex number with $|c| < |c+1|$, and let E_*, E^* and L be the disks given by (2.5), 2.6) and (2.7). Take

$$2\alpha = \arg c(1+c)$$

in P_α. Then the following holds if $E_* \subseteq P_\alpha$:

If all $a_n \in E_*$, then $K(\frac{a_n}{1}) \in L$.

Remarks:

(i) The conclusion does not hold if E_* is replaced by E^*.

(ii) Let for a fixed ρ $c \to \infty$ in such a way that $Re(c) \to +\infty$ and $Im[c(1+c)]$ is kept fixed (or tends to a finite limit). Then from a certain $|c|$ on E_* (and E^*) will be contained in the parabolic region (2.4) with $\alpha = 0$, i.e. in

$$P_0 : |z| - Re(z) \leqq \frac{1}{2}.$$

It is easy to see, that when $c \to \infty$ as described, then $|1+c| - |c| \to 1$, and hence the radii of E_* and E^* both tend to ρ. The common center of E_* and E^* is

$$a = c(1+c) - \frac{c}{1+\bar{c}}\rho^2 \quad ; \text{ where } \quad \frac{c}{1+\bar{c}} \to 1.$$

With the notations from (2.1) for E_* we find

(2.8) $$\begin{cases} a - c(1+c) \to -\rho^2 \\ r' \to \rho \end{cases}$$

The disks E_* and E^* may be described by

$$|z-[c(1+c)-\rho^2+\varepsilon]| \leqq \rho+\varepsilon_* ,$$
$$|z-[c(1+c)-\rho^2+\varepsilon]| \leqq \rho+\varepsilon^* ,$$

where

$$\varepsilon = \rho^2\left(1-\frac{c}{1+\bar{c}}\right) ,$$
$$\varepsilon_* = \rho\left(|1+c|-|c|-1+(|\tfrac{c}{1+c}|-1)\rho\right) ,$$
$$\varepsilon^* = \rho\left(|1+c|-|c|-1+(1-|\tfrac{c}{1+c}|)\rho\right) .$$

All three tend to 0 when $c\to\infty$ as described.

3. A test of the strategy

We have an element region (2.1) contained in the parabolic region P_0. If there exists a complex number c and a positive ρ, such that

$$(3.1)\qquad \begin{aligned} c(1+c)\left(1-\frac{\rho^2}{|1+c|^2}\right) &= a \\ \rho(|1+c|-|c|)\left(1-\frac{\rho}{|1+c|}\right) &= r' , \end{aligned}$$

we know that all $\overset{\infty}{\underset{n=1}{K}}\left(\frac{a_n}{1}\right)$; $a_n \in E$, given by (2.1), are in the disk (2.7).

For

$$(3.2)\qquad \begin{aligned} p &= a+r'e^{i\theta} \\ q &= a-r'e^{i\theta} \end{aligned}$$

we have

$$f = \frac{p}{1}+\frac{q}{1}+\frac{p}{1}+\frac{q}{1}+\cdots$$
$$= \frac{1}{2}\left[\sqrt{(1+p+q)^2-4pq}-1-q+p\right]$$
$$(3.3)\qquad = \frac{1}{2}\left[\sqrt{1+4a+4r'^2e^{i2\theta}}-1+2r'e^{i\theta}\right]$$

where the branch of $\sqrt{\ }$ is chosen such that

$$\sqrt{1+4a+4r'^2e^{i2\theta}} = 1+2c \quad \text{for} \quad r'=0 \ (\rho=0).$$

We find that

$$f-c = \frac{1}{2}\left[\sqrt{1+4a+4r'^2e^{i2\theta}} - (1+2c-2r'e^{i\theta})\right]$$

$$= \frac{4[a-c(1+c)] + 4r'(1+2c)e^{i\theta}}{2\left[\sqrt{1+4a+4r'^2e^{i2\theta}} + (1+2c-2r'e^{i\theta})\right]} = \frac{N}{D}.$$

$$N = 4(1+2c)\left[\frac{-1}{1+2c}\frac{c}{1+\bar{c}}\rho^2 + r'e^{i\theta}\right]$$

$$= 4(1+2c)[\rho e^{i\theta} + \eta_1(c)],$$

where $\eta_1(c) \to 0$ when $c \to \infty$ such that $\mathrm{Re}(c) \to \infty$ and $\mathrm{Im}[c(1+c)]$ has a finite limit.

$$D = 2(1+2c)\left\{\left[1+\frac{4}{(1+2c)^2}\left(r'^2e^{i2\theta} - \frac{c}{1+\bar{c}}\rho^2\right)\right]^{\frac{1}{2}} + 1 - \frac{2r'e^{i\theta}}{1+2c}\right\}$$

$$= 4(1+2c)[1+\eta_2(c)],$$

where $\eta_2(c) \to 0$ when $c \to \infty$ as above. From this follows that

(3.4) $\lim\limits_{c\to\infty}(f-c) = \rho e^{i\theta}$
as desribed

and thus

(3.5) $\lim\limits_{c\to\infty}|f-c| = \rho$
as described

Hence: For a fixed ρ and fixed $\mathrm{Im}[c(1+c)]$ there is to every $\varepsilon > 0$ an M, such that for $\mathrm{Re}(c) > M$ the set of points (3.3) is a closed curve between the circles

(3.6) $\quad |w-c| \leqq \rho \pm \varepsilon$

This is the precise meaning of the phrase "asymptotically exact" used about the strategy desribed. An illustration of this for three different locations of the center of (2.1), is given on page 44 .

Observation

If, instead of letting $a \to \infty$ essentially along a parallel to the positive real axis, we had followed a ray $\arg(a) = 2\alpha$, we would no longer have had "asymptotically exact" strategy, but still an interesting result. In this case we would have had

$$|1+c| - |c| \to \cos\alpha$$

and

$$\frac{c}{1+\bar{c}} \to e^{i2\alpha}$$

and in (3.4), (3.5) and (3.6) ρ would have been replaced by $\rho\cos\alpha$. For instance: (3.6) would have been replaced by

(3.7) $\quad |w-c| \leqq \rho\cos\alpha \pm \varepsilon$

An illustration of this with $\alpha = \frac{\pi}{3}$ and hence $\cos\alpha = \frac{1}{2}$, is given on page 45 for three locations of the center.
The illustrations may also be found in [1].

In the following illustrations the dots represent the closed curves (given by the strategy mentioned on page 39 , and fully described in [1]), and the drawn circles are the boundary of the (limit) regions (2.7).

On page 44 the element region (2.1) is centered on the positive real axis, with constant radius. We observe that by moving the element region to the right (increasing values of the center) the closed curve and the boundary of the (limit) region "coincide", (3.6).

On page 45 the element region is centered on the ray $\arg(a) = 2\alpha = \frac{2\pi}{3}$, still with constant radius. By increasing values of $|a|$ we observe that the closed curve "tends" to a circle with radius equal to the half of the radius of the (limit) region, (3.7).

On page 46 we have used both E_* and E^* (2.5), (2.6), as element region. We observe the same effect here as on page 44 , now with both inward and outward "coinciding" with the (limit) region.

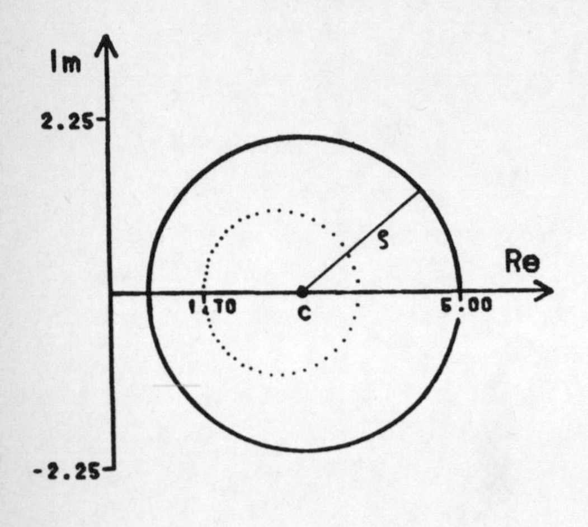

$E : |z-a| \leqq r'$

$a \in \mathbf{R}^+$; $r' = 1$

$L : |w-c| \leqq \rho$

$a = 9$

$c = 3$, $\rho = 2$

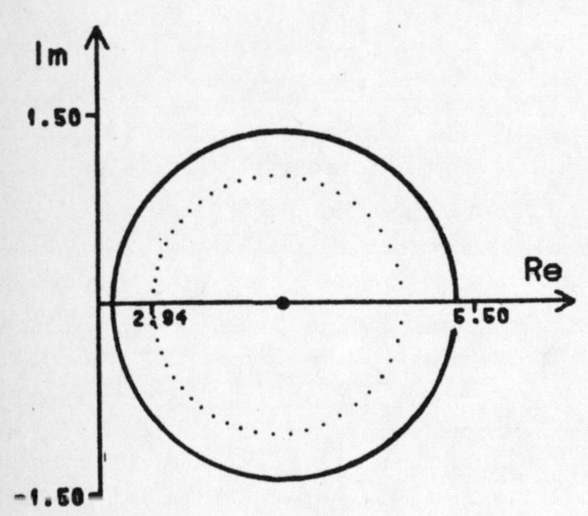

$a = 14 + 2\sqrt{5}$

$c = 4$, $\rho = \dfrac{5 - \sqrt{5}}{2}$

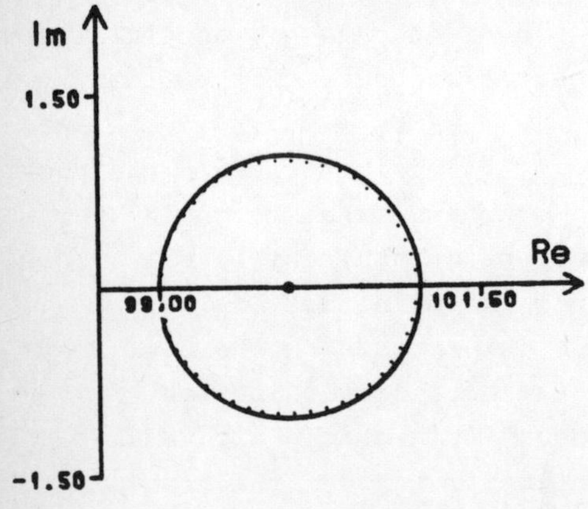

Im

1.50

Re

99.00 101.50

-1.50

$a = 10098.98...$

$c = 100$, $\rho = 1.01...$

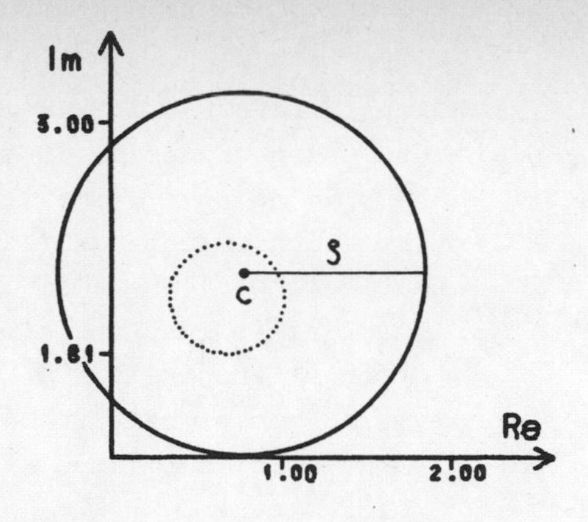

$E : |z-a| \leqq r'$

$\arg(a) = 2\alpha = \frac{2\pi}{3}$; $r' = \frac{1}{3}$

$L : |w-c| \leqq \rho$

a = (-2.60,4.50)

c = (0.77,2.10), $\rho = 1.09$

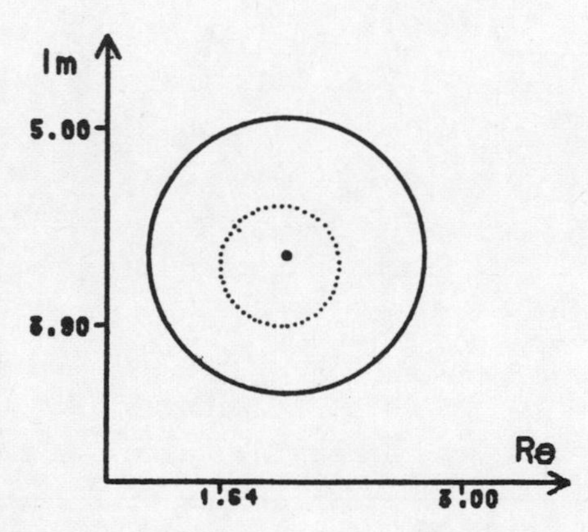

a = (-12.03,20.84)

c = (1.99,4.28), $\rho = 0.78$

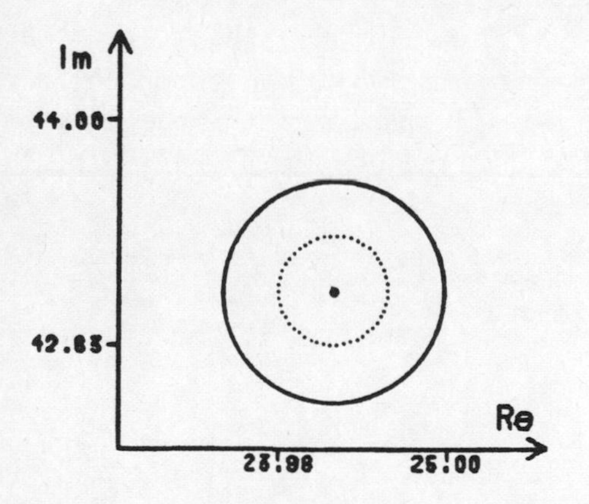

a = (-1230.82,2131.85)

c = (24.31,42.97), $\rho = 0.68$

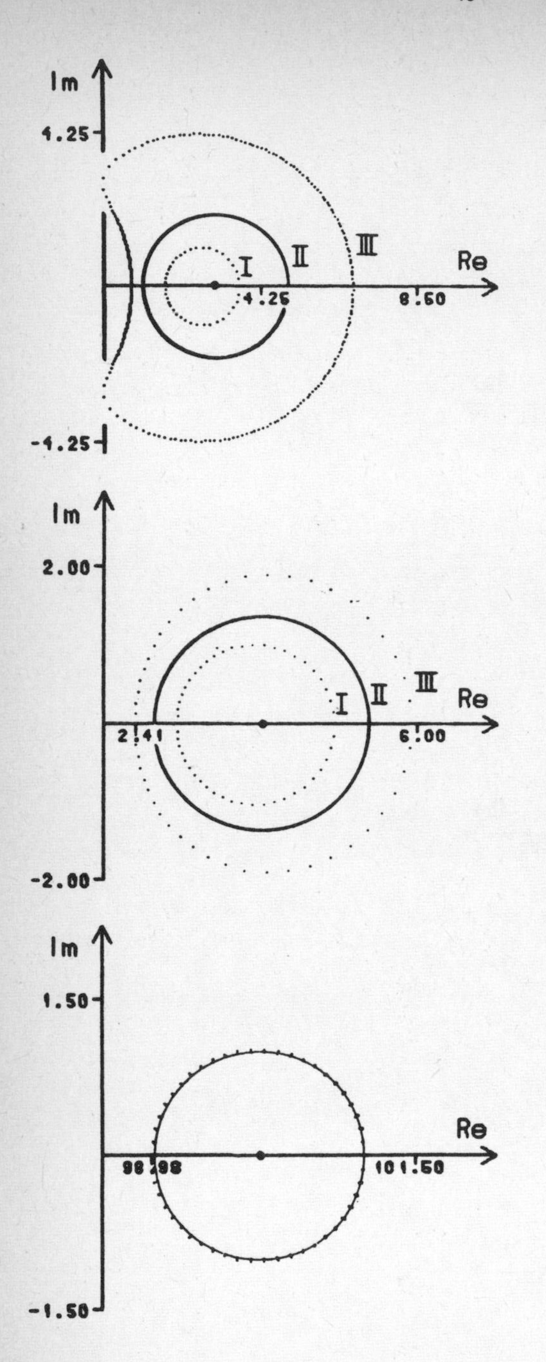

If we agree upon describing our strategy as using some kind of a "magician's hat", the illustrations may be explained as follows: Take a circle and put it into the hat (forget about all the mumbo-jumbo going on inside) and Voila! out jumps a closed curve.

In each of these three final illustrations we have taken both the circles ∂E_* and ∂E^* (on page 44 and 45 we only used ∂E_*) and "put them into the hat". The two sets of points coming out are shown as dotted curves. With increasing values of the common center of E_* and and E^* we observe (as on page 44) that the dotted curves may come arbitrarily close to the boundary of our (limit) region L.

I: ∂E_* as "input"
III: ∂E^* as "input"
II: $\partial L = \partial\{w \in \mathbb{C} \mid |w-c| \leqq \rho\}$

REFERENCES

1. Roy M. Istad, Om limitområder og numerisk bestemmelse av dem. Thesis for the degree cand.scient. at the University of Trondheim 1985. (Norwegian)

2. Lisa Jacobsen and W.J. Thron, Oval convergence regions and circular limit regions for continued fractions $K(\frac{a_n}{1})$. In preparation.

3. William B. Jones and W.J. Thron, Continued fractions: Analytic theory and applications. ENCYCLOPEDIA OF MATHEMATICS AND ITS APPLICATIONS, No. 11, Addison-Wesley, Reading, Mass. 1980.

4. Eigil Rye and Haakon Waadeland, Reflections on value regions, limit regions and truncation errors for continued fractions. Numerische Mathematik 47, 191-215 (1985).

5. W.J. Thron, On parabolic convergence regions for continued fractions, Math. Zeitschr. 69, 1958, 173-182.

6. Haakon Waadeland, Local properties of continued fractions. J. of Comp. a. Appl. Math. To appear.

ON THE CONVERGENCE OF LIMIT PERIODIC CONTINUED FRACTIONS $K(a_n/1)$, WHERE $a_n \to -1/4$. PART II

Lisa Jacobsen
Department of Mathematics and Statistics
University of Trondheim, AVH
N-7055 Dragvoll
Norway

Abstract. A previous divergence criterion by A. Magnus and the author is generalized. We show that $K(a_n/1)$ diverges (by oscillation) if $a_n = -1/4 - (c + \epsilon_n)/n(n+1)$, where $c > 1/16$ and $\epsilon_n \in \mathbb{R}$, $|\epsilon| = O(1/\log^\alpha n)$, $\alpha > 2$.

1. Introduction: A Continued Fraction

$$K\frac{a_n}{1} = \frac{a_1}{1} + \frac{a_2}{1} + \frac{a_3}{1} + \ldots = \cfrac{a_1}{1 + \cfrac{a_2}{1 + \cfrac{a_3}{1 + \ldots}}}, \quad 0 \neq a_n \in \mathbb{C}, \qquad [1.1]$$

is said to converge if its sequence of approximants

$$f_n = \frac{a_1}{1} + \frac{a_2}{1} + \ldots + \frac{a_n}{1}, \quad n = 1, 2, 3, \ldots \qquad [1.2]$$

converges in $\hat{\mathbb{C}} = \mathbb{C} \cup \{\infty\}$. The value of $K(a_n/1)$ is then $f = \lim f_n$.

It is of importance to establish convergence/divergence criteria for continued fractions $K(a_n/1)$. A particularly interesting class is formed by the limit periodic continued fractions; i.e., $\{a_n\}$ approaches a limit in $\hat{\mathbb{C}}$. Many interesting functions $f(z)$ have for instance useful continued fraction expansions of the form

$$K(\beta_n z/1), \text{ where } \beta_n \to \beta \in \hat{\mathbb{C}} \qquad [1.3]$$

(see [6, Sec. 6]). It is well known that if $\beta \neq \infty$, then [1.3] converges for all z such that $|\arg(\beta z + 1/4)| < \pi$, [7, Satz 2.43] (uniformly on compact subsets containing no poles of the limit function). On the interval $-\infty < \beta z \leq -1/4$ and for $\beta = \infty$ the continued fraction [1.3] may converge or diverge, depending on how $\beta_n z$ approaches βz [1, 4, 9].

In this paper we shall prove a divergence theorem for a family of continued fractions $K(a_n/1)$ where $a_n \to -1/4$. This may seem very special, since it only happens for the continued fractions [1.3] if $\beta \neq 0, \infty$, and then only at the point $z = -1/4\beta$, which for the examples in [6, Sec. 6] turns out to be a branch point for the limit function. However, let $K(\beta_n z/1)$ be a continued fraction with $\beta \to \infty$, and let $\{g_n(z)\}$ denote its sequence of approximants. If

$$1 + \beta_2 z \neq 0,\ 1 + (\beta_{2n-1} + \beta_{2n})z \neq 0 \text{ for } n = 2, 3, 4, \ldots \qquad [1.4]$$

then the even part of $K(\beta_n z/1)$; that is, a continued fraction with approximants $\{g_{2n}(z)\}_{n=1}^{\infty}$, can be written

$$\frac{\frac{\beta_1 z}{1 + \beta_2 z}}{1} - \frac{\frac{\beta_2\beta_3 z^2}{(1+\beta_2 z)(1+\beta_3 z+\beta_4 z)}}{1} - \frac{b_3(z)}{1} - \frac{b_4(z)}{1} - \ldots, \qquad [1.5]$$

where

$$b_n(z) = \frac{\beta_{2n-2}\,\beta_{2n-1} z^2}{(1 + \beta_{2n-3} z + \beta_{2n-2} z)(1 + \beta_{2n-1} z + \beta_{2n} z)}$$

$$= \frac{1}{\left[\frac{1}{\beta_{2n-2} z} + \frac{\beta_{2n-3}}{\beta_{2n-2}} + 1\right]\left[\frac{1}{\beta_{2n-1} z} + 1 + \frac{\beta_{2n}}{\beta_{2n-1}}\right]}, \quad n = 3,4,5,,,. \qquad [1.6]$$

(See for instance [7, p. 12 and p. 6].) If $\beta_n/\beta_{n-1} \to 1$ (which is often the case), then $b_n(z) \to 1/4$, independent of the value of z. That is, [1.5] is a continued fraction of the form $K(a_n/1)$ with $a_n \to -1/4$. Clearly, if [1.5] diverges, so also does $K(\beta_n z/1)$.

Similarly, if

$$1 + (\beta_{2n} + \beta_{2n+1})z \neq 0, \quad n = 1, 2, 3, \ldots, \qquad [1.7]$$

then the odd part of $K(\beta_n z/1)$; that is, a continued fraction with approximants $\{g_{2n+1}(z)\}_{n=1}^{\infty}$, can also be written as a limit periodic continued fraction $K(a_n/1)$ with $a_n \to -1/4$ (if $\beta_n/\beta_{n-1} \to 1$). (For further information, see [5].)

2. Some Basic Notation.

In accordance with [6], for instance, we introduce the notation

$$S_n(\omega) = \frac{a_1}{1} + \frac{a_2}{1} + \ldots + \frac{a_n}{1 + \omega}, \quad n \in \mathbb{N}. \qquad [2.1]$$

It is easy to show that S_n is a (nonsingular) linear fractional transformation. With this notation the approximants [1.2] can be written $f_n = S_n(0)$.

Clearly, $K(a_n/1)$ converges if and only if its <u>Nth tail</u>

$$\overset{\infty}{\underset{n=1}{K}} \frac{a_{N+n}}{1} = \frac{a_{N+1}}{1} + \frac{a_{N+2}}{1} + \frac{a_{N+3}}{1} + \ldots, \quad N \in \mathbb{N} \cup \{0\}, \qquad [2.2]$$

converges in $\hat{\mathbb{C}}$. Hence, it suffices to prove convergence/divergence of a tail of $K(a_n/1)$. We shall let $f^{(N)}$ denote the value of [2.2] if $K(a_n/1)$ converges. Then

$$f = f^{(0)} = S_N(f^{(N)}), \; f^{(N-1)} = a_N/(1 + f^{(N)}), \; N=1,2,3,\ldots . \qquad [2.3]$$

We say that $\{f^{(N)}\}$ is the sequence of <u>right tails</u> of $K(a_n/1)$, [11]. Similarly, every sequence $\{g^{(n)}\}_{n=0}^{\infty}$ of extended complex numbers satisfying the same recursion [2.3], is called a sequence of <u>right or wrong tails</u> of $K(a_n/1)$, regardless of whether $K(a_n/1)$ converges or not. They are called a sequence of <u>wrong tails</u> of $K(a_n/1)$ if they are not the sequence of right tails of $K(a_n/1)$ or if $K(a_n/1)$ diverges [11].

3. <u>Main Results</u>.

Let $K(a_n/1)$ be a continued fraction, where $a_n \to -1/4$ along a ray. Then it is well known (it follows for instance from [9]) that $K(a_n/1)$ always converges, except possibly if this ray is the negative real axis. It follows from one of Pringsheim's convergence criteria [8], [7 Satz 2.22, 2], that $K(a_n/1)$ converges if

$$|a_n + 1/4| \le 1/4(4n^2 - 1), \; n = 1, 2, 3, \ldots, \qquad [3.1]$$

[10]. Moreover, by [4] it follows that $K(a_n/1)$ diverges if

$$a_n = -1/4-c/(n+p)(n+p+1), \; n=1,2,3,\ldots, \; c>1/16, \; p\in\mathbb{R}\backslash\mathbb{Z}^-. \qquad [3.2]$$

In view of this, it seems reasonable to believe that $K(a_n/1)$ diverges if

$$a_n = -1/4 - (c+\epsilon_n)/(n+p)(n+p+1), \; n = 1, 2, 3, \ldots, \qquad [3.3]$$

where $c > 1/16$, $p \in \mathbb{R}\backslash\mathbb{Z}^-$, $\epsilon_n \in \mathbb{R}$ and $\epsilon_n \to 0$ fast enough. We shall prove that this is indeed so. But first we shall prove the following convergence/divergence criterion for real continued fractions.

THEOREM 3.1. Given the continued fraction $K(a_n/1)$, where

$$a_n = g^{(n-1)}(1 + g^{(n)}) \in \mathbb{R}, \quad n = 1, 2, 3, \ldots \qquad [3.4]$$

and

$$g^{(0)} \in \mathbb{C}\setminus\mathbb{R}, \ \frac{\mathrm{Im}\, g^{(n)}}{\mathrm{Re}\, g^{(n)}} > 0 \text{ for all } n, \ \frac{\mathrm{Im}\, g^{(n)}}{\mathrm{Re}\, g^{(n)}} \to 0. \qquad [3.5]$$

Then $K(a_n/1)$ converges if and only if

$$\sum_{n=1}^{\infty} \frac{\mathrm{Im}\, g^{(n)}}{\mathrm{Re}\, g^{(n)}} < \infty. \qquad [3.6]$$

Remarks 3.2.

(i) The conditions [3.4]-[3.5] are equivalent to requiring that $K(a_n/1)$ is a real continued fraction with a sequence $\{g^{(n)}\}$ of wrong tails satisfying [3.5]. ($\{g^{(n)}\}$ can not be a sequence of right tails since $K(a_n/1)$ is real whereas $g^{(0)} \notin \mathbb{R}$. Moreover, $g^{(n)} \neq 0, \infty$ since $g^{(n)} = -1 + a_n/g^{(n-1)}$ where $a_n \neq 0$ and similarly $\mathrm{Im}\, g^{(n)} \neq 0$.)

(ii) We could just as well require $(\mathrm{Im}\, g^{(n)})/(\mathrm{Re}\, g^{(n)}) < 0$ for all n, since the complex conjugates $\{\bar{g}^{(n)}\}$ are also a sequence of wrong tails for $K(a_n/1)$.

(iii) As mentioned in Section 2, it suffices that a tail of $K(a_n/1)$ satisfies the conditions.

Proof: Since $\{g^{(n)}\}$ and $\{\bar{g}^{(n)}\}$ are two distinct sequences of finite (wrong) tails for $K(a_n/1)$, it follows from [2] that

$$S_n(0) = g^{(0)} \frac{1 - \prod_{j=1}^{n} g^{(j)}/\bar{g}^{(j)}}{1 - \prod_{j=0}^{n} g^{(j)}/\bar{g}^{(j)}} = g^{(0)} \frac{1 - \exp(2i \sum_{j=1}^{n} \phi_j)}{1 - \exp(2i \sum_{j=0}^{n} \phi_j)}, \qquad [3.7]$$

where

$$\phi_j = \arg g^{(j)} = \mathrm{Arctan}\, \frac{\mathrm{Im}\, g^{(j)}}{\mathrm{Re}\, g^{(j)}} \in (0, \pi/2]. \qquad [3.8]$$

Since $\tan(\omega/2) < \omega < \tan\omega$ for $\omega \in (0,\pi/3)$ and $(\mathrm{Im}\, g^{(j)})/(\mathrm{Re}\, g^{(j)}) \to 0$, it follows that

$$\frac{\mathrm{Im}\, g^{(j)}}{2\mathrm{Re}\, g^{(j)}} < \phi_j < \frac{\mathrm{Im}\, g^{(j)}}{\mathrm{Re}\, g^{(j)}} \text{ for sufficiently large } j. \qquad [3.9]$$

It follows immediately that if [3.6] holds, then $\sum_{j=1}^{\infty}\phi_j$ and $\sum_{j=0}^{\infty}\phi_j$ converge to distinct values (modulo π), and therefore $\{S_n(0)\}$ converges.

Assume that the series in [3.6] diverge to ∞. We shall first prove that then $\{S_n(0)\}$ is no Cauchy sequence. Let $\epsilon > 0$ be arbitrarily chosen. For arbitrary $m, n \in \mathbb{N}$ we have by use of [3.7]:

$$|S_n(0)-S_{n+m}(0)| = \frac{c}{|1-\exp(2i\sum_{j=0}^{n}\phi_j)|}\left|\frac{1-\exp(2i\sum_{j=n+1}^{n+m}\phi_j)}{1-\exp(2i\sum_{j=0}^{n+m}\phi_j)}\right|, \qquad [3.10]$$

where

$$c = |g^{(0)}||e^{2i\phi_0} - 1| \neq 0 \qquad [3.11]$$

by [3.5]. Since $\phi_j \to 0$ and $\sum\phi_j \to \infty$, we can find some arbitrarily large $N \in \mathbb{N}$ such that

$$|1 - \exp(2i\sum_{j=0}^{N}\phi_j)| < \min\{1, c/2\epsilon\}. \qquad [3.12]$$

Let N be so chosen. Then we can always find some $m \in \mathbb{N}$ arbitrarily large such that

$$|1 - \exp(2i\sum_{j=N+1}^{N+m}\phi_j)|/|1 - \exp(2i\sum_{j=0}^{N+m}\phi_j)| > \tfrac{1}{2}. \qquad [3.13]$$

This means that

$$|S_N(0) - S_{N+m}(0)| > 2\epsilon\cdot\tfrac{1}{2} = \epsilon \qquad [3.14]$$

for some arbitrarily large N and m. That is, $\{S_n(0)\}$ is no Cauchy sequence. Hence, $\{S_n(0)\}$ either oscillates or converges to ∞.

If $S_n(0) \to \infty$, then the 1. tail $K(a_{1+n}/1)$ of $K(a_n/1)$ converges to $f^{(1)} = S_1^{-1}(\infty) = -1$. This leads to a contradiction since $\{g^{(1+n)}\}_{n=0}^{\infty}$ is a sequence of wrong tails for $K(a_{1+n}/1)$, and we can repeat the argument. Hence, $\{S_n(0)\}$ oscillates if [3.6] does not hold.

□

Unfortunately the conditions in Theorem 3.1 are on the wrong tails $\{g^{(n)}\}$ rather than on $\{a_n\}$. However, it is easy to show that the continued fractions $K(a_n/1)$ with a_n given by [3.2] satisfy the conditions since they have sequences

$$g^{(n)} = -\frac{1}{2} - \frac{1/4 + i\sqrt{c - 1/16}}{n + p + 1}, \quad n = 0, 1, 2, \ldots \qquad [3.15]$$

of wrong tails. In this case [3.6] does not hold. Hence, this represents a new and easier proof of the divergence of these continued fractions (see [4]).

More generally, the following theorem from [3] (here in a simplified version, both more special and slightly more general) can be helpful to obtain sufficient conditions on $\{a_n\}$ for $K(a_n/1)$ to have a sequence of right or wrong tails with certain properties.

THEOREM 3.3. Given the continued fraction $K(\tilde{a}_n/1)$ with a sequence $\{\tilde{g}^{(n)}\}$ of finite right or wrong tails. Let $\{t_n\}_{n=0}^{\infty}$, $t_n > 0$, and $\{\tilde{D}_n\}_{n=0}^{\infty}$, $\tilde{D}_n > 0$, be chosen such that

$$D_n = t_n|1+\tilde{g}^{(n)}| - t_{n-1}|\tilde{g}^{(n-1)}| \geq \tilde{D}_n \geq \tilde{D}_{n+1}, \; n=1,2,3,\ldots . \qquad [3.16]$$

Moreover, let

$$R_{n-1} = \frac{\tilde{D}_{n-1}(1 - \mu)}{2t_{n-1}}, \; r_n = R_{n-1} \cdot \frac{2D_n - \tilde{D}_n(1 - \mu)}{2t_n}, \; n=1,2,3,\ldots, \qquad [3.17]$$

where $-1 \leq \mu \leq 1$ is constant. Then the following hold.

A. $E_n = \{z \in \mathbb{C};\ |z - \tilde{a}_n| \leq r_n\}$, $n = 1, 2, 3, \ldots$
and
$V_n = \{z \in \mathbb{C};\ |z - \tilde{g}^{(n)}| \leq R_n\}$, $n = 0, 1, 2, \ldots$
are corresponding element and pre-value regions.

B. If

$$R_n \prod_{j=1}^{n} \frac{|\tilde{g}^{(j-1)}| + R_{j-1}}{|1 + \tilde{g}^{(j)}| - R_j} \to 0 \text{ as } n \to \infty, \qquad [3.18]$$

then $\{E_n\}$ is a (uniform) sequence of modified convergence regions with respect to $\{V_n\}$.

EXPLANATION. Part A means that $E_n/(1 + V_n) \subseteq V_{n-1}$. Part B means that if $a_n \in E_n$ and $\omega_n \in V_n$ for all n, then $S_n(\omega_n) \to F$, where $F \in V_0$ is independent of $\{\omega_n\}$. The important thing in this connection is that we then also have

$$S_n^{(N)}(\omega_{N+n}) = \frac{a_{N+1}}{1} + \frac{a_{N+2}}{1} + \ldots + \frac{a_{N+n}}{1 + \omega_{N+n}} \to F^{(N)} \text{ as } n \to \infty \qquad [3.19]$$

for all $N \in \mathbb{N} \cup \{0\}$, where $\{F^{(N)}\}$ is a sequence of right or wrong tails of $K(a_n/1)$, $F^{(N)} \in V_N$.

The idea is now to use $K(\tilde{a}_n/1)$ with $\tilde{a}_n$ given by [3.2] and wrong tails $\{\tilde{g}^{(n)}\}$ given by [3.15], as the auxiliary continued fraction in Theorem 3.3.

LEMMA 3.4. Let

$$t_n = (n + 1)q_n, \; n = 0, 1, 2, \ldots, \; 0 < q_n < q_{n+1}, \qquad [3.20]$$

and

$$\tilde{g}^{(n)} = -\frac{1}{2} - \frac{1 + i\gamma}{4(n + 1)}, \; n = 0, 1, 2, \ldots, \; \gamma > 0. \qquad [3.21]$$

Then

$$D_n = t_n|1+\tilde{g}^{(n)}|-t_{n-1}|\tilde{g}^{(n-1)}| = \frac{|2n+1+i\gamma|}{4}(q_n-q_{n-1}) > 0. \qquad [3.22]$$

If in addition $\sum_{n=1}^{\infty} 1/t_n < \infty$, $\{\tilde{D}_n\}$ is a sequence such that

$$D_n \geq \tilde{D}_n \geq \tilde{D}_{n+1} > 0, \; n = 1, 2, 3, \ldots, \; \tilde{D}_0 \geq \tilde{D}_1, \qquad [3.23]$$

and R_n is given by [3.17] then [3.18] holds.

Proof: [3.22] follows by straightforward verification. Let $\{\tilde{D}_n\}_{n=0}^{\infty}$ satisfy [3.23]. (Such sequences can always be found.) Then

$$\begin{aligned} R_n \prod_{j=1}^{n} \frac{|\tilde{g}^{(j-1)}| + R_{j-1}}{|1 + \tilde{g}^{(j)}| - R_j} &= \frac{\tilde{D}_n(1 - \mu)}{2(n+1)q_n} \cdot \prod_{j=1}^{n} \frac{j+1}{j} \cdot \frac{|2j+1+i\gamma| + 2\tilde{D}_{j-1}(1-\mu)/q_{j-1}}{|2j+1-i\gamma|-2\tilde{D}_j(1-\mu)/q_j} \qquad [3.24] \\ &\leq \frac{\tilde{D}_1(1-\mu)}{2q_n} \prod_{j=1}^{n} \left(1 + \frac{4\tilde{D}_0(1-\mu)/q_{j-1}}{|2j+1+i\gamma| - 2\tilde{D}_1(1-\mu)/q_j}\right), \end{aligned}$$

where the product converges to a finite value as $n \to \infty$ and $\tilde{D}_1(1-\mu)/2q_n \to 0$. □

COROLLARY 3.5. With the notation from Lemma 3.4, we have: If $\sum 1/t_n < \infty$ and

$$a_n \in \mathbb{C}\setminus\{0\}, \; |a_n - \tilde{a}_n| \le \tilde{D}_{n-1} \frac{|2n+1+i\gamma|}{16n(n+1)} \left[\frac{1}{q_{n-1}} - \frac{1}{q_n}\right] \qquad [3.25]$$

for $n = 1, 2, 3, \ldots$, then $K(a_n/1)$ has a sequence $\{g^{(n)}\}$ of right or wrong tails such that

$$|g^{(n)} - \tilde{g}^{(n)}| \le \frac{\tilde{D}_n}{2(n+1)q_n}, \; n = 0, 1, 2, \ldots . \qquad [3.26]$$

Proof: Using $K(\tilde{a}_n/1)$, $\{\tilde{g}^{(n)}\}$, $\{t_n\}$, $\{D_n\}$ and $\{\tilde{D}_n\}$ in Theorem 3.3 with $\mu = 0$, gives the implication by means of Lemma 3.4. □

We can now prove the following main result.

THEOREM 3.6. The continued fraction $K(a_n/1)$, where

$$a_n = -\frac{1}{4} - \frac{c + \epsilon_n}{n(n+1)}, \qquad c > 1/16, \qquad \epsilon_n \in \mathbb{R}, \qquad [3.27]$$

diverges if there exists a sequence $\{d_n\}$, $d_n > 0$ such that $|\epsilon_n| \le d_n$ for all n, where

$$\sum_{k=1}^{\infty} d_k/k < \infty, \qquad \sum_{n=1}^{\infty} \left[\frac{1}{n} \sum_{k=n}^{\infty} \frac{d_k}{k}\right] < \infty \qquad [3.28]$$

and

$$\liminf_{n\to\infty} d_n/(\sum_{k=n}^{\infty} d_k/k)^2 > 16. \qquad [3.29]$$

Proof: By Theorem 3.1 it suffices to prove that $K(a_n/1)$ has a sequence of finite wrong tails $\{g^{(n)}\}$ satisfying [3.5] from some n on, such that [3.6] does not hold. With the notation from Lemma 3.4 we have, with

$$\tilde{a}_n = \tilde{g}^{(n-1)}(1 + \tilde{g}^{(n)}), \qquad \gamma = \sqrt{16c - 1},$$

that

$$|a_n - \tilde{a}_n| = |\epsilon_n|/n(n+1), \; n = 1, 2, 3, \ldots .$$

Assume that $|\epsilon_n| \le d_n$ for all n, where $\{d_n\}$, $d_n > 0$ satisfies [3.28]-[3.29]. Let

$$q_n = 1/16 \sum_{k=n+1}^{\infty} d_k/|2k + 1 + i\gamma|, \qquad n = 0, 1, 2, \ldots .$$

Then D_n given by [3.22] becomes

$$D_n = \frac{|2n+1+i\gamma|}{4} \cdot \frac{d_n/|2n+1+i\gamma|}{16(\sum_{k=n+1}^{\infty} d_k/|2k+1+i\gamma|)(\sum_{k=n}^{\infty} d_k/|2k+1+i\gamma|)}$$

$$> \frac{d_n}{64(\sum_{k=n}^{\infty} d_k/|2k+1+i\gamma|)^2} \geq 1 \quad \text{from some } n \text{ on.}$$

Without loss of generality we assume that $D_n \geq 1$ for all n. (If this is not so, we can study a tail $\underset{n=1}{\overset{\infty}{K}} (a_{N+n}/1)$ of $K(a_n/1)$ for sufficiently large N.) This means that we can choose $\tilde{D}_n = 1$ for all n. Since

$$\frac{|\epsilon_n|}{n(n+1)} \leq \frac{d_n}{n(n+1)} = \frac{|2n+1+i\gamma|}{16n(n+1)}\left[\frac{1}{q_{n-1}} - \frac{1}{q_n}\right] \quad \text{for all } n$$

and $\sum_{n=1}^{\infty} 1/(n+1)q_n < \infty$ by [3.28], it then follows by Corollary 3.5 that $K(a_n/1)$ has a sequence $\{g^{(n)}\}$ of right or wrong tails such that

$$|g^{(n)}-\tilde{g}^{(n)}| \leq \frac{1}{2(n+1)q_n} = \frac{8}{n+1}\sum_{k=n+1}^{\infty} \frac{d_k}{|2k+1+i\gamma|} \quad \text{for all } n.$$

Moreover, there exists an $M \in \mathbb{N}$ such that $g^{(n)}$ satisfies [3.5] for $n \geq M$, because

$$\text{Re } g^{(n)} \leq \text{Re } \tilde{g}^{(n)} + \frac{8}{n+1}\sum_{k=n+1}^{\infty} \frac{d_k}{|2k+1+i\gamma|} \to -\frac{1}{2} \text{ as } n \to \infty$$

and

$$\text{Im } g^{(n)} \leq \text{Im } \tilde{g}^{(n)} + \frac{8}{n+1}\sum_{k=n+1}^{\infty} \frac{d_k}{|2k+1+i\gamma|}$$

$$= -\frac{1}{n+1}\left(\frac{\gamma}{4} - 8\sum_{k=n+1}^{\infty} \frac{d_k}{|2k+1+i\gamma|}\right) < 0 \quad \text{from some } n \text{ on.}$$

That $\{g^{(n)}\}$ does not satisfy [3.6] can be seen by the following:

$$\sum_{n=M+1}^{\infty} \frac{\operatorname{Im} g^{(n)}}{\operatorname{Re} g^{(n)}} \geq \sum_{n=M+1}^{\infty} \frac{\operatorname{Im} \tilde{g}^{(n)} + \frac{8}{n+1} \sum_{k=n+1}^{\infty} d_k/|2k+1+i\gamma|}{\operatorname{Re} \tilde{g}^{(n)} - \frac{8}{n+1} \sum_{k=n+1}^{\infty} d_k/|2k+1+i\gamma|}$$

$$= \sum_{n=M+1}^{\infty} \frac{\gamma - \mu_n}{4(n+1)(\frac{1}{2}+\nu_n)}, \text{ where } \mu_n \to 0,\ \nu_n \to 0.$$

□

The conditions on ϵ_n in Theorem 3.6 are rather complicated. For applications the following special case may be helpful.

<u>THEOREM 3.7</u>. The continued fraction $K(a_n/1)$, where

$$a_n = -\frac{1}{4} - \frac{c+\epsilon_n}{n(n+1)} \in \mathbb{R}, \quad c > 1/16, \quad \epsilon_n = O(1/\log^\alpha n) \qquad [3.30]$$

for $\alpha > 2$, diverges.

<u>Proof</u>: It suffices to prove that the conditions [3.28]-[3.29] hold if

$$d_n = M/\log^\alpha n \quad \text{for } n \geq 4,\ \alpha > 2$$

for some $M > 0$. This can be seen since then

$$\sum_{k=n}^{\infty} \frac{d_k}{k} < \int_{n-1}^{\infty} \frac{M\,dx}{x \log^\alpha x} = M\int_{\log(n-1)}^{\infty} \frac{du}{u^\alpha} = \frac{M}{(\alpha-1)\log^{\alpha-1}(n-1)}$$

and

$$\sum_{n=4}^{\infty} \left[\frac{1}{n} \sum_{k=n}^{\infty} \frac{d_k}{k}\right] < \frac{M}{\alpha-1} \sum_{n=4}^{\infty} \frac{1}{(n-1)\log^{\alpha-1}(n-1)} < \frac{M}{(\alpha-1)(\alpha-2)\log^{\alpha-2} 2}$$

and finally

$$\frac{d_n}{\left[\sum_{k=n}^{\infty} d_k/k\right]^2} > \frac{M/\log^\alpha n}{M^2/((\alpha-1)\log^{\alpha-1}(n-1))^2} > \frac{(\alpha-1)^2}{M} \log^{\alpha-2}(n-1) \to \infty.$$

□

<u>COROLLARY 3.8</u>. The continued fraction $K(a_n/1)$, where

$$a_n = -\frac{1}{4} - \frac{c+\epsilon_n}{(n+p)(n+q)} \in \mathbb{R},\ c > 1/16,\ \epsilon_n = O(1/\log^\alpha n) \qquad [3.32]$$

for $\alpha > 2$ and $p, q, \in \mathbb{R}\backslash\mathbb{Z}^-$ diverges.

Proof: The elements a_n can be written

$$a_n = -\frac{1}{4} - \frac{c + \epsilon_n}{n(n+1)} \cdot \frac{n}{n+p} \cdot \frac{n+1}{n+q}$$

$$= -\frac{1}{4} - \frac{c + \epsilon_n}{n(n+1)}\left[1 - \frac{p}{n+p} - \frac{q-1}{n+q} + \frac{p(q-1)}{(n+p)(n+q)}\right]$$

$$= -\frac{1}{4} - \frac{c + \epsilon'_n}{n(n+1)}$$

where $|\epsilon'_n| \leq |\epsilon_n| + K/n \leq M/\log^\alpha n + K/n \leq M_1/\log^\alpha n$. □

REFERENCES

1. Gill, J., Infinite Compositions of Möbius Transformations, Trans. Amer. Math. Soc. 176 (1973), 479-487.
2. Jacobsen, L., Composition of Linear Fractional Transformations in Terms of Tail Sequences, to appear in Proc. Amer. Math. Soc.
3. Jacobsen, L., Nearness of Continued Fractions, to appear in Math Scand.
4. Jacobsen, L. and Magnus, A., On the Convergence of Limit Periodic Continued Fractions $K(a_n/1)$, where $a_n \to -1/4$, Lecture Notes in Math. 1105 (1984), 243-248.
5. Jacobsen, L. and Waadeland, H., Even and Odd Parts of Limit Periodic Continued Fractions, to appear in J. CAM.
6. Jones, W. B. and Thron, W. J., Continued Fractions: Analytic Theory and Applications, *Encyclopedia of Math. and Its Appl.*, Addison-Wesley, 1980.
7. Perron, O., Die Lehre von den Ketterbrüchen, Band II, B. G. Teubner, 1957.
8. Pringsheim, I., Über die Konvergenz unendlicher Kettenbrüche, S.-B. Bayer. Akad. Wiss. Math. - Nat. Kl. 28 (1899), 295-324.
9. Thron, W. J., On Parabolic Convergence Regions for Continued Fractions, Math. Zeitschr. 69 (1958), 173-182.
10. Thron, W. J. and Waadeland, H., Accelerating Convergence of Limit Periodic Continued Fractions $K(a_n/1)$, Numer. Math. 34 (1980), 155-170.
11. Waadeland, H., Tales about Tails, Proc. Amer. Math. Soc. 90 (1984), 57-64.

A THEOREM ON SIMPLE CONVERGENCE REGIONS FOR CONTINUED FRACTIONS $K(a_n/1)$

Lisa Jacobsen
Department of Mathematics and Statistics
The University of Trondheim AVH
N-7055 Dragvoll, Norway

1. Introduction. Convergence criteria for continued fractions

$$(1.1)\quad K\left(\frac{a_n}{1}\right) = \frac{a_1}{1} + \frac{a_2}{1} + \frac{a_3}{1} + \cdots = \cfrac{a_1}{1 + \cfrac{a_2}{1 + \cfrac{a_3}{1} + \cdots}}\;, \quad a_n \in ¢ \sim \{0\}$$

are often stated in terms of simple convergence regions $E \subseteq ¢$, $E \sim \{0\} \neq \emptyset$, [, p. 78]; that is

$$(1.2)\quad a_n \in E \sim \{0\} \text{ for all } n \Rightarrow K(a_n/1) \text{ converges.}$$

A useful tool to obtain such convergence regions is the concept of value regions V belonging to E, introduced by Scott and Wall [8]; that is

$$(1.3)\quad V \subseteq \hat{¢},\ V \neq \emptyset \text{ and } E \subseteq V,\ E/(1+V) \subseteq V, \text{ [5, p. 67].}$$

The term region is here used loosely to mean any subset of the complex plane. In applications though, E and V usually are regions in the ordinary, strict sense; i.e., open connected sets, possibly including parts of their boundaries.

A direct consequence of (1.3) is that if $K(a_n/1)$ is a continued fraction from E, i.e., all $a_n \in E$, then its approximants

$$(1.4)\quad f_n = \frac{a_1}{1} + \frac{a_2}{1} + \cdots + \frac{a_n}{1}\;, \quad n \in N$$

satisfy $f_n \in V$. This means that if $K(a_n/1)$ converges to f (i.e., $\lim f_n = f$), then $f \in \overline{V}$ (the closure of V in $\hat{¢}$). We say that $L \subseteq \hat{¢}$ is a limit region for E if $E/(1+L) \subseteq L$ and L contains the values (limits) of every convergent continued fraction from E, [2,6].

In 1965 Hillam and Thron proved that if V is a circular disk containing the origin in its interior, then E, the corresponding element region, is a simple convergence region, [1]. In 1977 Roach proved that $-1 \notin \overline{V+V}$ and V bounded also are sufficient conditions for E to be a simple convergence region, [7].

In this paper we shall present a set of sufficient conditions based on *pre-value* regions W for E; that is

$$(1.5) \qquad W \subseteq \hat{\mathbb{C}}, \; W \neq \emptyset \; \text{ and } \; E/(1+W) \subseteq W,$$

a concept which was introduced by Thron [9] in 1943 for another purpose. Please observe that a pre-value region W for E is a value region for E only if E W, in many cases a severe restriction. Such pre-value regions can be regarded as value regions for *modified approximants*, $w \in W$

$$(1.6) \qquad S_n(w) = \frac{a_1}{1} + \frac{a_2}{1} + \cdots + \frac{a_n}{1+w}, \quad n \in \mathbb{N}$$

for continued fractions $K(a_n/1)$. If $w \in W$, then $S_n(w) \in W$.

We shall prove the following results.

Theorem 1.1. *Let* W *be a pre-value region for the set* $E \subseteq \hat{\mathbb{C}}$. *Let* $\overline{W}$ *contain at least two points*. *Then* W *is a limit region for* E.

Theorem 1.2. *Let* E *be an open connected subset of* $\mathbb{C}$, *and let* W *be a bounded pre-value region for* E. *Then every finitely connected*, *open subset* G *of* E *with* $\overline{G} \subseteq E$ *is such that* $\overline{G}$ *is a convergence region*.

The proofs of these results are found in Section 3. In Section 4 we extend these results to twin regions. Section 2 contains some basic definitions.

2. *Some basic concepts*. Working with modified approximants $S_n(w)$, it is inconvenient to be dependent on a convergence concept based on classical approximants $f_n = S_n(0)$. We say that $K(a_n/1)$ *converges generally* to a value $f \in \hat{\mathbb{C}}$ if there exist two sequences $\{v_n\}$ and $\{w_n\}$ from $\hat{\mathbb{C}}$ such that

$$(2.1) \quad \lim S_n(v_n) = \lim S_n(w_n) = f, \quad \liminf d(v_n, w_n) > 0, \; [4].$$

Here $d(x,y)$ denotes the usual chordal metric on the Riemann sphere. General convergence has the following properties, [4].

(i) The limit f is unique.

(ii) If $K(a_n/1)$ converges to f in the usual sense, then $K(a_n/1)$ converges generally to f. In particular this means that if W is a general limit region for E, then it is also a limit region for E.

(iii) $K(a_n/1)$ converges (generally) to f if and only if its *kth*

<u>tail</u>

(2.2) $$\overset{\infty}{\underset{n=1}{K}} \left(\frac{a_{k+n}}{1}\right) = \frac{a_{k+1}}{1} + \frac{a_{k+2}}{1} + \frac{a_{k+3}}{1} + \cdots, \quad k \in \mathbb{N} \cup \{0\}$$

converges (generally). Let $f^{(k)}$ denote the value of (2.2). Then $\{f^{(k)}\}$ is the <u>sequence of right (g-right) tails</u> for $K(a_n/1)$. It satisfies the recursion relations

(2.3) $$Y_{k-1} = a_k/(1+Y_k), \quad k = 1,2,3,\ldots .$$

A sequence $\{g^{(k)}\}$ satisfying (2.3) is called a sequence of wrong (g-wrong) tails for $K(a_n/1)$ if $K(a_n/1)$ diverges (generally) or $g^{(0)} \neq f$, [10].

(iv) If $K(a_n/1)$ converges generally to f and $\{g^{(n)}\}$ is a sequence of g-wrong tails for $K(a_n/1)$ with $g^{(0)} \neq f$, then

(2.4) $$\liminf d(w_n, g^{(n)}) > 0 \Rightarrow \lim S_n(w_n) = f.$$

(v) If $K(a_n/1)$ converges generally to f and $\{g^{(n)}\}$ $\{p^{(n)}\}$ are two sequences of g-wrong tails for $K(a_n/1)$ with $g^{(0)}$, $p^{(0)} \neq f$, then

$$\lim d(g^{(n)}, p^{(n)}) = 0 .$$

A set $X \subseteq \hat{\mathbb{C}}$, $X \neq \emptyset$ is a <u>region for right or wrong tails</u> for $E \subseteq \mathbb{C}$ if

(2.5) $$-1 + E/X \subseteq X, \text{ [3]}.$$

It has the property that if an element $g^{(N)}$ of a sequence $\{g^{(n)}\}$ of right or wrong tails for a continued fraction from E is contained in X, then $g^{(n)} \in X$ for all $n \geq N$. It is straight forward to prove that if W is a pre-value region for E and $X = \hat{\mathbb{C}} - W \neq \emptyset$, then X is a region for right or wrong tails for E.

3. <u>Proofs</u>. Proof of Theorem 1.1. Let $K(a_n/1)$ be a convergent continued fraction from E with value $f \in \hat{\mathbb{C}}$. (If no such continued fraction exists, then the result holds trivially. Then $K(a_n/1)$ converges generally to f. Let $g^{(0)} \neq f$. Then $g^{(n)} = S_n^{-1}(g^{(0)})$ gives a sequence of g-wrong tails. Since W contains at least two points, we can always find a sequence $\{w_n\}$, $w_n \in W$ such that

$$\liminf_{n\to\infty} d(w_n, g^{(n)}) > 0 .$$

By (2.4) follows then that $S_n(w_n) \to f$. Since $w_n \in W$ for all n, we have $S_n(w_n) \in W$ for all n, and hence $f = \lim S_n(w_n) \in \overline{W}$. □

Indeed, we have proved a stronger result than Theorem 1.1. We have proved that $\overline{W}$ is a general limit region for E; that is that $\overline{W}$ contains the values of all generally convergent continued fractions from E.

To prove Theorem 1.2 we shall use some lemmas.

Lemma 3.1. *Let* $E \subseteq \mathbb{C}$, $E^0 \neq \emptyset$ *be given. Then there exist circular disks* $D \subseteq E^0$ E *and* L *such that* D *is a convergence region,* L *is a limit region for* D, *and* $\operatorname{diam} S_n(L) \to 0$ *for every continued fraction from* D. E^0 *denotes the interior of* E.

Proof. Since $E^0 \neq \emptyset$ is open, it follows that $E^0 \sim (-\infty, -1/4] \neq \emptyset$. Let $a \in E^0 \sim (-\infty, -1/4]$. From the convergence neighborhood theorem [5, Theorem 4.45] follows that there exists a $\rho > 0$ such that the disk $B(a; \rho)$ is a convergence region. Since E^0 is open, this ρ can be chosen small enough to ensure that $D = B(a; \rho) \subseteq E^0$. From [2, Result 9] follows the existence of L for ρ sufficiently small. □

Lemma 3.2. *Let* W *be a pre-value region for* E. *Then* $X = -1 - W$ *is a region for right or wrong tails for* E.

Proof. Let $a \in E$ be chosen arbitrarily. Then

$$-1 + \frac{a}{X} = -1 + \frac{a}{-1-W} = -\left(1 + \frac{a}{1+W}\right) \subseteq -(1+W) = X\ . \ \square$$

Proof of Theorem 1.2. Let G be an N-connected open set such that $\overline{G} \subseteq E$. Let D be a convergence disk in G with limit region L such that $\operatorname{diam} S_n(L) \to 0$ for every continued fraction from D. Then we can find N simply connected, open subsets G_n; $n = 1,2,\dots,N$ such that

$$D \subseteq G_n \quad \text{for} \quad n = 1,2,\dots,N \quad \text{and} \quad \overline{G} = \bigcup_{n=1}^{N} \overline{G}_n\ ,$$

and we can find N simply connected, open subsets $E_n \subseteq E$ such that $\overline{G}_n \subseteq E_n$ for $n = 1,2,\dots,N$. Let $\alpha_n\colon E_n \to U = \{z;\ |z| < 1\}$ be a Riemann mapping function for E_n with $\alpha_n(C_D) = 0$, where C_D is the center of D, for $n = 1,2,..,N$. Then there exists an $\varepsilon > 0$ such that

$$\alpha_n(\overline{G}_n) \subseteq \overline{U}_{1-\varepsilon} = \{z;\ |z| \leq 1 - \varepsilon\} \quad \text{for} \quad n = 1,2,\dots,N\ .$$

Let $K(a_n/1)$ be a continued fraction from $\overline{G}$ and let $w_1, w_2 \in W \cap L$, $w_1 \neq w_2$. (Two such points can always be found since by Theorem 1.1 W is a limit region for E, and since we can easily find two continued fractions from D converging to distinct values.) We shall first prove that $\lim S_n(w_1) = \lim S_n(w_2)$; that is, that $K(a_n/1)$ converges generally.

We can always find a sequence $\{k_n\}_{n=1}^{\infty}$ from $\{1,2,\ldots,N\}$ such that $a_n \in \overline{G}_{k_n}$ for all n. Let us define $\tilde{\alpha}_n: E_{k_n} \to \mathbb{C}$ for all $n \in \mathbb{N}$ by

$$\tilde{\alpha}_n(z) = \frac{1-\varepsilon}{\alpha_{k_n}(a_n)} \alpha_{k_n}(z) \quad \text{if} \quad a_n \neq C_D ,$$

$$\tilde{\alpha}_n^{-1}(z) = a_n \quad \text{if} \quad a_n = C_D .$$

Then $\tilde{\alpha}^{-1}$ is an analytic function on U (or more precisely on a region containing U) for all n with

$$\tilde{\alpha}_n^{-1}(0) = C_D, \quad \tilde{\alpha}^{-1}(1-\varepsilon) = a_n .$$

Moreover there exists a $\delta > 0$ such that

$$\text{(3.1)} \qquad \tilde{\alpha}_n^{-1}(U_\delta) \subseteq D \quad \text{for all} \quad n .$$

because by Schwarz' lemma

$$|\tilde{\alpha}_n^{-1}(r\, e^{i\theta})| \leq r \cdot \operatorname{diam} E,$$

where $\operatorname{diam} E < \infty$ since W is bounded.

Since $\tilde{\alpha}_n^{-1}(u) \subseteq E$ for all n, it follows that the functions

$$F_{n,j}(z) = \frac{\tilde{a}_1^{-1}(z)}{1} + \frac{\tilde{a}_2^{-1}(z)}{1} + \cdots + \frac{\tilde{a}_n^{-1}(z)}{1+w_j}, \quad n \in \mathbb{N}, \quad j = 1,2$$

map U into W which is bounded. Hence $\{F_{n,j}\}_{n=1}^{\infty}$ is a sequence of uniformly bounded analytic functions on U which by (3.1) converges to a function $f(z)$ on U_δ, independent of j. By Stieltjes-Vitali's theorem then follows that also $\{F_{n,j}(1-\varepsilon)\}_{n=1}^{\infty}$ converges to a limit independent of j. That is

$$\lim_{n\to\infty} F_{n,j}(1-\varepsilon) = \lim_{n\to\infty} S_n(w_j) = f(1-\varepsilon) \quad \text{for} \quad j = 1,2.$$

That is, $K(a_n/1)$ converges generally.

To see that $K(a_n/1)$ also converges in the classical sense, we shall use Lemma 3.2 and (2.4). Since $f_n = S_n(0) = S_{n+1}(\infty)$ and ∞ has a positive chordal distance from $-1-w$, the result follows. □

4. *Twin regions*. The pair (E_1, E_2) is a set of *twin convergence regions* if all continued fractions from (E_1, E_2) converge; i.e., all continued fractions $K(a_n/1)$ with $a_{2n-1} \in E_1$ and $a_{2n} \in E_2$ for all n. (W_0, W_1) is a set of *pre-value* regions for (E_1, E_2) if

(4.1) $W_0, W_1 \subseteq \hat{\mathbb{C}}$, $W_0, W_1 \neq \emptyset$ and $E_1/(1+W_1) \subseteq W_0$, $E_2/(1+W_0) \subseteq W_1$.

Clearly, simple convergence regions are special cases of twin convergence regions. We shall see that Theorem 1.1 and Theorem 1.2 also hold for twin regions.

Theorem 4.1. *Let* $E_1, E_2 \subseteq \mathbb{C}$ *and* W_0, W_1 *satisfy* (4.1). *Let* W_0 *contain at least two points. Then* $(\overline{W}_0, \overline{W}_1)$ *is a set of general limit regions for the twin regions* (E_1, E_2).

By that we shall logically mean that every generally convergent continued fraction from (E_1, E_2) has a value in $\overline{W}_0$, and that every generally convergent continued fraction from (E_2, E_1) has a value in $\overline{W}_1$. The proof follows the proof of Theorem 1.1 with $w_{2n} \in W_0$, w_{2n+1} W_1. (Clearly W_1 must also contain at least two points if $E_1 \sim \{0\} \neq \emptyset$ and $E_2 \sim \{0\} \neq \emptyset$.

Theorem 4.2. *Let* E_1, E_2 *be open connected subsets of* $\mathbb{C}$, *and let* (W_0, W_1) *be a set of bounded pre-value regions for the twin regions* (E_1, E_2). *Then every twin set* $(\overline{G}_1, \overline{G}_2)$ *with* $\overline{G}_1 \subseteq E_1$, $\overline{G}_2 \subseteq E_2$ *where* G_1, G_2 *are open and finitely connected, is a set of twin convergence regions*.

The proof of this theorem is also a copy of the proof of Theorem 1.2. We only need to prove that Lemma 3.1 and 3.2 hold for twin regions, that is:

Lemma 4.3. *Let* $E_1, E_2 \subseteq \mathbb{C}$, $E_1^0, E_2^0 \neq \emptyset$ *be given. Then there exist circular disks* $D_1 \subseteq E_1^0$, $D_2 \subseteq E_2^0$ *and* L_0, L_1 *such that* (D_1, D_2) *is a set of twin convergence regions*, (L_0, L_1) *is a set of pre-value regions for* (D_1, D_2) *and*

$$\operatorname{diam} S_{2n}(L_0) \to 0 \ , \quad \operatorname{diam} S_{2n-1}(L_1) \to 0$$

for every continued fraction from (D_1, D_2).

Lemma 4.4. *Let* (W_0, W_1) *be pre-value regions for the twin set* (E_1, E_2). *Then* (X_0, X_1), *where*

$$X_0 = -1 - W_1, \quad X_1 = -1 - W_0$$

<u>is a set of regions for right or wrong tails for</u> (E_1, E_2).

By this we shall mean that

$$-1 + E_1/X_0 \subseteq X_1 \quad \text{and} \quad -1 + E_2/X_1 \subseteq X_0 ,$$

such that if $g^{(0)} \in X_0$, then $g^{(2n)} = S_{2n}^{-1}(g^{(0)}) \in X_0$ and $g^{(2n-1)} = S_{2n-1}^{-1}(g^{(0)}) \in X_1$ for all $n \geq 1$.

<u>Proof</u> of Lemma 4.3. A linear fractional transformation S_2 is elliptic or parabolic if and only if its trace satisfies $\operatorname{tr} S \in [0,4]$. Since $E_1^0, E_2^0 \neq \emptyset$ we can therefore always find $a_1 \in E_1^0$, $a_2 \in E_2^0$ such that the linear fractional transformations

$$S_2(w) = \frac{a_1}{1} + \frac{a_2}{1+w} , \quad S_2^{(1)}(w) = \frac{a_2}{1} + \frac{a_1}{1+w}$$

are loxodromic or hyperbolic with finite fixed points. The existence of D_1, D_2, L_0, L_1, then follows from [2, Result 9]. □

<u>Proof</u> of Lemma 4.4. We have

$$-1 + \frac{E_1}{X_0} = -1 + \frac{E_1}{-1-W_1} = -(1 + \frac{E_1}{1+W_1}) \subseteq -(1+W_0) = W_1 .$$

The computation of $-1 + E_2/X_1$ goes similarly. □

References

1. K. L. Hillam and W. J. Thron, "A general convergence criterion for continued fractions $K(a_n/b_n)$," <u>Proc. Amer. Math. Soc.</u> 16 (1965), 1256-1262.

2. L. Jacobsen, "Modified approximants for continued fractions. Construction and applications," <u>Det. Kgl. Norske Vid. Selsk. Skr.</u> (1983), No. 3, 1-46.

3. L. Jacobsen, "Regions for right or wrong tails of continued fractions," <u>Det Kgl. Norske Vid. Selsk. Skr.</u> (1985), No. 7, 1-3.

4. L. Jacobsen, "General convergence for continued fractions," <u>Trans. Amer. Math. Soc.</u>, to appear.

5. W. B. Jones and W. J. Thron, <u>Continued fractions</u>. <u>Analytic theory and Applications</u>, Encyclopedia of Math and its Appl., Addison-Wesley, Reading, Mass., (1980). Now available from Cambridge Univ. Press.

6. M. Overholt, "The values of continued fractions with complex elements," <u>Det. Kgl. Norske Vid. Selsk. Skr</u>. (1983), No. 1, 109-116.

7. F. A. Roach, "Boundedness of value regions and convergence of continued fractions," <u>Proc. Amer. Math. Soc</u>. 62 (1977), 299-304.

8. W. T. Scott and H. S. Wall, "Value regions for continued fractions," Bull. Amer. Math. Soc. 47 (1941), 580-585.

9. W. J. Thron, "Two families of twin convergence regions for continued fractions," Duke Math. J. 10 (1943), 677-685.

10. H. Waadeland, "Tales about tails," Proc. Amer. Math. Soc. 90 (1984), 57-64.

FURTHER RESULTS ON THE COMPUTATION OF INCOMPLETE GAMMA FUNCTIONS

Lisa Jacobsen
Dept. of Mathematics
and Statistics
University of Trondheim, AVH
N-7055 Dragvoll, NORWAY

William B. Jones*
Dept. of Mathematics
Campus Box 426
University of Colorado
Boulder, CO 80309-0426 U.S.A.

Haakon Waadeland
Dept. of Mathematics
and Statistics
University of Trondheim, AVH
N-7055 Dragvoll, NORWAY

1. Introduction. A continued fraction of the form

$$K(\alpha_n z/1), \quad z \in S_\pi := [z: |\arg z| < \pi], \quad \alpha_n > 0$$

is called a Stieltjes fraction (S-fraction). Thus every S-fraction can be expressed in the form

$$K(a_n/1), \quad \arg a_n = 2\alpha, \quad -\frac{\pi}{2} < \alpha < \frac{\pi}{2} .$$

In this paper we investigate the subclass $S(\mu, M, \alpha)$ of S-fractions $K(a_n/1)$ such that

$$\arg a_n = 2\alpha, \quad -\frac{\pi}{2} < \alpha < \frac{\pi}{2}, \quad n = 1,2,3,\ldots, \tag{1.1a}$$

$$\mu \leq |a_{n+1}| - |a_n| \leq M, \quad n = 1,2,3,\ldots, \tag{1.1b}$$

and

$$\lim_{n\to\infty} a_n = \infty . \tag{1.1c}$$

It follows from (1.1b) that

$$(n-1)\mu + |a_1| \leq |a_n| \leq (n-1)M + |a_1|, \quad n = 1,2,3,\ldots, \tag{1.2a}$$

and hence

$$\frac{1}{|a_n|} \geq \frac{1}{(n-1)M+|a_1|} \quad \text{and} \quad \sum_{n=1}^{\infty} \frac{1}{|a_n|} = \infty . \tag{1.2b}$$

One of the main reasons for our interest in the class $S(\mu, M, \alpha)$ is its relation to continued fraction expansion of the complementary incomplete gamma function

*W.B.J. was supported in part by the U.S. National Science Foundation under Grant DMS-8401717 and by grants from the U.S. Educational Foundation in Norway (Fulbright-Hays Grant), The Norwegian Marshall Fund and the University of Colorado Council on Research and Creative Work.

(1.3) $\Gamma(a,z) := \int_z^\infty e^{-t} t^{a-1} dt, \quad a \in \mathbb{C}, \quad z \in S_\pi := [z: |\arg z| < \pi]$

where the path of integration is $t = z + \tau, \ 0 \le \tau < \infty$. It is well known [6, p. 348, (9.6.17)] that

(1.4a)
$$\frac{e^z \Gamma(a,z)}{z^a} = \frac{z^{-1}}{1} + \frac{(1-a)z^{-1}}{1} + \frac{1 \cdot z^{-1}}{1} + \frac{(2-a)z^{-1}}{1} + \frac{2z^{-1}}{1} + \cdots$$
$$= \mathop{K}_{n=1}^{\infty} \left(\frac{\alpha_n z^{-1}}{1}\right), \quad z \in S_\pi, \quad a \notin [1,2,3,\ldots] ,$$

where

(1.4b) $\alpha_1 := 1, \quad \alpha_{2n} := n-a, \quad \alpha_{2n+1} := n, \quad n = 1,2,3,\ldots$.

(see also [7]). In the case where

(1.5) $0 \le a < 1,$

the continued fraction in (1.4) belongs to $S(\mu, M, \alpha)$ where

(1.6) $\mu := -a|z|^{-1}, \quad M := |z|^{-1}\max[a,(1-a)], \quad 2\alpha := \mathrm{Arg}\,(z^{-1})$.

Some important special cases of this include the <u>complementary error function</u>

(1.7) $\mathrm{erfc}\,(z) := \frac{2}{\sqrt{\pi}} \int_z^\infty e^{-t^2} dt = \frac{1}{\sqrt{\pi}} \Gamma(\frac{1}{2}, z^2), \quad a := \frac{1}{2}$

and the <u>exponential integral</u>

(1.8) $E_1(z) := \int_z^\infty e^{-t} t^{-1} dt = \Gamma(0,z), \quad a := 0$.

Thus we obtain from (1.4) that, for $\mathrm{Re}\, z > 0$,

(1.9a)
$$\frac{\sqrt{\pi}\, e^{z^2}}{z} \mathrm{erfc}\,(z) = \frac{z^{-2}}{1} + \frac{\frac{1}{2} z^{-2}}{1} + \frac{\frac{2}{2} z^{-2}}{1} + \frac{\frac{3}{2} z^{-2}}{1} + \frac{\frac{4}{2} z^{-2}}{1} + \cdots$$
$$= \mathop{K}_{n=1}^{\infty} \left(\frac{\alpha_n z^{-2}}{1}\right) \in S\left(\frac{-1}{2|z|^2}, \frac{1}{2|z|^2}, \frac{\mathrm{Arg}\, z^{-2}}{2}\right) ,$$

where

(1.9b) $\alpha_1 := 1, \quad \alpha_n := \frac{n-1}{2}, \quad n = 2,3,4,\ldots;$

for $z \in S_\pi$,

(1.10a)
$$e^z E_1(z) = \frac{z^{-1}}{1} + \frac{1 \cdot z^{-1}}{1} + \frac{1 \cdot z^{-1}}{1} + \frac{2z^{-1}}{1} + \frac{2z^{-1}}{1} + \cdots$$

$$= \underset{n=1}{\overset{\infty}{K}} \left(\frac{\alpha_n z^{-1}}{1}\right) \in S(0, |z|^{-1}, \frac{\text{Arg } z^{-1}}{2}) ,$$

where

(1.10b) $\alpha_1 := 1, \quad \alpha_{2n} := \alpha_{2n+1} := n, \quad n = 1,2,3,\ldots$.

In Section 2 we use results from the parabola theorem of Thron [8] to investigate convergence, truncaton error bounds and speed of convergence of S-fractions in $S(\mu, M, \alpha)$. A priori truncation error bounds are given in Theorem 2.3 for $S(\mu, M, \alpha)$ and the results are interpreted in Theorem 2.4 and Corollaries 2.5 and 2.6, respectively, for the complementary incomplete gamma functions, the complementary error funcion and the exponential integral. Section 3 is used to describe results on modified continued fractions $K(a_n,1;w_n)$ whose reference continued fraction $K(a_n/1) \in S(\mu, M, \alpha)$. Theorem 3.1 gives sufficient conditions to ensure that $K(a_n/1)$ and $K(a_n,1;w_n)$ converge to the same (finite) limit; also shown is an inequality (3.4b) that can be used to obtain truncation error bounds for the nth approximant g_n of $K(a_n,1;w_n)$. We discuss the choice of modifying factors w_n and, for one such choice, we given sequences of element and value regions (Theorem 3.2). As one application of Theorem 3.2 we state a result on convergence acceleration in Theorem 3.7. To illustrate the convergence behavior of $K(a_n/1)$ and $K(a_n,1;w_n)$ we give at the end of Section 3 some contour maps of the number of significant digits (SD) obtained by approximants of $K(a_n/1)$ and $K(a_n,1;w_n)$ for the complementary error funcion erfc (z). We note that other results of the type described in Section 2 could be obtained by the inclusion region method and results of Henrici and Pfluger [4]. The a priori truncation error estimates for Stieltjes fractions of Thron [9] give similar results to those given here in Theorem 2.3(A).

We conclude this introduction by summarizing some definitions and terminology that are used. For further details the reader can refer to [6], [1]. Associated with a continued fraction

(1.11) $$K\left(\frac{a_n}{1}\right) = \underset{n=1}{\overset{\infty}{K}} \left(\frac{a_n}{1}\right) = \frac{a_1}{1} + \frac{a_2}{1} + \frac{a_3}{1} + \cdots$$

are the linear fractional transformations

(1.12a) $s_n(w) := \frac{a_n}{1+w}, \quad n = 1,2,3,\ldots$

and

(1.12b) $\quad S_1(w) := s_1(w), \quad S_n(w) := S_{n-1}(s_n(w)), \quad n = 2,3,\ldots$.

It is readily seen that

$$(1.13) \qquad S_n(w) = \frac{a_1}{1} + \frac{a_2}{1} + \cdots + \frac{a_{n-1}}{1} + \frac{a_n}{1+w}, \quad n = 1,2,3,\ldots$$

and the nth approximant f_n of $K(a_n/1)$ is

$$(1.14) \qquad f_n := S_n(0) = \frac{a_1}{1} + \frac{a_2}{1} + \cdots + \frac{a_n}{1}, \quad n = 1,2,3,\ldots .$$

The continued fraction $K(a_n/1)$ is said to converge to $f \in \hat{\mathbb{C}}$ if $\lim f_n = f$. The mth tail of $K(a_n/1)$ is the continued fraction

$$(1.15) \qquad \mathop{K}_{n=m+1}^{\infty} \left(\frac{a_n}{1}\right) = \frac{a_{m+1}}{1} + \frac{a_{m+2}}{1} + \frac{a_{m+3}}{1} + \cdots, \quad m = 0,1,2,\ldots .$$

If we define

$$(1.16) \qquad S_n^{(m)}(w) = \frac{a_{m+1}}{1} + \frac{a_{m+2}}{1} + \cdots + \frac{a_{m+n-1}}{1} + \frac{a_{m+n}}{1+w}$$

then the nth approximant $f_n^{(m)}$ of the mth tail is

$$(1.17) \qquad f_n^{(m)} := S_n^{(m)}(0) = \frac{a_{m+1}}{1} + \frac{a_{m+2}}{1} + \cdots + \frac{a_{m+n}}{1} .$$

Hence

$$(1.18) \qquad f_{n+m} = S_{n+m}(0) = S_m(f_n^{(m)}).$$

It follows that $\lim_{n\to\infty} f_n$ exists iff $\lim_{n\to\infty} f_n^{(m)}$ exists for $m = 0,1,2,\ldots$. Therefore a continued fraction converges iff all of its tails converge, and when this happens we obtain

$$(1.19) \qquad f = S_m(f^{(m)}), \quad m = 0,1,2,\ldots \ (S_0(w) := w)$$

Given a continued fraction $K(a_n/1)$ and a sequence of modifying factors $W = \{w_n\}$, we obtain a <u>modified continued fraction</u> $K(a_n,1;w_n)$ (see [1] for further details). The nth approximant of $K(a_n,1;w_n)$ is defined by

$$(1.20) \qquad g_n := S_n(w_n) = \frac{a_1}{1} + \frac{a_2}{1} + \cdots + \frac{a_{n-1}}{1} + \frac{a_n}{1+w_n} .$$

The modified continued fraction is said to converge to a value $g \in \hat{\mathbb{C}}$ if $\{g_n\}$ converges to g. The continued fraction $K(a_n/1)$ is called the <u>reference continued fraction</u> for $K(a_n,1;w_n)$. We have $K(a_n/1) = K(a_n,1;0)$, so that a modified continued fraction can be thought of as a generalization of a continued fraction. We obtain the reference continued fraction when all modifying factors $w_n = 0$. One advantage of modified continued fractions is that if w_n is close to $f^{(n)}$, then (see (1.19)) we may have

$$(1.21) \qquad \left|\frac{f-g_n}{f-f_n}\right| = \left|\frac{S_n(f^{(n)}) - S_n(w_n)}{S_n(f^{(n)}) - S_n(0)}\right| = \delta_n$$

where $\delta_n \to 0$ as $n \to \infty$. When this occurs we say that the modified continued fraction accelerates the convergence of its reference continued fraction.

A sequence $E = \{E_n\}$ of subsets of $\mathbb{C}$ is called a <u>sequence of element regions for</u> $K(a_n,1;w_n)$ if

$$(1.22) \qquad 0 \neq a_n \in E_n, \quad n = 1,2,3,\ldots .$$

A sequence $V = \{V_n\}$ of subsets of $\hat{\mathbb{C}}$ is called a <u>sequence of value regions corresponding to a sequence of element regions</u> $E = \{E_n\}$ <u>at</u> $W = \{w_n\}$ if

$$(1.23) \qquad \frac{E_n}{1+w_n} \in V_{n-1} \quad \text{and} \quad \frac{E_n}{1+V_n} \subseteq V_{n-1}, \quad n = 1,2,3,\ldots .$$

We denote by $\mathcal{V}(E,W)$ the family of all sequences of value regions corresponding to E at W. The following are readily verified:

$$(1.24) \qquad \left[w_{n-1} \in V_{n-1} \text{ and } \frac{E_n}{1+V_n} \subseteq V_{n-1},\ n \geq 1\right] \Rightarrow \{V_n\} \in (E,W)$$

and

$$(1.25) \qquad \left[\{V_n^{(\alpha)}\} \in \mathcal{V}(E,W),\ \alpha \in A\right] \Rightarrow \left\{\bigcap_{\alpha \in A} V_n^{(\alpha)}\right\} \in (E,W) .$$

In case $w_n = 0$ for all $n \geq 1$ and (1.23) holds, we say that $V = \{V_n\}$ is a <u>sequence of value regions corresponding to</u> $E = \{E_n\}$.

2. <u>Convergence of Continued Fractions in</u> $S(\mu, M, \alpha)$.

Use is made here of a parabola theorem due to Thron [8]. For completeness we state this result as Theorem 2.1. Involved are regions $P_n(\alpha)$ and $H(\alpha)$ defined, for $n = 1,2,3,\ldots$ and

$$(2.1a) \qquad -\frac{\pi}{2} < \alpha < \frac{\pi}{2}, \quad 0 < g_n < 1, \quad 0 < k_n \leq 1,$$

by

(2.1b) $\quad P_n := [w: |w| - \text{Re}(w\, e^{-i2\alpha}) \le 2k_n g_n(1-g_{n+1})\cos^2\alpha]$

and

(2.1c) $\quad H_n := [v: \text{Re}(v\, e^{-i\alpha}) \ge -g_n \cos\alpha]$.

It is readily seen that the boundary ∂P_n of P_n is a parabola with focus at the origin $w = 0$, axis through the ray $\arg w = 2\alpha$ and passing through the point $w = 2k_n g_n(1-g_{n+1})\cos^2\alpha$. The set H_n is a half-plane with boundary ∂H_n perpendicular to the line passing through the ray $\arg v = \alpha$ such that ∂H_n passes through $v = -g_n$. We also define

$$(2.2) \qquad d_n := \frac{\prod_{j=1}^{n}\left(\frac{1}{g_{j+1}} - 1\right)}{\sum_{k=0}^{n-1}\prod_{j=1}^{k}\left(\frac{1}{g_{j+1}} - 1\right)}, \quad n = 1,2,3,\ldots .$$

Theorem 2.1. (Thron) *Let* $\{P_n\}$, $\{H_n\}$ *and* $\{d_n\}$ *be defined by* (2.1) *and* (2.2). *Let* $K(a_n/1)$ *be a continued fraction satisfying*

$$(2.3) \qquad 0 \ne a_n \in P_n , \quad n = 1,2,3,\ldots .$$

Let $\{S_n\}$ *be defined by* (1.12) *and let* f_n *denote the* n^{th} *approximant*. *Then*:

(A) $\{S_n(H_n)\}$ *is a nested sequence of closed circular disks and* (*since* $0 \in H_n$ *and* $f_n = S_n(0)$)

$$(2.4a) \qquad f_{n+m} = S_{n+m}(0) \in S_{n+m}(H_{n+m}) \subseteq S_n(H_n),$$

and

$$(2.4b) \qquad |f_{n+m} - f_n| \le 2R_n := 2 \text{ rad } S_n(H_n) ,$$

for $n = 1,2,3,\ldots,\ m = 0,1,2,\ldots$.

(B) *For* $n = 1,2,3,\ldots$.

$$(2.5) \qquad R_n \le \frac{|a_1|}{2(1-g_2)\cos\alpha \prod_{j=2}^{n}\left(1 + \frac{(1-k_j+d_{j-1})\cos^2\alpha\, g_j(1-g_{j+1})}{|a_j|}\right)}$$

By taking $g_j = 1/2$ for all j we obtain from (2.2) $d_n = 1/n$, $n \ge 1$. Next we make the restriction $0 < k_j = 1 - \varepsilon < 1$ for $j \ge 1$. The regions P_n and H_n of (2.16) then reduce, respectively, to

$$(2.6a) \qquad P(\alpha,\varepsilon) := [w: |w| - \text{Re}(w\, e^{-i2\alpha}) \le \tfrac{1}{2}(1-\varepsilon)\cos^2\alpha] ,$$

and

(2.6b) $\quad H(\alpha) := [v\colon \operatorname{Re}(v\, e^{-i\alpha}) \geq -\frac{1}{2}\cos\alpha]$.

The following result is an immediate consequence of Theorem 2.1.

Corollary 2.2. Let $P(\alpha,\varepsilon)$ and $H(\alpha)$ be defined by (2.6) where $0 < \varepsilon < 1$ and $-\frac{\pi}{2} < \alpha < \frac{\pi}{2}$. Let $K(a_n/1)$ be a continued fraction with n^{th} approximant f_n and with elements satisfying

(2.7) $\quad 0 \neq a_n \in P(\alpha,\varepsilon), \quad n = 1,2,3,\ldots$.

Let $\{S_n\}$ be defined by (1.12). Then:

(A) $\{S_n(H(\alpha)\}$ is a nested sequence of closed circular disks and, for $n = 1,2,3,\ldots$ and $m = 0,1,2,\ldots,$

(2.8a) $\quad f_{n+m} = S_{n+m}(0) \in S_{n+m}(H(\alpha)) \subseteq S_n(H(\alpha)),$

and

(2.8b)
$$|f_{n+m}-f_n| \leq \frac{2|a_1|}{(\cos\alpha)\prod_{j=2}^{n}\left[1+\frac{\varepsilon\cos^2\alpha}{4|a_j|}\right]} .$$

(B) If, in addition

(2.9)
$$\sum_{n=1}^{\infty} \frac{1}{|a_n|} = \infty$$

then $K(a_n/1)$ converges to a finite value $f \in S_n(H(\alpha)), n \geq 1$ and in (2.8b) f_{n+m} can be replaced by f thus giving a truncation error bound for f_n.

Proof. (A) follows directly from theorem 2.1. (B) follows from (A) and the fact that an infinite product $\Pi(1+b_j)$, $b_j > 0$, diverges to 0 iff $\Sigma\, b_j = \infty$.

In the special case in which $\operatorname{Arg} a_n = 2\alpha$, $n \geq 1$, we obtain

$$0 \neq a_n \in P(\alpha,\varepsilon) \quad \text{for all} \quad 0 < \varepsilon < 1 .$$

Therefore (2.8b) holds for all $0 < \varepsilon < 1$ and hence it holds in the limiting case $\varepsilon = 1$. We state this result in

Corollary 2.2'. Let $K(a_n/1)$ be a continued fraction such that

(2.10) $\quad \operatorname{Arg} a_n = 2\alpha \,, \quad -\frac{\pi}{2} < \alpha < \frac{\pi}{2} \,, \quad n = 1,2,3,\ldots$.

Then: (A) If f_n denotes the nth approximant, then, for $n = 1,2,3,$ $\ldots,$ $m = 0,1,2,\ldots,$

$$(2.11)\qquad |f_{n+m}-f_n| \le \frac{2|a_1|}{(\cos\alpha)\prod_{j=1}^{n}\left[1+\frac{\cos^2\alpha}{4|a_j|}\right]} .$$

(B) <u>If in addition</u>

$$(2.12)\qquad \sum_{n=1}^{\infty} \frac{1}{|a_n|} = \infty$$

<u>then</u> $K(a_n/1)$ <u>converges to a finite limit</u> f <u>and</u>, <u>in</u> (2.11), f_{n+m} <u>can be replaced by</u> f <u>to give a truncation error bound for</u> f_n.

We now apply Corollary 2.3 to the class $S(\mu, M, \alpha)$.

<u>Theorem 2.3.</u> <u>Let</u> $K(a_n/1)$ <u>belong to</u> $S(\mu, M, \alpha)$ <u>and let</u> f_n <u>denote its</u> n^{th} <u>approximant</u>. <u>Then</u>:

(A) $K(a_n/1)$ <u>converges to a finite value</u> f <u>and</u>

$$(2.13)\qquad |f-f_n| \le \frac{2|a_1|}{(\cos\alpha)\prod_{j=2}^{n}\left[1+\frac{\cos^2\alpha}{4|a_j|}\right]} , \quad n = 2,3,4,\ldots .$$

(B) <u>Let</u> m_0 <u>be chosen such that</u>

$$(2.14a)\qquad b_n := \frac{\cos^2\alpha}{4|a_n|} \le \frac{1}{2} , \quad \text{for } n \ge m_0 \ge 2 ,$$

<u>and let</u>

$$(2.14b)\qquad A(m_0) := \frac{2|a_1|}{(\cos\alpha)\prod_{j=2}^{m_0-1}\left[1+\frac{\cos^2\alpha}{4|a_j|}\right]} , \quad K := \frac{|a_1|+M}{2} .$$

<u>Then</u>

$$(2.14c)\qquad |f-f_n| \le A(m_0)\left(\frac{m_0}{n+1}\right)^{\frac{\cos^2\alpha}{8k}} , \quad n = m_0,\ m_0+1,\ m_0+2,\ \ldots .$$

<u>Proof</u>. (A) follows from Corollary 2.2 and (1.2b).

(B): First we note that

$$(2.15)\qquad \frac{1}{2}|z| \le |\text{Log}(1+z)| \le \frac{3}{2}|z|, \quad \text{if } |z| \le \frac{1}{2} .$$

To verify (2.15) we write the Taylor series expansion

$$\text{Log}(1+z) = z - \frac{1}{2}z^2 + \frac{1}{3}z^3 - \frac{1}{4}z^4 + \cdots$$
$$= z[1 + z(v(z))]$$

where

$$|v(z)| = \frac{1}{2}\left|1 - \frac{2}{3}z + \frac{2}{4}z^2 - \frac{2}{5}z^3 + \cdots\right|$$

$$\leq \frac{1}{2}\left(1 + \frac{2}{3}|z| + \frac{2}{4}|z|^2 + \frac{2}{5}|z|^3 + \cdots\right) < \frac{1}{2(1-|z|)} \leq 1 .$$

Hence, if $|z| \leq \frac{1}{2}$,

$$|\mathrm{Log}(1+z)| \leq |z|[1 + |z|\cdot|v(z)|] \leq |z|(1+|z|) \leq \frac{3}{2}|z|$$

and

$$|\mathrm{Log}(1+z)| \geq |z|[1 - |z|\cdot|v(z)|] \geq |z|(1-|z|) \geq \frac{1}{2}|z| .$$

It follows from (2.11a) and (2.15) that, for $n \geq m_0 \geq 2$,

$$P_{m_0}^n := \prod_{j=m_0}^{n} (1+b_j) = e^{\sum_{j=m_0}^{n}(1+b_j)}$$

$$\geq e^{\frac{1}{2}\sum_{j=m_0}^{n} b_j} = \left(e^{\sum_{m_0}^{n}\frac{1}{|a_j|}}\right)^{\frac{\cos^2\alpha}{8}} .$$

It follows from (2.14b) and (1.2b) that

$$\frac{1}{|a_j|} > \frac{1}{(j-1)M+|a_1|} > \frac{1}{Kj}, \quad \text{for } j \geq m_0 \geq 2 . \tag{2.17}$$

Since $\sum_{j=m_0}^{n} \frac{1}{j} \geq \int_{m_0}^{n+1} \frac{dx}{x} = \ell n\left(\frac{n+1}{m_0}\right)$, we obtain from (2.16) and (2.14)

$$P_{m_0}^n \geq e^{\left(\sum_{j=m_0}^{n}\frac{1}{j}\right)\frac{\cos^2\alpha}{8K}} \geq \left(\frac{n+1}{m_0}\right)^{\frac{\cos^2\alpha}{8K}} .$$

The assertions in (2.14) follow from this and Corollary 2.3.

We consider now the application of Theorem 2.3 to incomplete gamma functions.

Theorem 2.4. Let

$$F(a,z) := \frac{e^z\Gamma(a,z)}{z^a}, \quad 0 \leq a < 1 \tag{2.18}$$

<u>and let</u> $K(\alpha_n z^{-1}/1)$ <u>denote the continued fraction in</u> (1.4). <u>Let</u> f_n <u>denote the</u> n^{th} <u>approximant</u>. <u>Then</u> :

(A) <u>For</u> $n = 2,3,4,\ldots,$

$$(2.19)\qquad |F(a,z)-f_n| \leq \frac{2}{|z|(\cos\alpha)\prod_{j=2}^{n}\left[1+\frac{|z|\cos^2\alpha}{4\alpha_j}\right]} .$$

(B) <u>Let</u>

$$(2.20)\qquad m_0 := [\![|z|\cos^2\alpha + \max(1,2a)]\!] + 1.$$

<u>Then</u>

$$(2.21a)\qquad |F(a,z)-f_n| \leq A(m_0)\left(\frac{m_0}{n+1}\right)^{B|z|},\quad n = m_0,\ m_0+1,\ m_0+2,\ldots,$$

<u>where</u>

$$(2.21b)\qquad A(m_0) := \frac{2}{|z|(\cos\alpha)\prod_{j=2}^{m_0-1}\left[1+\frac{|z|\cos^2\alpha}{4\alpha_j}\right]},\quad B := \frac{\cos^2\alpha}{2(1+\max(a,1-a))} .$$

<u>Proof</u>. This result follows readily from Theorem 2.3 since by (1.6) $M = |z|^{-1}\max(a,1-a)$ and hence $K := (|a_1|+M)/2 = (1+\max(a,1-a))/2|z|$. Moreover, it is easily shown from (2.14a) and (1.4b) that

$$b_n \leq \frac{1}{2}\quad \text{if}\quad n \geq m_0 := [\![|z|\cos^2\alpha + (\max(1,2a)]\!] + 1 .$$

In the following two corollaries we state explicitly the results of Theorem 2.4 for the complementary error function (1.7) and the exponential integral (1.8).

<u>Corollary</u> 2.5. <u>Let</u>

$$(2.22)\qquad F(\tfrac{1}{2},z^2) := \frac{\sqrt{\pi}}{z}e^{z^2}\operatorname{erfc}(z)$$

<u>and let</u> f_n <u>denote the</u> n^{th} <u>approximant of</u> (1.9). <u>Then</u>:

(A) <u>For</u> $n = 2,3,4,\ldots,$

$$(2.23)\qquad |F(\tfrac{1}{2},z^2) - f_n| \leq \frac{2}{|z|(\cos\alpha)\prod_{j=2}^{n}\left[1+\frac{2|z|\cos^2\alpha}{j-1}\right]} .$$

(B) <u>Let</u>

$$(2.24a)\qquad m_0 := [\![|z|\cos^2\alpha]\!] + 2,\quad B := \frac{\cos^2\alpha}{3},$$

<u>and</u>

(2.24b) $$A(m_0) = \frac{2}{|z|(\cos\alpha)\prod_{j=2}^{m_0-1}\left[1+\frac{2|z|\cos^2\alpha}{j-1}\right]}.$$

Then

(2.24c) $$\left|F(\tfrac{1}{2},z^2) - f_n\right| \le A(m_0)\left(\frac{m_0}{n+1}\right)^{B|z|}, \quad n = m_0, m_0+1, m_0+2, \ldots .$$

Corollary 2.6. Let

(2.25) $$F(0,z) := e^z E_1(z)$$

and let f_n denote the nth approximant of (1.10). Then:

(A) For $m = 1,2,3,\ldots,$

(2.26a) $$|F(0,z) - f_{2m}| \le \frac{2}{|z|(\cos\alpha)\prod_{j=1}^{m-1}\left[1+\frac{|z|\cos^2\alpha}{4j}\right]^2\left[1+\frac{|z|\cos^2\alpha}{4m}\right]},$$

(2.26b) $$|F(0,z) - f_{2m+1}| \le \frac{2}{|z|(\cos\alpha)\prod_{j=1}^{m}\left[1+\frac{|z|\cos^2\alpha}{4j}\right]^2}.$$

(B) Let

(2.27a) $$m_0 := [\![|z|\cos^2\alpha]\!] + 2, \quad B := \frac{\cos^2\alpha}{4}$$

and let $A(m_0)$ be defined by (2.21b) where $\alpha_{2m} = \alpha_{2m+1} = m$, $m \ge 1$. Then

(2.27b) $$|F(0,z) - f_n| \le A(m_0)\left(\frac{m_0}{n+1}\right)^{B|z|}, \quad n = m_0, m_0+1, m_0+2, \ldots .$$

3. Modified Continued Fractions. We begin this section by giving sufficient conditions to insure that a modified continued fraction converges to the same finite limit as its reference continued fraction. Our result also provides an approach for obtaining truncation error bounds for the approximants g_n.

Theorem 3.1. Let $K(a_n/1)$ be a continued fraction such that, for some $-\frac{\pi}{2} < \alpha < \frac{\pi}{2}$ and $0 < \varepsilon < 1$,

(3.1) $$0 \ne a_n \quad P(\alpha,\varepsilon), \quad n = 1,2,3,\ldots \quad \text{and} \quad \sum_{n=1}^{\infty} \frac{1}{|a_n|} = \infty,$$

where $P(\alpha,\varepsilon)$ and $H(\alpha)$ are defined by (2.6). Let $K(a_n,1;w_n)$ be a modified continued fraction with n^{th} approximant g_n and let

$V = \{V_n\}$ _be a sequence of value regions corresponding to a sequence of element regions_ $E = \{E_n\}$ _at_ $W = \{w_n\}$ _such that_

(3.2a) $\qquad a_n \in E_n, \quad n = n_0+1, n_0+2, n_0+3, \ldots,$

and

(3.2b) $\qquad w_n \in V_n \cap H(\alpha), \quad n = n_0+1, n_0+2, n_0+3, \ldots .$

Then: (A) $\{S_n(V_n)\}_{m_0}^{\infty}$ _and_ $\{S_n(H(\alpha))\}_1^{\infty}$ _are nested sequences of subsets of_ $\hat{\mathbb{C}}$ _and, for_ $n \geq n_0 + 1$ _and_ $m \geq 0$,

(3.3) $\quad g_{n+m} := S_{n+m}(w_{n+m}) \in S_{n+m}(V_{n+m}) \cap S_{n+m}(H(\alpha)) \subseteq S_n(V_n) \cap S_n(H(\alpha)),$

(B) $K(a_n/1)$ _and_ $K(a_n,1;w_n)$ _converge to the same finite value_ f,

(3.4a) $\qquad f \in S_n(V_n) \cap S_n(H(\alpha)), \quad n = n_0+1, n_0+2, \ldots,$

and

(3.4b) $\qquad |f-g_n| \leq \text{diam } S_n(V_n), \quad n = n_0+1, n_0+2, \ldots .$

(C) _If_ $f^{(m)}$ _denotes the_ m^{th} _tail of_ $K(a_n/1)$, _then_,

(3.5) $\qquad f^{(m)} \in V_m, \quad m = n_0+1, n_0+2, \ldots .$

Proof. (A): It follows from (3.2), (1.12), and (1.23) that

$$g_{n+m} := S_{n+m}(w_{n+m}) \in S_{n+m}(V_{n+m}) = S_{n+m-1}\left(\frac{a_{n+m}}{1+V_{n+m}}\right) \subseteq$$

$$S_{n+m-1}(V_{n+m-1}) \subseteq \cdots \subseteq S_n(V_n) \subseteq S_{n-1}(V_{n-1}),$$

for $n \geq n_0 + 1$ and $m \geq 0$. We obtain (3.3) from this, Corollary 2.2(A) and (3.2b).

(B): It follows from (3.1) and Corollary 2.3 that $K(a_n/1)$ converges to a finite limit f and

$$f \in S_n(H(\alpha)), \quad n = 1,2,3,\ldots .$$

Moreover, by Theorem 2.1, $\lim_{n\to\infty} R_n = \lim_{n\to\infty} \text{rad } S_n(H(\alpha)) = 0$. It follows from this and (3.3) that $\{g_n\}$ converges also to f and that (3.4) holds.

(C) is an immediate consequence of the fact that

$$f = S_m(f^{(m)}) \in S_m(V_m), \quad m = n_0+1, n_0+2, \ldots . \quad \square$$

The first question about modified continued fractions is: How do we choose the modifying factors w_n? From (1.19) we see that, if possible, one wishes to have w_n near the tail $f^{(n)}$. For limit-periodic continued fractions $K(a_n/1)$ where $\lim_{n\to\infty} a_n = a \in \mathbb{C} - (-\infty, -\frac{1}{4}]$, one may wish to choose

$$(3.6) \qquad w_n = K\left(\frac{a}{1}\right) = \hat{f} := \sqrt{a + \frac{1}{4}} - \frac{1}{2}, \quad \mathrm{Re}\sqrt{} > 0,$$

since it is well known [6, p. 113-114] that

$$(3.7) \qquad \lim_{n\to\infty} f^{(n)} = \hat{f}$$

(See also [10]).

A continued fraction $K(a_n/1) \in S(\mu, M, \alpha)$ is also limit-periodic, but in this case

$$\lim_{n\to\infty} a_n = a = \infty .$$

Therefore, guided by the choice of w_n in (3.6) for finite a, we consider modifying factors

$$(3.8) \qquad w_n := w_n^{(0)} := \sqrt{a_{n+1} + \frac{1}{4}} - \frac{1}{2}, \quad \mathrm{Re}\sqrt{} > 0, \quad n = 0,1,2,\ldots .$$

Thus w_n is a fixed point of the transformation $w = a_{n+1}/(1+w)$. Other choices that we have considered include the following

$$(3.9a) \qquad w_n^{(1)} := \frac{\sqrt{a_{n+1} + \frac{1}{4}} + \sqrt{a_n + \frac{1}{4}}}{2} - \frac{1}{2},$$

$$(3.9b) \qquad w_n^{(2)} := \sqrt{\frac{a_n + a_{n+1}}{2} + \frac{1}{4}} - \frac{1}{2},$$

$$(3.9c) \qquad w_n^{(3)} := \frac{a_{n+1}}{1 + \left(\sqrt{a_{n+2} + \frac{1}{4}} - \frac{1}{2}\right)} = \frac{a_{n+1}}{a_{n+2}}\left(\sqrt{a_{n+2} + \frac{1}{4}} - \frac{1}{2}\right),$$

$$(3.9d) \qquad w_n^{(4)} := \sqrt{a_{n+1}} - \frac{1}{2}$$

It can be readily shown that if $\{\rho_n\}$ and $\{\Gamma_n\}$ satisfy

$$(3.10a) \qquad 0 < \rho_n < |1+\Gamma_n|, \quad \Gamma_n \in \mathbb{C}, \quad n = 0,1,2,\ldots$$

and if $\{V_n\}$ and $\{E_n\}$ are defined by

$$(3.10b) \qquad V_n := [v: |v-\Gamma_n| \leq \rho_n], \quad n = 0,1,2,\ldots,$$

(3.10c) $E_n := [w: |w(1\mp\Gamma_n) - \Gamma_{n-1}(|1+\Gamma_n|^2-\rho_n)| + \rho_n|w| \leq \rho_{n-1}(|1+\Gamma_n|^2-\rho_n^2)]$

$n = 1,2,3,\ldots,$

then

(3.11) $\Gamma_{n-1} \in V_{n-1}$ and $\frac{E_n}{1+V_n} \subseteq V_{n-1}$, $n = 1,2,3,\ldots$.

(for a proof see, for example, [1, Lemma 3.2]). We shall prove:

Theorem 3.2. Let $K(a_n/1) \in S(\mu, M, \alpha)$ and let $\{w_n\}$ be defined by (3.8). Let

(3.12a) $E_n := [w: |w(1+\bar{w}_n) - w_{n-1}(|1+w_n|^2-\rho^2)| + \rho|w| \leq \rho(|1+w_n|^2-\rho^2)]$,

$n = 1,2,3,\ldots,$

(3.12b) $V_n := [v: |v-w_n| \leq \rho]$, $n = 0,1,2,\ldots,$

where

(3.12c) $0 < \rho < \cos\alpha$,

and

(3.12d) $-2\rho(\cos\alpha - \rho) < \mu \leq M < 2\rho(\cos\alpha + \rho)$

Then there exists $n_0 \geq 1$ such that: (A)

(3.13) $w_{n-1} \in V_{n-1}$ and $\frac{E_n}{1+V_n} \subseteq V_{n-1}$, $n = n_0, n_0+1, n_0+2,\ldots$.

(B)

(3.14) $a_n \in E_n$, $n = n_0, n_0+1, n_0+2,\ldots$.

Proof. (A) follows from (3.10) by setting $\rho_n = \rho$, $\Gamma_n = w_n$ and noting that, by (3.8) and (1.1c), $\lim_{n\to\infty} w_n = \infty$ so that (3.10a) holds eventually. To prove (B) we first prove the following lemma, in each of which we assume the hypotheses of Theorem 3.2.

Lemma 3.3.

(A)

(3.15) $\lim_{n\to\infty} (w_n - w_{n-1}) = 0,$

(B)

(3.16) $\lim_{n\to\infty} \frac{w_{n-1}}{1+\bar{w}_n} = e^{i2\alpha}$,

(C)

(3.17) $$\lim_{n\to\infty}(|1 + w_n| - |w_n|) = \cos\alpha .$$

Proof. (A): By (3.8)

$$w_n - w_{n-1} = \left(\sqrt{a_{n+1}+\frac{1}{4}} - \sqrt{a_n+\frac{1}{4}}\right)\frac{\sqrt{a_{n+1}+\frac{1}{4}} + \sqrt{a_n+\frac{1}{4}}}{\sqrt{a_{n+1}+\frac{1}{4}} + \sqrt{a_n+\frac{1}{4}}}$$

$$= \frac{a_{n+1} - a_n}{|a_{n+1}|e^{i\alpha}\sqrt{1+\frac{1}{4a+1}} + |a_n|e^{i\alpha}\sqrt{1+\frac{1}{4a_n}}} .$$

Therefore (3.15) holds since $|a_{n+1}-a_n| = \big||a_{n+1}| - |a_n|\big| \le M$ and $\lim |a_n| = \infty$.

(B):

$$\frac{w_{n+1}}{1+\bar{w}_n} = \frac{|a_n|e^{i\alpha}\sqrt{1+\frac{1}{4a_n}} - \frac{1}{2}}{|a_{n+1}|e^{-i\alpha}\sqrt{1+\frac{1}{4a_{n+1}}} + \frac{1}{2}}$$

$$= e^{i2\alpha}\frac{\sqrt{1+\frac{1}{4a_n}} - \frac{1}{2|a_n|}}{\left|\frac{a_{n+1}}{a_n}\right|\sqrt{1+\frac{1}{4a_{n+1}}} + \frac{1}{2|a_n|}} .$$

Therefore (3.16) holds since $\lim a_n = \infty$ and

$$\left|\frac{a_{n+1}}{a_n}\right| - 1 = \frac{|a_{n+1}|-|a_n|}{|a_n|} \le \frac{M}{|a_n|} \to 0 \quad \text{as} \quad n \to \infty$$

so that $\lim |a_{n+1}/a_n| = 1$.

(C): Let $\alpha_n := \text{Arg}(w_n + \frac{1}{2}) = \text{Arg}\sqrt{a_{n+1} + \frac{1}{4}} = \alpha + \text{Arg}\sqrt{1+\frac{1}{4a_{n+1}}}$ (see Figure 1). It follows that $\lim \alpha_n = \alpha$. Let β_n and γ_n be defined as in Figure 1 so that $\alpha_n = \beta_n + \gamma_n$. It follows that $\lim \beta_n = \alpha$ and $\lim \gamma_n = 0$. by the law of cosines $|1 + w_n|^2 - |w_n|^2 = 2|1 + w_n|\cos\beta_n - 1$; hence

$$\lim_{n\to\infty}(|1+w_n|-|w_n|) = \lim_{n\to\infty}\frac{2|1+w_n|\cos\beta_n - 1}{|1+w_n|+|w_n|}$$

$$= \lim_{n\to\infty} \frac{2\cos\beta_n}{1 + \left|\frac{w_n}{1+w_n}\right|} - \lim_{n\to\infty} \frac{1}{|1+w_n|+|w_n|} = \cos\alpha \ .\square$$

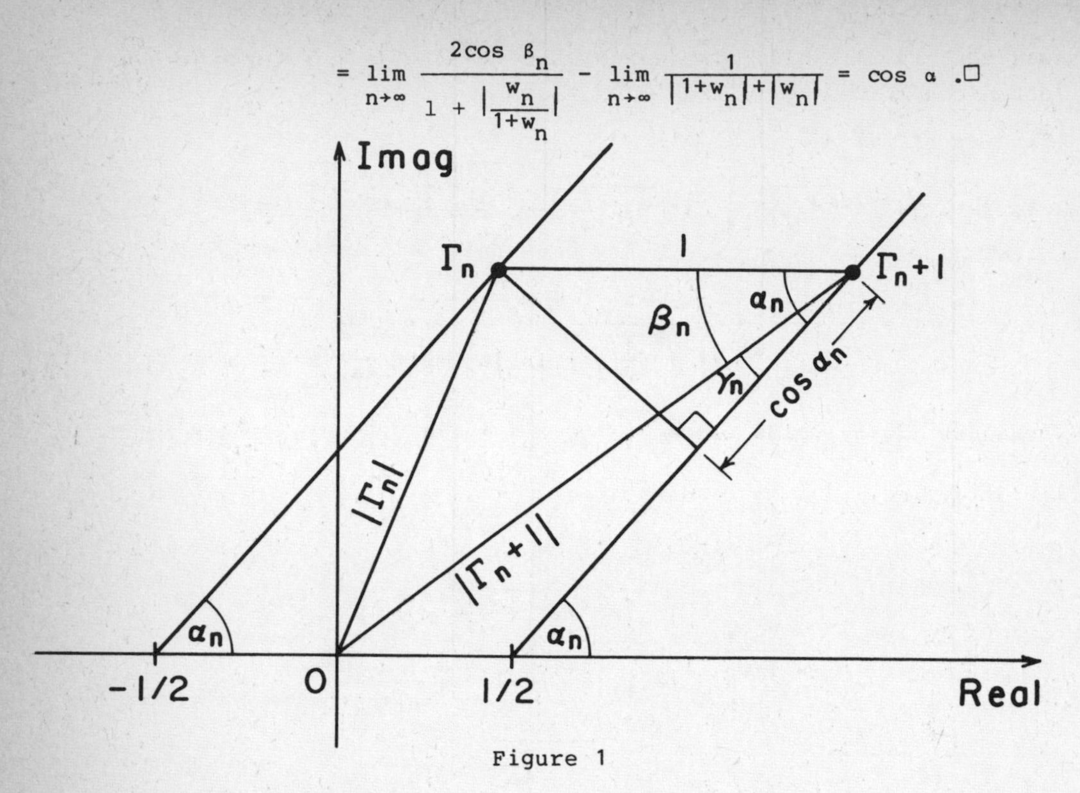

Figure 1

<u>Lemma</u> 3.4. <u>Let</u>

$$a_n' := w_{n-1}(1+w_n), \quad n = 1,2,3,\ldots \ . \tag{3.18}$$

<u>Then</u>: (A) <u>There exists</u> n_1 <u>such that</u>

$$a_n' \in E_n \quad \text{for} \quad n = n_1, n_1+1, n_1+2,\ldots \ . \tag{3.19}$$

(B) <u>If</u> ε_n <u>is defined by</u>

$$a_n - a_n' = -\frac{1}{2}(a_{n+1}-a_n) + \varepsilon_n, \quad n = 1,2,3,\ldots, \tag{3.20a}$$

<u>then</u>

$$\lim_{n\to\infty} \varepsilon_n = 0 \ . \tag{3.20b}$$

<u>Proof</u>. (A): It follows from (3.12a) that $a_n' \in E_n$ iff

$$\left|\frac{w_{n-1}}{1+w_n}\right|\rho^2 + |w_{n-1}| - |w_n| \le (|1+w_n|-|w_n|) - \frac{\rho^2}{|1+w_n|} \ . \tag{3.21}$$

By Lemma 3.3, $\left|\frac{w_{n-1}}{1+w_n}\right| \to 1$, $\big||w_{n-1}| - |w_n|\big| \le |w_{n-1}-w_n| \to 0$, $\rho^2/|1+w_n| \to 0$ and $|1+w_n| - |w_n| \to \cos\alpha > 0$, as $n \to \infty$. Hence there exists an n_1

such that (3.21) holds for $n > n_1$ provided that $\rho^2 < \cos\alpha$; but this latter condition is implied by (3.12c).

(B): By (3.18) and (3.8)

$$a_n - a_n' = a_n - \sqrt{(a_n+\frac{1}{4})(a_{n+1}+\frac{1}{4})} + \frac{1}{2}\left(\sqrt{a_{n+1}+\frac{1}{4}} - \sqrt{a_n+\frac{1}{4}}\right) + \frac{1}{4}\,.$$

By the method of rationalization

$$a_n - \sqrt{(a_n+\frac{1}{4})(a_{n+1}+\frac{1}{4})} = \frac{a_n^2 - a_n a_{n+1} - \frac{1}{4}a_n - \frac{1}{4}a_{n+1} - \frac{1}{16}}{a_n + \sqrt{a_n a_{n+1} + \frac{1}{4}(a_n+a_{n+1}) + \frac{1}{16}}}$$

$$= \frac{a_n - a_{n+1} - \frac{1}{4}\left(1+\frac{a_{n+1}}{a_n}\right) - \frac{1}{16\,a_n}}{1 + \sqrt{\frac{a_{n+1}}{a_n} + \frac{1}{4}\left(\frac{1}{a_n} + \frac{a_{n+1}}{a_n^2}\right) + \frac{1}{16\,a_n^2}}}\,.$$

Also by Lemma 3.3

$$\frac{1}{2}\left(\sqrt{a_{n+1}+\frac{1}{4}} - \sqrt{a_n+\frac{1}{4}}\right) = \frac{1}{2}(w_n - w_{n-1}) \to 0 \quad \text{as} \quad n \to \infty$$

From (1.1) we obtain

$$\left|\frac{a_{n+1}}{a_n} - 1\right| = \frac{|a_{n+1}-a_n|}{|a_n|} \leq \frac{\max(M,-\mu)}{|a_n|} \to 0 \quad \text{as} \quad n \to \infty\,,$$

so that

$$\text{(3.22)} \qquad \lim_{n\to\infty} \frac{a_{n+1}}{a_n} = 1.$$

Combining these results with (3.20a) yields

$$\varepsilon_n = (a_{n+1}-a_n)\left[\frac{1}{2} - \frac{1}{1 + \sqrt{\frac{a_{n+1}}{a_n} + \frac{1}{4}\left(\frac{1}{a_n} + \frac{a_{n+1}}{a_n^2}\right) + \frac{1}{16\,a_n^2}}}\right]$$

$$- \frac{\frac{1}{4}\left(1+\frac{a_{n+1}}{a_n}\right) - \frac{1}{16\,a_n}}{1 + \sqrt{\frac{a_{n+1}}{a_n} + \frac{1}{4}\left(\frac{1}{a_n} + \frac{a_{n+1}}{a_n^2}\right) + \frac{1}{16\,a_n^2}}} + \frac{1}{2}(w_n - w_{n-1}) + \frac{1}{4}\,,$$

from which (3.20b) follows.

<u>Lemma</u> 3.5. <u>The set of all limit points of</u> $\{a_n - a_n'\}$ <u>is contained in the interval</u>

(3.23) $\quad D := e^{i2\alpha}[-\frac{M}{2}, -\frac{\mu}{2}] = [xe^{i2\alpha}: -\frac{M}{2} \leq x \leq -\frac{\mu}{2}]$.

Proof. It follows from (1.1) that

$$-\frac{1}{2}(a_{n+1}-a_n) = -\frac{1}{2}e^{i2\alpha}(|a_{n+1}|-|a_n|) \in D, \quad n = 1,2,3,\ldots .$$

Thus the assertion of this lemma is a consequence of Lemma 3.4.

Lemma 3.6. *If* D *is defined by* (3.23) *and if*

(3.24) $\quad D \in C := [w \in ¢: |w + \rho^2 e^{i2\alpha}| < \rho \cos \alpha]$,

Then there exists n_0 *such that*

(3.25) $\quad a_n \in E_n$, *for* $n = n_0, n_0+1, \ldots$.

Proof. It follows from (3.12) and (3.18) that $a_n \in E_n$ iff

(3.26) $\quad |a_n - a_n' + \frac{w_{n-1}}{1+\overline{w}_n}\rho^2| \leq \rho(|1+w_n| - \frac{|a_n|}{|1+w_n|}) - \frac{\rho^3}{|1+w_n|}$.

Since $\lim_{n\to\infty} \rho^3/|1+w_n| = 0$ and

$$|1+w\;| - \frac{|a_n|}{|1+w_n|} = |1+w_n| - \frac{|a_n'+(a_n-a_n')|}{|1+w_n|}$$

$$= (|1+w_n| - |w_n|) + |w_n| - |w_{n-1} + \frac{a_n-a_n'}{1+w_n}| \to \cos\alpha ,$$

as $n \to \infty$, there exists a sequence $\{\eta_n\}$ such that $\lim \eta_n = 0$ and for $n \geq 1$

$$\rho(|1+w_n| - \frac{|a_n|}{|1+w_n|}) - \frac{\rho^3}{|1+w_n|} = \rho\cos\alpha + \eta_n .$$

If $\{\delta_n\}$ is defined by $w_{n-1}/(1+\overline{w}_n) = \delta_n e^{i2\alpha}$, then by Lemma 3.3(B), $\lim \delta_n = 1$. It follows from this and (3.26) that $a_n \in E_n$ iff

(3.27) $\quad |a_n - a_n' + \delta_n \rho^2 e^{i2\alpha}| \leq \rho\cos\alpha + \eta_n$.

Let K be a compact subset of C such that

(3.28) $\quad D \subset K \subset C$ and $\operatorname{dist}(K, \partial C) > 0$.

Then by Lemma 3.5 there exists n_2 such that

$$a_n - a'_n \in K \quad \text{if} \quad n \geq n_2 .$$

By (3.28) there exists n_3 such that, for $n \geq n_3$,

$$K \subset C_n := [w \in \mathbb{C}: |w + \delta_n \rho^2 e^{i2\alpha}| \leq \rho \cos \alpha + \eta_n] .$$

Let $n_0 = \max(n_2, n_3)$. It follows that

$$a_n - a'_n \in K \subset C_n \quad \text{for} \quad n \geq n_0 .$$

Hence (3.25) holds.

Proof of Theorem 3.2. By Lemma 3.6 it suffices to show that (3.24) holds. But (3.24) holds iff

$$(3.29) \qquad \left|\frac{M}{2} - \rho^2\right| < \rho \cos \alpha \quad \text{and} \quad \left|\frac{\mu}{2} - \rho^2\right| < \rho \cos \alpha .$$

It can be readily shown that (3.29) is implied by (3.12c,d).

Theorem 3.2 can be used in conjunction with (3.4b) of Theorem 3.1 to obtain truncation error bounds for the approximants g_n of $K(a_n, 1; w_n)$. One can also apply Theorem 3.2 to investigate the numerical stability of the backward recurrence algorithm to compute g_n. These investigations will be dealt with in a subsequent paper. We conclude this paper with an application of Theorem 3.2 to convergence acceleration. The following result has recently been proved by the authors [5, (3.29)].

Theorem 3.7. *Let* $K(a_n/1) \in S(\mu, M, \alpha)$ *and let* $\{w_n\}$ *be defined by* (3.8). *Let* f_n *and* g_n *denote the* n^{th} *approximants of* $K(a_n/1)$ *and* $K(a_n, 1; w_n)$, *respectively. Then there exists* n_0 *such that*

$$(3.30) \qquad |f - g_n| \leq |f - f_n| \frac{2\rho}{|w_n| - \rho}, \quad n = n_0, n_0+1, n_0+2, \ldots .$$

It follows from Theorem 3.7 that the modified continued fraction $K(a_n, 1; w_n)$ yields convergence acceleration in the sense of (1.21). To illustrate this phenomenon we consider the function

$$(3.31) \qquad w(z) := e^{-z^2} \operatorname{erfc}(-iz) ,$$

first introduced and tabulated by Faddeeva and Terent'ev [2]. This function is represented, for $z \in Q_1 := [z: \operatorname{Re} z \geq 0, \operatorname{Im} z > 0]$, by the continued fraction

$$(3.32a) \quad \frac{-i\sqrt{\pi}}{z} w(z) = \frac{z^{-2}}{1} + \frac{-\frac{1}{2} z^{-2}}{1} + \frac{-\frac{2}{2} z^{-2}}{1} + \frac{-\frac{3}{2} z^{-2}}{1} + \frac{-\frac{4}{2} z^{-2}}{1} + \cdots$$

$$= \operatorname*{K}_{n=1}^{\infty} \left(\frac{a_n}{1}\right) = \operatorname*{K}_{n=1}^{\infty} \left(\frac{-\alpha_n z^{-2}}{1}\right)$$

where

(3.32b) $\alpha_1 := 1, \quad \alpha_n = \frac{n-1}{2}, \quad n = 2,3,4,\ldots$

(3.32c) $\mathrm{Arg}\ a_n := 2\alpha := \mathrm{Arg}\ (-z^{-2})$

where

(3.32d) $a_n := -\alpha_n z^{-2}, \quad n = 1,2,3,\ldots$.

Thus $K(a_n/1) \in S(\frac{-1}{2|z|^2}, \frac{1}{2|z|^2}, \frac{\mathrm{Arg}(-z^{-2})}{2})$.

Let f_n and g_n denote the nth approximants of $K(a_n/1)$ and $K(a_n,1;w_n)$, respectively, where w_n is the modifying factor (3.8). Figures 2 and 3 show level curves of $\widetilde{SD}(f_5)$ and $\widetilde{SD}(g_5)$ where

$$\widetilde{SD}(f_n) := -\log_{10} \left|\frac{w(z)-f_n}{w(z)}\right| - .301$$

is approximately the number (SD) of significant digits in the approximation of $w(z)$ by f_n. We used a library subroutine MERRCZ developed by Gautschi [3] to compute $w(z)$ in single precision, with a 13-decimal digit machine. It can be seen that, in much of the domain of the graphs $\widetilde{SD}(g_5) \geq \widetilde{SD}(f_5) + 1$; thus the convergence has been accelerated. In future studies even faster acceleration will be sought, by using other modifying factors such as those given by (3.9) and those given by Wynn [11].

Acknowledgement. The authors wish to thank Robert B. Jones for preparing the contour maps in Figures 2 and 3.

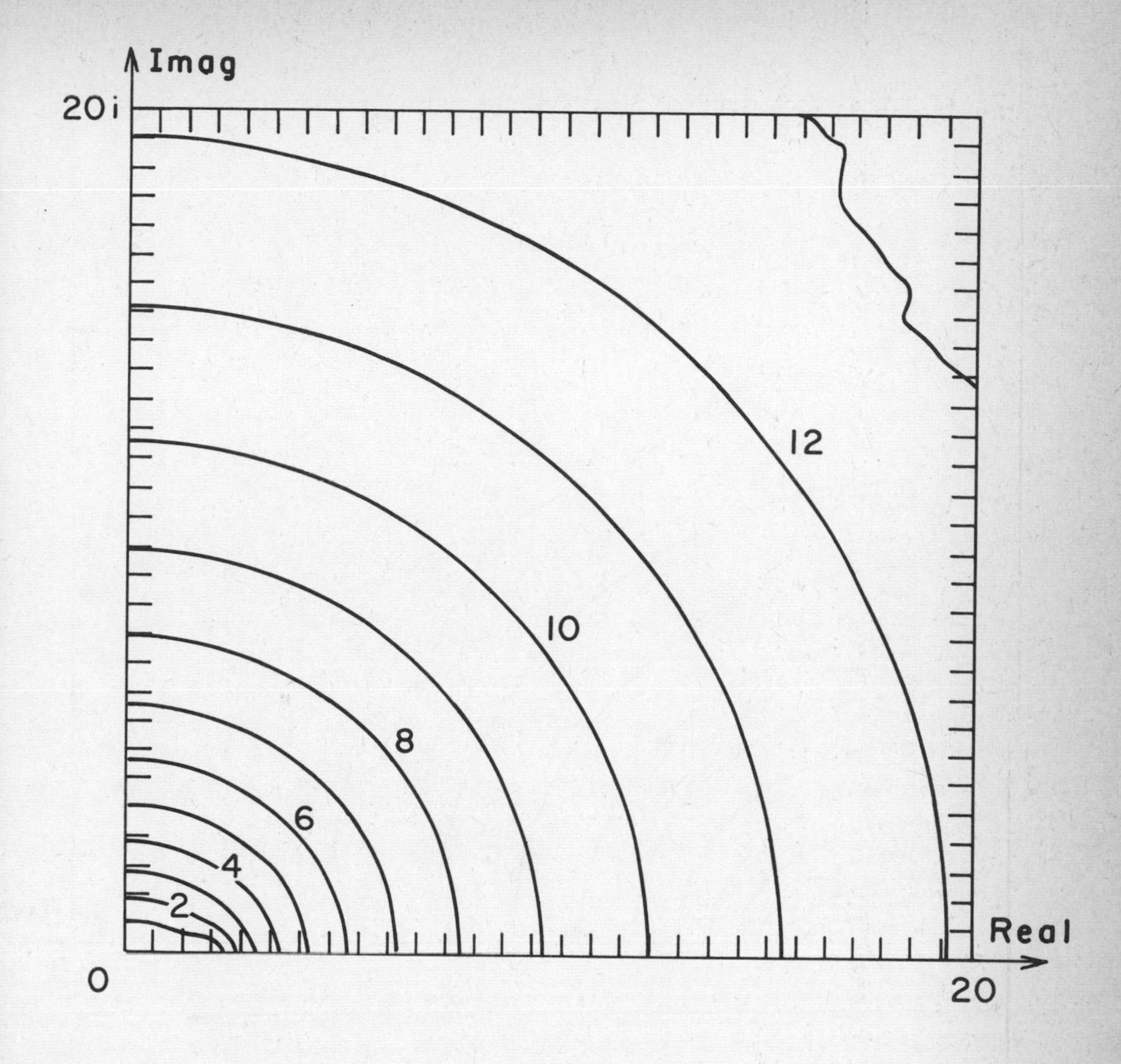

Figure 2. Level curves of $SD(f_5(z))$, where $f_5(z)$ is the 5th approximant of the Stieltjes continued fraction representation of $w(z)$: e^{-z^2} erfc $(-iz)$.

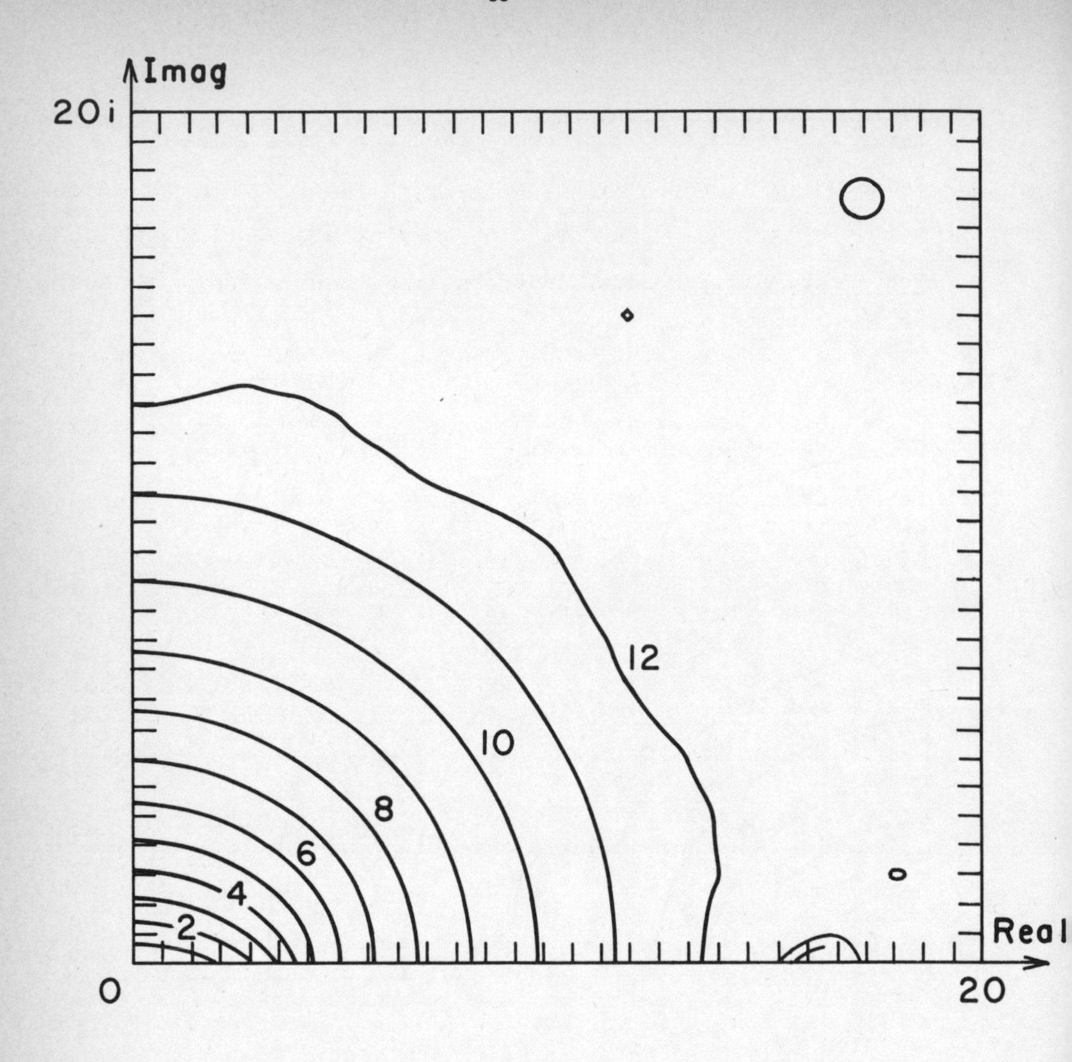

Figure 3. Level curves of $SD(g_5(z))$, where $g_5(z)$ is the 5th approximant of the modified continued fraction representation of $x(z) := e^{-z^2} \operatorname{erfc}(-iz)$.

References

1. Baltus, Christopher and William B. Jones, A family of best value regions for modified continued fractions, this Volume.

2. Faddeeva, V.N. and N.N. Terent'ev, Tables of values of the function $w(z) = e^{z^2} [1 + \frac{2i}{\sqrt{\pi}} \int_0^z e^{t^2} dt]$ for complex argument, Gosud. Izdat. Teh.-Teor. Lit., Moscow, 1954; English transl., Pergamon Press, New York, 1961.

3. Gautschi, Walter, Efficient computation of the complex error function, SIAM J. Numer. Anal. 7(1970), 187-198.

4. Henrici, P. and Pia Pfluger, Truncation error estimates for Stieltjes fractions, Numer. Math. 9 (1966), 120-138.

5. Jacobsen, Lisa, William B. Jones and Haakon Waadeland, Convergence acceleration for continued fractions $K(a_n/1)$, where $a_n \to \infty$, Proceedings of a Conference on Rational Approximation and Its Applications to Theoretical Physics, (Lancut, Poland, July 1-4, 1985), Lecture Notes in Mathematics, Springer-Verlag, New York, to appear.

6. Jones, William B. and W. J. Thron, Continued Fractions: Analytic Theory and Applications, Encyclopedia of Mathematics and its Applications, 11, Addison-Wesley Publishing Company, Reading, Mass. (1980). Distributed now by Cambridge University Press, New York.

7. Jones, William B. and W. J. Thron, On the computation of incomplete gamma functions in the complex domain, J. Comp. and Applied Math. 12 and 13 (1985), 401-417.

8. Thron, W. J., On parabolic convergence regions for continued fractions, Math Zeitschr. 69 (1958), 173-182.

9. Thron, W. J., A priori truncation error estimates for Stieltjes fractions, in E. B. Christoffel (Ed., P.L. Butzer and F. Fehér) Aachen Birkhäuser Verlag, Basel, (1981), 203-211.

10. Thron, W. J. and Haakon Waadeland, Accelerating convergence of limit periodic continued fractions $K(a_n/1)$, Numer. Math. 34, (1980), 155-170.

11. Wynn, P., Converging factors for continued fractions, Numer. Math. 1 (1959), 272-320.

Oval convergence regions and circular limit regions for continued fractions $K(a_n/1)$

L. Jacobsen
Department of Mathematics
and Statistics
The University of Trondheim AVH
N-7055 Dragvoll, Norway

W. J. Thron*
Department of Mathematics
Campus Box 426
University of Colorado
Boulder, CO 80309-0426 U.S.A.

1. Introduction. In the early nineteen forties it was realized that value regions could play an important role in the convergence theory of continued fractions [21]. More recently it became clear that value regions can also be useful in truncation error analysis [10, 12] and stability analysis for computation of convergent continued fractions [9]. As modified approximants gained in interest the fact that pre-value regions can be regarded as value regions for modified continued fractions [4], became significant. In particular pre-value regions and limit regions are often advantageous to use in truncation error analysis [16], and stability analysis [7].

In 1945 Lane [13], with slightly different notation and in a somewhat more general setting, proved the following.

Theorem A. *If*

(1.1) $$|\Gamma| \leq |1+\Gamma| \quad \text{(or equivalently } \operatorname{Re}\Gamma \geq -1/2)$$

and

(1.2) $$0 < \rho < |1+\Gamma| ,$$

then

(1.3) $$V(\Gamma,\rho) := [v: |v-\Gamma| \leq \rho]$$

*W.J.T.'s research was supported in part by the U.S. National Science Foundation under Grant No. DMS 8401717.

is a pre value region for the element region

(1.4) $E(\Gamma,\rho):=[u: |u(1+\overline{\Gamma}) - \Gamma(|1+\Gamma|^2-\rho^2)| + \rho|u| \leq \rho(|1+\Gamma|^2-\rho^2)]$.

(For the definitions of element, value, pre value and limit regions, see the next section.)

The main purpose of this article is to determine when the regions $E(\Gamma,\rho)$ (which are bounded by Cartesian ovals, see for example [15]) are convergence regions or uniform convergence regions. (For definitions we refer to the next section.)

In Section 2 we introduce basic definitions and notation, mainly in accordance with [11]. In Section 3 we prove some convergence results for $E(\Gamma,\rho)$. Section 4 contains different useful representations of $E(\Gamma,\rho)$ and in Section 5 we describe some geometric properties of $E(\Gamma,\rho)$. In Section 6 we use $V(\Gamma,\rho)$ to derive truncation error estimates for continued fractions $K(a_n/1)$ with all $a_n \in E(\Gamma,\rho)$.

The well known Worpitzky theorem, [23] and the uniform parabola theorem, [14, 17, 11 Theorem 4.40] give very useful and frequently used simple uniform convergence regions. In Section 7 we find necessary and sufficient conditions for $E(\Gamma,\rho)$ to be contained in the Worpitzky disk or in a parabolic uniform convergence region. The advantage of the ovals is that $V(\Gamma,\rho)$ is often much smaller than even the best (smallest) value region corresponding to the parabolic region in which $E(\Gamma,\rho)$ is contained, and therefore may give better truncation error estimates and stability properties. Since not all $E(\Gamma,\rho)$ are contained in previously known simple convergence regions, some of them also represent new convergence results.

In Section 8 a number of special classes of ovals $E(\Gamma,\rho)$ are described, in particular a one-parameter family in which the Worpitzky disk and the parabolas, from the uniform parabola theorem, are the extreme cases.

2. Notation and basic concepts. Continued fractions $K(a_n/1)$ can be defined in terms of sequences of linear fractional transformations as follows [17]. Set

$$(2.1) \qquad s_n(w) := \frac{a_n}{1+w}, \quad a_n \neq 0, \quad n \geq 1.$$

Then define, inductively,

$$(2.2) \qquad S_n(w) := S_{n-1}(s_n(w)), \quad n \geq 1, \quad S_0(w) := (w).$$

The quantities $S_n(0)$ are called the nth *approximants* of $K(a_n/1)$, $S_n(w_n)$ is called a *modified* approximant. If $\{S_n(0)\}$ converges to a limit $\mathcal{L}$, then the continued fraction is said to *converge to the limit* $\mathcal{L}$, or to have the *value* $\mathcal{L}$.

Let E be a set in $\mathbb{C}$, $E \sim \{0\} \neq \emptyset$. Then $V \subseteq \hat{\mathbb{C}} = \mathbb{C} \cup \{\infty\}$ is a *value region* for the element region E if

$$(2.3) \qquad \frac{E}{1+V} \subset V$$

and

$$(2.4) \qquad E \subset V \quad [21].$$

The second condition is satisfied if $0 \in V$. If only condition (2.3) holds for a pair of sets $\langle E,V \rangle$, then V is called a *pre–value region* [8] (or a *modified value region* [4]) for E. We note that (2.3) implies

$$(2.5) \qquad S_n(w_n) \in V \text{ for } w_n \in V, \ a_k \in E, \ k = 1, \cdots, n,$$

and if (2.4) also holds, then

$$(2.6) \qquad S_n(0) \in V \text{ for } a_k \in E, \ k = 1, \cdots, n.$$

A *limit region* for an element region E is a pre-value region for E which contains all limits of convergent continued fractions $K(a_n/1)$ with all $a_n \in E$.

The significance of these various types of regions for continued fractions can be seen from the following. Let V be a pre-value region for the element region E, and let $K(a_n/1)$ be a continued fraction with all $a_n \in E$. If V is also a value region for E (i.e. (2.4) holds), then $a_{n+1} \in V$. From property (2.3) we therefore get

$$(2.7) \qquad S_m^{(n)}(0) = s_{n+1} \circ s_{n+1} \circ \cdots \circ s_{n+m}(0)$$
$$= s_{n+1} \circ s_{n+1} \circ \cdots \circ s_{n+m-1}(a_{n+m}) \in V, \quad m = 1,2, \cdots .$$

Since clearly $S_{n+m}(w) = S_n \circ S_m^{(n)}(w)$, we have

$$(2.8) \qquad |S_{n+m}(0) - S_{n+1}(0)| \leq \text{diam}\, S_n(V), \quad n,m = 1,2, \cdots .$$

This means: if $K(a_n/1)$ converges to f, then

$$(2.9) \qquad |f - S_{n+1}(0)| \leq \text{diam}\, S_n(V), \quad n = 0,1,2, \cdots .$$

If diam $S_n(V) \to 0$, then $K(a_n/1)$ converges.

Similarly, we always have

$$(2.10) \quad |S_{n+m}(w_{n+m}) - S_n(w_n)| \leq \text{diam}\, S_n(V) \text{ for } w_{n+m}, w_n \in V,$$

It follows that if diam $S_n(V) \to 0$, then $\{S_n(w_n)\}$ converges; and if $\{S_n(w_n)\}$ converges to a limit $\hat{f}$, then

$$(2.11) \qquad |\hat{f} - S_n(w_n)| \leq \text{diam}\, S_n(V).$$

From [6] follows that if $K(a_n/1)$ converges to f and diam $V > 0$, then

$$(2.12) \quad [\lim S_n(w_n) = \hat{f} \text{ for all sequences } \{w_n\},\ w_n \in V] \Rightarrow \hat{f} = f .$$

In particular, the closure $\bar{V}$ is a limit region for E if (2.12) holds for every continued fraction $K(a_n/1)$ with all $a_n \in E$.

Since value regions may be much larger than limit regions (or pre value regions), the truncation error estimates (2.10) based on pre value regions are often better than the ones based on value regions (2.8).

If every continued fraction $K(a_n/1)$, with all $a_n \in E$, converges, we say that E is a convergence region. If in addition there exists a sequence $\{\lambda_n\}$ of positive numbers converging to 0, such that

(2.13) $|f - S_n(0)| \leq \lambda_n$, $n = 1,2,3, \cdots$ for all $K(a_n/1)$, $a_n \in E$,

then E is called a uniform convergence region. Similarly, if there exists a sequence $\{\lambda_n\}$ of positive numbers converging to 0 such that

(2.14) $|\hat{f} - S_n(w_n)| \leq \lambda_n$, $n = 1,2,3, \cdots$ for all $K(a_n/1)$, $a_n \in E$ and all $w_n \in V$,

then E is said to be a uniform modified convergence region with respect to V.

3. Some convergence theorems for ovals. In this paper we are concerned with the corresponding element – pre- value regions given by Lane, as stated in Theorem A in Section 1. Using the notation

$$a := \Gamma(1+\Gamma)(1-\rho^2/|1+\Gamma|^2), \tag{3.1}$$

the expression for $E(\Gamma,\rho)$ can be written

$$E(\Gamma,\rho) = [u: |u-a| + \frac{\rho}{|1+\Gamma|}|u| \leq \frac{\rho}{|1+\Gamma|}(|1+\Gamma|^2 - \rho^2)] \tag{3.2}$$

If $\operatorname{Re}\Gamma = -1/2$, then $E(\Gamma,\rho) = [a]$ and $a < 0$. It is then well known that $E(\Gamma,\rho)$ is a simple uniform convergence region if $a \geq -1/4$, and that $K(a/1)$ diverges otherwise. (See for instance [11, Theorem 3.1].) In the remainder of this article we shall therefore let $E(\Gamma,\rho)$ denote the set given by (3.2) with $\operatorname{Re}\Gamma > -1/2$ and $0 < \rho < |1+\Gamma|$.

It is clear that $\partial E(\Gamma,\rho)$ is a Cartesian oval with foci at 0 and a (see for example [15]). It will be convenient to refer to the whole region $E(\Gamma,\rho)$ as an oval. We shall do so throughout this article.

For these ovals we can prove the following results.

Theorem 3.1. *If* $\rho > |\Gamma|$, *then* $E(\Gamma,\rho)$ *is a convergence region and* $V(\Gamma,\rho)$ *is a corresponding limit region.*

Proof. That $E(\Gamma,\rho)$ is a convergence region follows directly from a result of Hillam and

Thron [3] (see also [11, Th. 4.27]). Since $0 \in V(\Gamma,\rho)$, this disk is a value region for $E(\Gamma,\rho)$. Since $V(\Gamma,\rho)$ is closed, it is also a limit region for $E(\Gamma,\rho)$.

If $\rho \leq |\Gamma|$, then further work needs to be done to determine when $E(\Gamma,\rho)$ is a convergence region. A partial answer for which we are able to give a simple direct proof, is the following.

Theorem 3.2. *Let* $0 < \rho < \mathrm{Re}(\Gamma+1/2)$. *Then* $E(\Gamma,\rho)$ *is a convergence region and a uniform modified convergence region with respect to its limit region* $V(\Gamma,\rho)$.

Proof. If $a_n \in E(\Gamma,\rho)$ and $w_n \in V(\Gamma,\rho)$, then

$$w_{n-1} = \frac{a_n}{1+w_n} \in V(\Gamma,\rho).$$

Now consider the function $y = a_n/(1+w)$. Its derivative is

$$\frac{dy}{dw} = \frac{-a_n}{(1+w)^2}.$$

Set $w = w_n$ and recall that $a_n = w_{n-1}(1+w_n)$. Thus

$$\frac{dy}{dw}\bigg|_{w=w_n} = -\frac{w_{n-1}}{1+w_n},$$

and

$$\frac{dS_n(w)}{dw}\bigg|_{w=w_n} = \prod_{k=1}^{n} \frac{-w_{k-1}}{1+w_k}.$$

This follows from the chain rule of differentiation. The product can be rewritten as

$$\left|\frac{dS_n(w)}{dw}\right|_{w=w_n} = \left|\frac{w_0}{1+w_n}\right| \prod_{k=1}^{n-1} \left|\frac{w_k}{1+w_k}\right|.$$

Since $\rho < \mathrm{Re}(\Gamma+1/2)$ insures that

$$\left|\frac{w_k}{1+w_k}\right| < 1 - \epsilon(\Gamma,\rho) \quad \text{for } w_k \in V(\Gamma,\rho),$$

it follows that

(3.3) $\pi \operatorname{diam} S_n(V(\Gamma,\rho)) = \int_{|w-\Gamma|=\rho} |S_n'(w)|\,|dw| < K(\Gamma,\rho)(1-\epsilon(\Gamma,\rho))^{n-1}$.

This proves that $E(\Gamma,\rho)$ is a uniform modified convergence region with respect to $V(\Gamma,\rho)$. That $E(\Gamma,\rho)$ also is a convergence region and $V(\Gamma,\rho)$ is its limit region follows from [6].

Corollary 3.3. *The oval* $E(\Gamma,\rho)$ *with*

(3.4) $\rho < \min\{|1+\Gamma|-|\Gamma|,\ (|1+\Gamma|^2-|\Gamma|^2)/2\}$

is a convergence region. It contains the point $\Gamma(1+\Gamma)$ *in its interior and thus is a convergence neighborhood of* $\Gamma(1+\Gamma)$.

Proof. We have $\Gamma(1+\Gamma) \in \operatorname{Int} E(\Gamma,\rho)$ if and only if

$$|\Gamma(1+\Gamma)-\Gamma(1+\Gamma)(1-\frac{\rho^2}{|1+\Gamma|^2}| + \frac{\rho}{|1+\Gamma|}|\Gamma(1+\Gamma)| < \frac{\rho}{|1+\Gamma|}(|1+\Gamma|^2-\rho^2),$$

i.e.,

$$\rho^2 + \rho|\Gamma| + |\Gamma(1+\Gamma)| - |1+\Gamma|^2 < 0,$$

which holds for $\rho < |1+\Gamma| - |\Gamma|$. Moreover, by Theorem 3.2, $E(\Gamma,\rho)$ is a convergence region if

$$\rho < \operatorname{Re}(\Gamma+1/2) = (|1+\Gamma|^2 - |\Gamma|^2)/2. \quad \square$$

From (3.3) we can also get *a priori* truncation error estimates for $S_n(w_n)$, $w_n \in V$. This is done in Section 6.

The case which is not covered by these two theorems, is the case

(3.5) $\operatorname{Re}(\Gamma+1/2) \leq \rho \leq |\Gamma|$,

which can occur only if

$$|\Gamma| - \operatorname{Re}\Gamma \geq 1/2 .$$

We therefore have the following corollary:

Corollary 3.4. *If*

$$(3.6) \qquad |\Gamma| - \mathrm{Re}\Gamma < 1/2$$

then $E(\Gamma,\rho)$ *is a simple convergence region.*

Condition (3.6) holds if Γ is an interior point of the Scott and Wall parabola [20].

For Γ and ρ satisfying (3.5), the following result may be of help.

Theorem 3.5. *Any compact subset of the interior of an oval* $E(\Gamma,\rho)$ *is a simple convergence region, and* $V(\Gamma,\rho)$ *is a corresponding limit region.*

Proof. This follows directly from [6]. □

Another helpful fact is that if $E(\Gamma,\rho)$ is contained in a (uniform) convergence region, then it is a (uniform) convergence region. Or more generally, if A is a (uniform) convergence region and $A \cap E(\Gamma,\rho) \sim \{0\} \neq \emptyset$, then $A \cap E(\Gamma,\rho)$ is a (uniform) convergence region. In Section 7 we compare $E(\Gamma,\rho)$ to the simple uniform convergence regions given by the Worpitzky theorem [23] and the uniform parabola theorem [14, 17].

4. Representations of $E(\Gamma,\rho)$. The representation (3.2) of the Cartesian oval $E(\Gamma,\rho)$ with $\Gamma \in \mathbb{C}$, $\mathrm{Re}\Gamma > -1/2$ $0 < \rho < |1+\Gamma|$ is simple and useful. But for analyzing $E(\Gamma,\rho)$ geometrically it is often better to use polar coordinates. An application of the law of cosines to the triangle with vertices u, a, 0 leads to the formula for $\partial E(\Gamma,\rho)$ given below.

$$(4.1) \quad \partial E(\Gamma,\rho) = [r\, e^{i(\theta+2\alpha)}:\ r := r(\theta) := Y(\theta) \pm \sqrt{Y^2(\theta)+D}\ ],$$

where

$$(4.2) \qquad 2\alpha := \begin{cases} \arg(\Gamma(1+\Gamma)) & \text{if } \Gamma \neq 0, \\ 0 & \text{if } \Gamma = 0, \end{cases}$$

and

(4.3) $Y(\theta) := |\Gamma(1+\Gamma)|\cos\theta - \rho^2, \quad D := (|1+\Gamma|^2 - \rho^2)(\rho^2 - |\Gamma|^2).$

The set in (4.1) is taken over all $\theta \in (-\pi,\pi]$ for which $r(\theta)$ is well defined and positive. Whether both signs in front of the square root in (4.1) or just one have to be used, depends on whether $0 \in \mathrm{Int}(E(\Gamma,\rho))$ or not; i.e., on whether $|\Gamma| < \rho$ or not. One has

(4.4) $r(\theta) = Y(\theta) + \sqrt{Y^2(\theta)+D}\,, \quad -\pi < \theta \leq \pi, \text{ if } |\Gamma| < \rho\,.$

(4.5) $r_{+,-}(\theta) = Y(\theta) \pm \sqrt{Y^2(\theta)+D}\,, \quad |\theta| \leq \theta_0 \leq \pi/2, \text{ if } |\Gamma| \geq \rho\,,$

where

(4.5) (a) $\cos\theta_0 := (\rho^2 + \sqrt{|D|})/|\Gamma(1+\Gamma)|\,.$

Since $0 \in \mathrm{Int}(E(\Gamma,\rho))$ only if $|\Gamma| < \rho$, whereas the other focus $a = \Gamma(1+\Gamma)(1-\rho^2/|1+\Gamma|^2)$ of $E(\Gamma,\rho)$ is always contained in $\mathrm{Int}(E(\Gamma,\rho))$, the oval is more easily described by using a as the polar origin.

(4.6) $E(\Gamma,\rho) = [a+Re^{i(\phi+2\alpha)}:\ 0 \leq R \leq R(\phi) := \frac{\rho}{|1+\Gamma|}\,(y(\phi) - \sqrt{y^2(\phi)-d})],$

where

(4.7) $y(\phi) := |1+\Gamma|^2 + \rho|\Gamma|\cos\phi, \quad d := (|1+\Gamma|^2 - \rho^2)(|1+\Gamma|^2 - |\Gamma|^2).$

The sign in front of the square root in (4.6) must always be the same. Checking shows the "$-$" to be correct.

Regions of the type we are considering can sometimes be more conveniently expressed in terms of $w^2 = u$. The first to observe this phenomenon for continued fractions were Paydon and Wall [17] who realized that the parabola

(4.8) $P_\alpha := [u = re^{i(\theta+2\alpha)}:\ r \leq \dfrac{\frac{1}{2}\cos^2\alpha}{1-\cos\theta}], \quad |\alpha| < \pi/2$

can also be written as

(4.9) $$P_\alpha = [u = w^2:\ |\mathrm{Im}(we^{-i\alpha})| \le \frac{1}{2}\cos\alpha].$$

Introducing $u = w^2$ where $we^{-i\alpha} = \xi + i\eta$ and $r = |u| = \xi^2 + \eta^2$, the curve $\partial E(\Gamma,\rho)$ with $\Gamma \neq 0$ can be described by

(4.10) $$\eta^2 = \frac{\rho^2(|1+\Gamma|-|\Gamma|)^2-(r-|\Gamma(1+\Gamma)|+\rho^2)^2}{4|\Gamma(1+\Gamma)|},$$

but not all points satisfying (4.10) are on the image of the oval in the w-plane. We also need

$$r = \xi^2 + \eta^2 \ge \eta^2 \text{ and } \eta^2 \ge 0.$$

Checking shows that for $\rho \neq 0$,

(4.11) $$E(\Gamma,\rho) = [u = w^2:\ |\mathrm{Im}(we^{-i\alpha})| \le (\rho^2(|1+\Gamma|-|\Gamma|)^2 - (r-|\Gamma(1+\Gamma)|+\rho^2)^2)^{1/2}/2\sqrt{|\Gamma(1+\Gamma)|}\,,\ \text{where } |v_1| \le r \le |v_2|]$$

where $|v_1| = r(0)$ if $\rho \le |\Gamma|$ and $|v_1| = r(\pi)$ if $\rho > |\Gamma|$ and $|v_2| = r(\pi)$. (See also (5.2)-(5.3).)

Finally, let us describe $\partial E(\Gamma,\rho)$ in terms of Cartesian coordinates. By setting $r\cos\theta = x$, $r\sin\theta = y$ in (4.1) we get after squaring twice

(4.12) $$(x^2+y^2)^2 - 4|\Gamma(1+\Gamma)|x(x^2+y^2) + 2(2|\Gamma(1+\Gamma)|^2-2\rho^4-D)x^2 - 2(2\rho^4+D)y^2 + 4D|\Gamma(1+\Gamma)|x + D^2 = 0.$$

The expression (4.1) is equivalent to (4.12) together with

(4.13) $$2|\Gamma(1+\Gamma)|x + D \ge x^2 + y^2$$

and if $D > 0$,

(4.14) $$r - Y(\theta) = \sqrt{x^2+y^2} - |\Gamma(1+\Gamma)|\frac{x}{\sqrt{x^2+y^2}} + \rho^2 \ge 0\,.$$

Hence,

(4.15) $$E(\Gamma,\rho) = [(x+iy)e^{i2\alpha}:\ (x,y) \text{ satisfies } (4.12)\text{–}(4.14)].$$

5. Geometric properties of the Cartesian oval $E(\Gamma,\rho)$. As mentioned earlier, the regions $E(\Gamma,\rho)$ were introduced into the theory of continued fractions by Lane [13] already in 1945. But the geometry of these regions was not taken up at that time. Later, Hillam and Thron, [3], and Jones and Thron [8] obtained such oval convergence regions, but again without discussing the geometry.

More interest was shown by Reid [18, 19]. He proved that our ovals are always convex (in addition to the obvious facts that they are closed simply connected domains symmetric about a line through $\Gamma(1+\Gamma)$ and 0 (if $\Gamma \neq 0$)).

If $\Gamma = 0$, then (3.2) reduces to the disk

$$(5.1) \qquad E(0,\rho) = [u \in \mathbb{C}: |u| \leq \rho(1-\rho)], \quad 0 < \rho < 1.$$

For $\rho = 1/2$ this is exactly the Worpitzky disk, [23]. In the following we assume $\Gamma \neq 0$.

Let us consider the representation (4.1) of the boundary of $E(\Gamma,\rho)$. Then we firstly see that $\partial E(\Gamma,\rho)$ intersects its axis of symmetry at the points

$$(5.2) \qquad v_1 := \begin{cases} (|\Gamma|-\rho)(|1+\Gamma|+\rho)e^{i2\alpha} & \text{if } \rho \leq |\Gamma|, \\ (|\Gamma|-\rho)(|1+\Gamma|-\rho)e^{i2\alpha} & \text{if } \rho > |\Gamma|, \end{cases}$$

and

$$(5.3) \qquad v_2 := (|\Gamma|+\rho)(|1+\Gamma|-\rho)e^{i2\alpha}\,.$$

Since $|v_1| < |v_2|$, we shall call v_1 the <u>lower vertex</u> and v_2 the <u>upper vertex</u> of $E(\Gamma,\rho)$.

We also have $Y^2(\theta) + D \geq 0$ for all permissible θ where

$$Y^2(\theta) + D = 0 \text{ only if } \rho < |\Gamma| \text{ and } |\theta| = \theta_0\,.$$

For $\rho \leq |\Gamma|$ and $|\theta| < \theta_0$ we therefore have

$$\frac{dr(\theta)}{d(\theta)} = Y'(\theta)\Big(1 \pm \frac{Y(\theta)}{\sqrt{Y^2(\theta)+D}}\Big) = -|\Gamma(1+\Gamma)|\sin\theta\, \frac{\pm r(\theta)}{\sqrt{Y^2(\theta)+D}}\,.$$

Thus $r(\theta)$ increases monotonely as the boundary of $E(\Gamma,\rho)$ is traced from v_1 to v_2.

For $\rho > |\Gamma|$ we have

$$\frac{dr(\theta)}{d\theta} = -|\Gamma(1+\Gamma)|\sin\theta \frac{r(\theta)}{\sqrt{Y^2(\theta)+D}},$$

so that again $r(\theta)$ increases monotonely as θ goes from $\pm\pi$ to 0; that is, as the boundary of $E(\Gamma,\rho)$ is traced from v_1 to v_2.

Using the representation (4.6) of $E(\Gamma,\rho)$, we always have

$$y^2(\phi) - d \geq 0,$$

where

$$y^2(\phi) - d = 0 \text{ only if } \rho = |\Gamma| \text{ and } \phi = \pi.$$

For all (ρ,ϕ) except $(|\Gamma|,\pi)$ we therefore have

$$\frac{dR(\phi)}{d\phi} = \frac{\rho}{|1+\Gamma|} y'(\phi)\left(1 - \frac{y(\phi)}{\sqrt{y^2(\phi)-d}}\right) = \frac{-y'(\phi)R(\phi)}{\sqrt{y^2(\phi)-d}}$$
$$= \frac{R(\phi)\rho|\Gamma|\sin\phi}{\sqrt{y^2(\phi)-d}}.$$

This means firstly that $\partial E(\Gamma,\rho)$ is a smooth Jordan curve, except possibly at $z = 0$ if $0 \in \partial E(\Gamma,\rho)$. Secondly, since $dR(\phi)/d\phi > 0$ for positive $\phi < \pi$, we have that $R(\phi)$ increases monotonely as $\partial E(\Gamma,\rho)$ is traced from v_2 to v_1. This could also have been seen from the representation (3.2) of $E(\Gamma,\rho)$, according to which $z = r(\theta)e^{i\theta} = R(\phi)e^{i\phi}$ is a point on $\partial E(\Gamma,\rho)$ if and only if

$$R(\phi) + \frac{\rho}{|1+\Gamma|} r(\theta) = \frac{\rho}{|1+\Gamma|} (|1+\Gamma|^2 - \rho^2).$$

Since $r(\theta)$ increases monotonely as $z = r(\theta)e^{i\theta} = R(\phi)e^{i\phi}$ moves from v_1 to v_2, $R(\phi)$ must decrease monotonely.

As a consequence of this we get, thirdly, that

$$V(a,R(0)) \subseteq E(\Gamma,\rho) \subseteq V(a,R(\pi)) \tag{5.4}$$

and

$$V(0,r(\pi)) \subseteq E(\Gamma,\rho) \subseteq V(0,r(0)) \quad \text{if} \quad |\Gamma| < \rho\,, \tag{5.5}$$

where $V(C,R)$ is understood to mean the disk

$$V(C,R) = [v\colon |v-C| \leq R]\,.$$

The relations (5.4) and (5.5) also hold if $\Gamma = 0$, since then all the regions involved coincide. We have proved the following.

Theorem 5.5. *The following inclusion relations are valid.*

(A) *If* $\rho \leq |\Gamma|$, *then*

$$V(a,\rho(|1+\Gamma|-|\Gamma|)(1-\frac{\rho}{|1+\Gamma|})) \subseteq E(\Gamma,\rho) \subseteq V(a,\rho(|1+\Gamma|-|\Gamma|)(1+\frac{\rho}{|1+\Gamma|}))$$

and

$$E(\Gamma,\rho) \subseteq V(0,|v_2|) = V(0,(|\Gamma|+\rho)(|1+\Gamma|-\rho)).$$

(B) *If* $\rho > |\Gamma|$, *then*

$$V(a,\rho(|1+\Gamma|-|\Gamma|)(1-\frac{\rho}{|1+\Gamma|})) \subseteq E(\Gamma,\rho) \subseteq V(a,\rho(|1+\Gamma|+|\Gamma|)(1-\frac{\rho}{|1+\Gamma|}))$$

and

$$V(0,|v_1|) \subseteq E(\Gamma,\rho) \subseteq V(0,|v_2|).$$

(C) *If* $\Gamma \neq 0$, *then*

$$\partial V(a,\rho(|1+\Gamma|-|\Gamma|)(1-\frac{\rho}{|1+\Gamma|})) \cap \partial E(\Gamma,\rho) = [v_2]$$

and

$$\partial V(a,\rho(|1+\Gamma|-|\Gamma|)(1+\frac{\rho}{|1+\Gamma|}) \cap \partial E(\Gamma,\rho) = [v_1] \quad \text{if} \quad \rho \leq |\Gamma|,$$

$$\partial V(a,\rho(|1+\Gamma|+|\Gamma|)(1-\frac{\rho}{|1+\Gamma|}) \cap \partial E(\Gamma,\rho) = [v_1] \quad \text{if} \quad \rho > |\Gamma|.$$

6. Truncation error bounds. An extensive discussion of truncation error questions is beyond the scope of this paper. We shall however give a few results which come rather easily. From the proof of Theorem 3.2 we get the following uniform *a priori* truncation error bounds if $\rho < \text{Re}(\Gamma+1/2)$.

Theorem 6.1. *If* $\rho < \text{Re}(\Gamma+1/2)$ *and*

$$a_n \in E(\Gamma,\rho) \sim [0], \quad w_n \in V(\Gamma,\rho) \text{ for } \textit{all} \text{ n},$$

then $K(a_n/1)$ *converges to a value* $f \in V(\Gamma,\rho)$, *and*

$$\text{(6.1)} \quad |f-S_n(w_n)| \leq 2\rho \frac{|\Gamma|+\rho}{|1+\Gamma|-\rho} \left(1-\frac{2(\text{Re}\Gamma+1/2-\rho)}{(|1+\Gamma|+\rho)^2}\right)^{(n-1)/2}, \quad n = 1,2,3,\cdots.$$

Proof. We have

$$|f-S_n(w_n)| \leq \text{diam } S_n(V(\Gamma,\rho)) \leq \rho\cdot \max_{w_k\in V(\Gamma,\rho)} \left\{ \left|\frac{w_0}{1+w_n}\right| \prod_{k=1}^{n-1} \left|\frac{w_k}{1+w_k}\right| \right\}$$

$$= 2\rho\cdot\frac{|\Gamma|+\rho}{|1+\Gamma|-\rho} \left(\max \left|\frac{w_k}{1+w_k}\right|\right)^{n-1},$$

where we have for $w_k = \Gamma + \rho e^{i\theta} \in \partial V(\Gamma,\rho)$

$$\text{(6.2)} \qquad \left|\frac{w_k}{1+w_k}\right|^2 = 1 - \frac{1+2\text{Re}\Gamma+2\rho\cos\theta}{|1+\Gamma+\rho e^{i\theta}|^2}$$

$$\leq 1 - \frac{2(\text{Re}\Gamma+1/2-\rho)}{(|1+\Gamma|+\rho)^2}.$$

We can also use $V(\Gamma,\rho)$ to derive *a posteriori* truncation error bounds for $S_n(w_n)$. *A posteriori* bounds are often sharper than *a priori* bounds. The following method is an easy generalization of a result by Overholt, [16].

Theorem 6.2. *If* $K(a_n/1)$ *converges to* $f \in \mathbb{C}$, *and*

$$a_n \in E(\Gamma,\rho) \sim [0], \quad w_n \in V(\Gamma,\rho) \text{ for all } n,$$

then

$$\text{(6.3)} \quad |f-S_n(w_n)| \leq \frac{2\rho|h_n||a_1|}{(|h_n+\Gamma|-\rho)|h_n+w_n|} \prod_{k=2}^{n} \left|\frac{h_k-1}{h_k}\right|, \quad n = 1,2,3, \cdots,$$

where

$$\text{(6.4)} \quad h_k := -S_k^{-1}(\infty) = 1 + a_k/h_{k-1}, \quad h_1 = 1.$$

Remark. The quantity h_k can be computed recursively using (6.4). This is usually a stable process.

If $S_n(w_n)$ is computed by the forward recurrence algorithm, that is by using the well known relation

$$\text{(6.5)} \qquad S_n(w_n) = \frac{A_n+A_{n-1}w_n}{B_n+B_{n-1}w_n}, \quad n = 1,2,3, \cdots,$$

where $\{A_n\}$ and $\{B_n\}$ are computed recursively by

$$\text{(6.6)} \qquad Y_n = Y_{n-1} + a_n Y_{n-2}, \quad n = 1,2,3, \cdots$$

with initial conditions

$$\text{(6.7)} \qquad A_{-1} = B_0 = 1, \quad A_0 = B_{-1} = 0,$$

then $h_k = B_k/B_{k-1}$ can be computed directly.

Proof of Theorem 6.2. By use of (6.5) we find that $f_n = A_n/B_n$, $h_n = B_n/B_{n-1}$ and

$$f - S_n(w_n) = \frac{h_n(w_n-f^{(n)})}{(h_n+f^{(n)})(h_n+w_n)} (f_n-f_{n-1}).$$

Overholt, [16], proved that

$$f_n - f_{n-1} = S_{n-1}(a_n) - S_{n-1}(0) = \frac{f_{n-2}-f_{n-1}}{1+h_{n-1}/a_n} = \frac{h_n-1}{h_n}(f_{n-2}-f_{n-1}) .$$

Since $f_0 = S_0(0) = 0$ and $-h_n \notin V(\Gamma,\rho)$ (because $S_n(-h_n) = \infty \notin V(\Gamma,\rho)$), this gives the result. □

A particularly interesting case arises if $K(a_n/1)$ is limit periodic; that is $\{a_n\}$ converges. If all a_n except possibly the very first ones, are contained in $E(\Gamma,\rho)$ for a suitable choice of Γ and ρ, the truncation error estimates above can be helpful. However, if $c = \lim a_n \neq 0,\infty$ and $c \notin (-\infty,-\frac{1}{4}]$, it is well known that with

$$\tilde{f} := K(c/1) = (\sqrt{1+4c} - 1)/2, \ \operatorname{Re}\sqrt{\cdots} > 0,$$

the sequence $\{S_n(\tilde{f})\}$ or $\{S_n(w_n)\}$ where $w_n \to \tilde{f}$ converges faster to the value f of $K(a_n/1)$ in the sense that if $f \neq \infty$, then

$$\frac{f-S_n(w_n)}{f-S_n(u_n)} \to 0 \ \text{ if } \ w_n \to \tilde{f} \ \text{ and } \ \liminf |u_n-\tilde{f}| > 0, \ [4,22].$$

Therefore we get better estimates if we can choose $\Gamma = \tilde{f}$ and let ρ vary with n, ρ_n, such that $\rho_n \to 0$. This idea was used by Baltus and Jones, [1]. Using our results we get

Theorem 6.3. *Let* $K(a_n/1)$ *be a limit periodic continued fraction with*

$$c = \lim a_n \in \hat{\mathbb{C}} \sim (-\infty,-\frac{1}{4}], \ c \neq 0,\infty .$$

Set

$$\Gamma_c := (\sqrt{1+4c} - 1)/2, \ \operatorname{Re}\sqrt{\cdots} > 0.$$

If there exists a $0 < \rho_1 < \operatorname{Re}(\Gamma_c+1/2)$ *such that* $a_n \in E(\Gamma_c,\rho_1)$, $n \geq 1$, *then*

$$(6.9) \quad |f-S_n(\Gamma_c)| \leq \rho_{n+1} \frac{|\Gamma_c|+\rho_0}{|1+\Gamma_c|} \prod_{k=2}^{n} (1-2\frac{\operatorname{Re}\Gamma_c+1/2-\rho_k}{(|1+\Gamma_c|+\rho_k)^2})^{1/2}, \ n = 1,2,\cdots,$$

where

(6.10) $\rho_k = \min\{\rho > 0\colon\ a_m \in E(\Gamma_c,\rho) \text{ for all } m \geq k\}$.

Remarks. The conditions on c imply that $K(a_n/1)$ converges. Since $\rho_n \to 0$, $c \in E(\Gamma_c,\rho_n)$ from some n on. (We have by straight forward computation that $c = \Gamma_c(1+\Gamma_c) \in E(\Gamma_c,\rho)$ if and only if $\rho \leq |1+\Gamma_c| - |\Gamma_c|$.)

Proof of Theorem 6.2. Since $a_m \in E(\Gamma_c,\rho_{n+1})$ for all $m \geq n+1$, it follows that the value $f^{(n)}$ of $\overset{\infty}{\underset{k=1}{K}}(a_{n+k}/1)$ is contained in $V(\Gamma_c,\rho_{n+1})$. Moreover, if $w_n \in V(\Gamma_c,\rho_n)$, then $s_n(w_n) = a_n/(1+w_n) \in V(\Gamma_c,\rho_{n-1})$.

From the proof of Theorem 3.2 it follows that

$$|f - S_n(w_n)| \leq \operatorname{diam} S_n(V(\Gamma_c,\rho_n)) = \frac{1}{\pi} \int_{|w-\Gamma_c|=\rho_n} |S_n'(w)|\,|dw|$$

$$\leq 2\rho_n \left|\frac{w_0}{1+w_n}\right| \prod_{k=1}^{n-1} \left|\frac{w_k}{1+w_k}\right|,$$

where $w_k = s_{k+1} \circ \cdots \circ s_n(w_n) \in V(\Gamma_c,\rho_k)$ for all $k < n$. From (6.2) it follows that

$$\left|\frac{w_k}{1+w_k}\right| \leq \sqrt{1 - 2\,\frac{\operatorname{Re}\Gamma_c + 1/2 - \rho_k}{(|1+\Gamma_c| + \rho_k)^2}} \quad \text{for } 1 \leq k < n.$$

Combining this and setting $w_n = \Gamma_c$ gives (6.9). □

Another interesting type of continued fractions where the oval theorems are useful are the S-fractions $K(\beta_n z/1)$, $\beta_n > 0$, $z \in \mathbb{C}$. If $\arg z = 2\alpha$, $|\alpha| < \pi/2$, and $|\beta_n z|$ is bounded, we can choose Γ and ρ such that $E(\Gamma,\rho)$ is a "thin" oval with axis along the ray $\arg z = 2\alpha$, "long" enough to contain all $\beta_n z$. We have the following.

Theorem 6.4. *Let* $K(\beta_n z/1)$, $\beta_n > 0$, $z \in \mathbb{C} \sim [0]$, $|\arg z| < \pi$ *be an S-fraction such that* $|\beta_n z| \leq M$ *for an* $M > 0$ *for all* n. *Then*

(6.11) $$V_\alpha := V(\Gamma,|\Gamma|) \cap V_\alpha^*,$$

where

$$(6.12)\quad |\Gamma| \geq M/2\cos(\frac{1}{2}\arg z) \text{ and } \arg(\Gamma(1+\Gamma)) = \arg z = 2\alpha$$

and

$$(6.13)\quad V_\alpha^* = \begin{cases} \mathbb{R}^+ \text{ if } \arg z = 0 \\ [v \in \mathbb{C}\colon\ 0 < (\arg v)(\operatorname{sgn}\arg z) \leq |\arg z|], \text{ otherwise} \end{cases}$$

is a value region for $K(\beta_n z/1)$.

Remarks. By saying that V_α is a value region for $K(\beta_n z/1)$ we mean that V_α is a value region for the element region $E := [\beta_n z\colon\ n \in \mathbb{N}]$.

In order to prove this result we shall use the following lemma.

Lemma 6.5. $|1+\Gamma| - |\Gamma| \geq \cos\alpha$.

This clearly holds for $\alpha = 0$. For $0 < |\alpha| < \pi/2$ we refer to Lemma 7.4, formula (7.8) in the next section.

Proof of Theorem 6.4. The oval $E(\Gamma,|\Gamma|)$ has vertices at $v_1 = 0$ and $v_2 = |v_2|e^{i2\alpha}$, where

$$|v_2| = 2|\Gamma|(|1+\Gamma|-|\Gamma|) \geq 2|\Gamma|\cos\alpha \geq M,\ \ 2\alpha = \arg z.$$

This means that $\beta_n z \in E(\Gamma,|\Gamma|)$ for all n. Since $0 \in V(\Gamma,|\Gamma|)$, we have that $V(\Gamma,|\Gamma|)$ is a value region for $E(\Gamma,|\Gamma|)$. From [2] it follows that the sector V_α^* is a value region for the ray $L_\alpha = [u \in \mathbb{C}\colon\ \arg u = 2\alpha]$. Hence $V_\alpha = V(\Gamma,|\Gamma|) \cap V_\alpha^*$ is a value region for $E(\Gamma,|\Gamma|) \cap L_\alpha$. □

To obtain truncation error estimates for $K(\beta_n z/1)$ we can also use $V(\Gamma,|\Gamma|)$ as a value region for $K(\beta_n z/1)$. If $\rho = |\Gamma| < \mathrm{Re}\Gamma + 1/2$; that is Γ is an inner point in the Scott and Wall parabola (3.6), then we can use Theorem 6.1. Theorem 6.2 can always be used.

7. Intersection of $E(\Gamma,\rho)$ and known simple uniform convergence regions. The best known simple uniform convergence regions are the Worpitzky disk

$$(7.1)\qquad W := [u\colon\ |u| \le 1/4],\ [23].$$

and bounded subsets of the parabolic region

$$(7.2)\quad P_\alpha := [u\colon\ |u| - \mathrm{Re}(ue^{-i2\alpha}) \le \frac{1}{2}\cos^2\alpha],\ |\alpha| < \pi/2,\ [14,17].$$

(Other representations of P_α are given by (4.8) and (4.9).) The parabola ∂P_α has its focus at 0, vertex at $-\frac{1}{4}\cos^2\alpha e^{i2\alpha}$ and axis along the ray $L_\alpha := [u\colon\ \arg(u) = 2\alpha]$, and it passes through $-\frac{1}{4}$. The Worpitzky disk W is contained in P_0, and $\partial W \cap \partial P_0 = [-\frac{1}{4}]$. (See [11, Theorem 4.45B].)

In view of the last comments in Section 3, it is of interest to study when $E(\Gamma,\rho)$ is contained in one of these convergence regions.

Let us introduce the notation

$$(7.3)\qquad \Sigma := |1+\Gamma| + |\Gamma|,\quad \Delta := |1+\Gamma| - |\Gamma|.$$

We then have

Theorem 7.1. $E(\Gamma,\rho) \subseteq W$ *if and only if*

$$(7.4)\quad \rho \ge (\Delta + \sqrt{\Sigma^2 - 1})/2 \quad or \quad \rho \le (\Delta - \sqrt{\Sigma^2 - 1})/2.$$

Proof. From Theorem 5.1 it follows that $E(\Gamma,\rho) \subseteq W$ if and only if $|v_2| \le 1/4$, which is equivalent to (7.4). □

Comparing $E(\Gamma,\rho)$ to the parabolic regions is more intricate. But because of the symmetry properties of $E(\Gamma,\rho)$ and P_α, we only have to compare $E(\Gamma,\rho)$ to P_α and P_β, where, for $\Gamma \neq 0$,

$$(7.5) \qquad \alpha = \frac{1}{2}\arg(\Gamma(1+\Gamma)), \quad \beta = \begin{cases} \alpha - \pi/2 & \text{if } \alpha > 0, \\ \alpha + \pi/2 & \text{if } \alpha \leq 0. \end{cases}$$

($P_{\pm\pi/2} := \emptyset$.) Indeed, we can even forget about P_β because of the following result.

Theorem 7.2. *If* $\Gamma \notin \mathbb{R}^+ \cup \{0\}$ *and* $E(\Gamma,\rho) \subseteq P_\beta$, *then* $E(\Gamma,\rho) \subseteq W$.

Proof. From Theorem 5.1 follows that $E(\Gamma,\rho) \subseteq W$ if and only if $|v_2| \leq 1/4$. Since $v_2 \in P_\beta$, we have

$$|v_2| \leq \frac{1}{4}\cos^2\beta \leq 1/4. \qquad \square$$

So we are left with the comparison of $E(\Gamma,\rho)$ to P_α if $|\alpha| < \pi/2$. For $\alpha = \pm\pi/2$ or $\Gamma = 0$ we have the following simple result.

Theorem 7.3. *If* $-1/2 < \Gamma \leq 0$, *then* $E(\Gamma,\rho) \subseteq W \subseteq P_0$. *Moreover,* $E(0,\frac{1}{2}) = W$ *and* $\partial E(\Gamma,\Gamma+1/2) \cap \partial W = [-1/4]$, if $\Gamma \neq 0$ and for $\rho \neq \Gamma + \frac{1}{2}$, $E(\Gamma,\rho) \subset \operatorname{Int} W$.

Proof. If $\Gamma = 0$, the result follows from (5.1).

For $\Gamma < 0$ we have

$$(7.6) \quad |v_2| = (|\Gamma|+\rho)(|1+\Gamma|-\rho) \leq (|\Gamma|+\rho)(|1+\Gamma|-\rho)\,\big|_{\rho=\Delta/2} = \frac{\Sigma^2}{4} = \frac{1}{4},$$

where the equality sign only holds for $\rho = \Delta/2 = \Gamma + 1/2$. That $\partial E(\Gamma,\rho) \cap \partial W = [-1/4]$ for this value of ρ follows from the monotonicity of $r(\theta)$. (See Section 5.) $\square$

To study when $E(\Gamma,\rho) \subseteq P_\alpha$, we shall use the following lemma.

Lemma 7.4. *If* $\Gamma \in \mathbb{C} \sim \mathbb{R}$, $\operatorname{Re}\Gamma > -1/2$ *and we let*

$$(7.7) \qquad \delta := \frac{1}{2} \arg \frac{\Gamma}{1+\Gamma},$$

then $\alpha \neq 0,\ \delta \neq 0,\ 0 < \Delta < 1 < \Sigma$, *and the following relations hold.*

$$(7.8) \quad \Delta = \frac{\cos\alpha}{\cos\delta} = \frac{\cos\alpha}{\sin\delta}\tan\delta,\ \Sigma = \frac{\sin\alpha}{\sin\delta} = \frac{\cos\alpha}{\sin\delta}\tan\alpha,$$

$$(7.9) \quad |\Gamma(1+\Gamma)| = \frac{1}{4}(\Sigma^2-\Delta^2) = \frac{\cos^2\alpha}{4\sin^2\delta}(\tan^2\alpha - \tan^2\delta),$$

$$(7.10) \quad |1+\Gamma|^2 + |\Gamma|^2 = \frac{\Sigma^2+\Delta^2}{2} = \frac{\cos^2\alpha}{2\sin^2\delta}(\tan^2\alpha+\tan^2\delta),$$

$$(7.11) \quad \cos^2\alpha = \frac{\Delta^2(\Sigma^2-1)}{\Sigma^2-\Delta^2},$$

Proof. Consider the figure

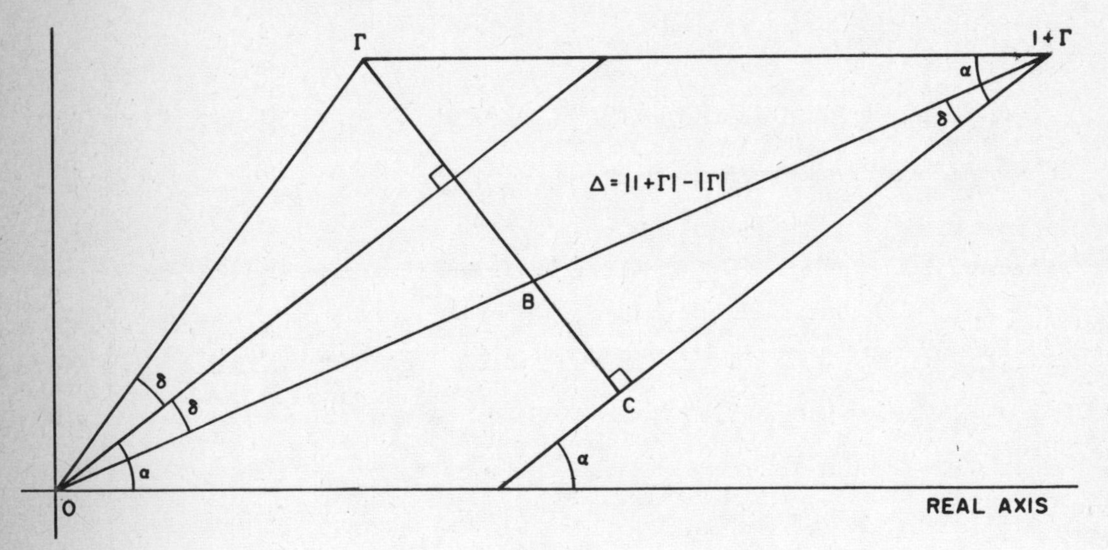

Figure 1.

From triangle $\Gamma C(1+\Gamma)$ one obtains that $\cos\alpha$ is equal to the length of $C(1+\Gamma)$. From triangle $BC(1+\Gamma)$ one has $\cos\delta = \cos\alpha/\Delta$. This proves (7.7)(a). Most of the remaining formulas are straight forward. To establish (7.11) we use the cosine law to obtain

$$\cos 2\delta = \frac{|\Gamma|^2+|1+\Gamma|^2-1}{2|\Gamma(1+\Gamma)|}$$

Hence

$$\cos^2\delta = \frac{1}{2}(\cos 2\delta+1) = \frac{|\Gamma|^2+|1+\Gamma|^2+2|\Gamma(H\Gamma)|-1}{4|\Gamma(1+\Gamma)|} = \frac{\Sigma^2-1}{\Sigma^2-\Delta^2}$$

from which the result follows. □

Using this we get the following theorem which restricts the possibilities for how $E(\Gamma,\rho)$ can "stick out" of P_α.

Theorem 7.5. *For* $\operatorname{Re}\Gamma > -1/2$, $0 < \rho < |1+\Gamma|$, $E(\Gamma,\rho)$ *does not contain any points on the ray* $z < -1/4$. *The point* $-1/4 \in E(\Gamma,\rho)$ *if and only if* $\rho = |\Gamma+1/2|$.

Proof. If $\Gamma \leq 0$, the result follows from Theorem 7.3. Assume that $\Gamma \notin \mathbb{R}^- \cup \{0\}$. We use the representation (4.1) of $\partial E(\Gamma,\rho)$. We are interested in $r_+(-2\alpha\pm\pi)e^{i\pi}$ which exists only if $|\Gamma| < \rho$ or if $|\Gamma| \geq \rho$ and $\pi-|2\alpha| < \theta_0$. Then

$$\begin{aligned} Y(-2\alpha\pm\pi) &= |\Gamma(1+\Gamma)|\cos(\pi-2\alpha) - \rho^2 \\ &= |\Gamma(1+\Gamma)|(1-2\cos^2\alpha) - \rho^2 \\ &= \frac{\Sigma^2+\Delta^2}{4} - \frac{\Sigma^2\Delta^2}{2} - \rho^2, \end{aligned}$$

and hence

$$r_+(-2\alpha\pm\pi) = \frac{\Sigma^2+\Delta^2}{4} - \frac{\Sigma^2\Delta^2}{2} - \rho^2 + \Sigma\Delta\sqrt{\rho^2-\frac{1}{4}(\Sigma^2-1)(1-\Delta^2)}$$

which attains its maximum for

$$\rho^2 = \frac{1}{4}(\Sigma^2+\Delta^2-1) = |\Gamma+1/2|^2$$

with

$$\max r(-2\alpha\pm\pi) = 1/4. \qquad \square$$

This result also follows from Theorem 3.5 since no simple convergence region can contain points from the ray $z < -1/4$.

There are two ways in which $E(\Gamma,\rho)$ can "stick out" of the parabolic region P_α: at the vertex and at the sides (and not the vertex). Let us first consider when $E(\Gamma,\rho)$ "sticks out" of P_α at the vertex; that is, when $v_1 \notin P_\alpha$, which is equivalent to $|v_1| > \frac{1}{4}\cos^2\alpha$ and $\rho > |\Gamma|$.

Theorem 7.6. *Let* v_1 *be the vertex of* $E(\Gamma,\rho)$ *given by* (5.2) *and let* $\Gamma \in ¢ \sim (-\infty,0]$, $\mathrm{Re}\Gamma > -1/2$. *Then* $v_1 \notin P_\alpha$ *if and only if*

$$(7.12)\qquad \frac{\Sigma}{2} - \frac{\Delta}{2}\sqrt{\frac{1-\Delta^2}{\Sigma^2-\Delta^2}} < \rho < \frac{\Sigma}{2} + \frac{\Delta}{2}\sqrt{\frac{1-\Delta^2}{\Sigma^2-\Delta^2}}\,.$$

Remarks. We always have

$$(7.13)\qquad |\Gamma| < \frac{\Sigma}{2} - \frac{\Delta}{2}\sqrt{\frac{1-\Delta^2}{\Sigma^2-\Delta^2}} \quad\text{and}\quad \frac{\Sigma}{2} + \frac{\Delta}{2}\sqrt{\frac{1-\Delta^2}{\Sigma^2-\Delta^2}} < |1+\Gamma|\,.$$

For $\Gamma > 0$ we have $1-\Delta^2 = 0$, and no ρ satisfies (7.12). In view of Theorem 7.3 this means that $v_1 \in P_\alpha$ if $\Gamma > -1/2$.

Proof of Theorem 7.6. We have from (5.2) and (7.11) that if $\rho > |\Gamma|$, then $|v_1| > \frac{1}{4}\cos^2\alpha$ if and only if

$$(\rho-|\Gamma|)(|1+\Gamma|-\rho) > \frac{1}{4}\,\frac{\Delta^2(\Sigma^2-1)}{\Sigma^2-\Delta^2}\,,$$

which is equivalent to (7.12).

Since for each $\Gamma \in ¢ \sim \mathbb{R}$ there exists $\rho > |\Gamma|$ such that (7.12) holds, this proves that not every oval is contained in some parabolic region P_ψ, $|\psi| < \pi/2$. It is easy to prove that we always have

$$(7.14)\qquad |v_1| \le 1/4\,.$$

A natural question is therefore, will every $E(\Gamma,\rho)$ where (7.12) holds be contained in W?

Comparing with Theorem 7.1 we see that this is not so. For instance if $\rho = \Sigma/2$ which satisfies (7.12), then $E(\Gamma,\rho) \sim W \neq \emptyset$ if

$$(\Delta-\sqrt{\Sigma^2-1})/2 < \Sigma/2 < (\Delta+\sqrt{\Sigma^2-1})/2$$

which holds if and only if $\Delta(2\Sigma-\Delta) > 1$, which is not difficult to satisfy. In view of Theorem 3.5, this means that some of the ovals $E(\Gamma,\rho)$ represent new convergence results.

Also Corollary 3.4 presents some new convergence results since there exists Γ and ρ such that $\Delta(2\Sigma-\Delta) > 1$ but $|\Gamma| - \mathrm{Re}\Gamma < 1/2$. This is illustrated by the following example.

Example 7.7. Choosing $\Gamma = 0.49i$ and $\rho = 0.9$ gives $|\Gamma| - \mathrm{Re}\Gamma = 0.49 < 1/2$, $\Gamma \notin \mathbb{R}$ and $|\Gamma| < \rho < |1+\Gamma| = 1.114$. From Theorem 7.1 it follows that $E(\Gamma,\rho) \sim W \neq \emptyset$ since

$$\Delta = |1+\Gamma| - |\Gamma| = 0.6235977, \quad \Sigma = |1+\Gamma| + |\Gamma| = 1.6035978$$

and thus

$$(\Delta+\sqrt{\Sigma^2-1})/2 = 0.9386014, \quad (\Delta-\sqrt{\Sigma^2-1})/2 < 0.$$

To see that $E(\Gamma,\rho) \sim P_\alpha \neq \emptyset$ we note that

$$\frac{\Sigma}{2} - \frac{\Delta}{2}\sqrt{\frac{1-\Delta^2}{\Sigma^2-\Delta^2}} = 0.6368127, \quad \frac{\Sigma}{2} + \frac{\Delta}{2}\sqrt{\frac{1-\Delta^2}{\Sigma^2-\Delta^2}} = 0.9667851 \qquad \square$$

Next we shall analyze the possibilities for $E(\Gamma,\rho)$ to "stick out at the sides" of P_α. The following result reveals that then we only have to consider ρ such that $0 < \rho < (\Sigma^2-\Delta^2)/2\Sigma$, where $|\Gamma| < (\Sigma^2-\Delta^2)/2\Sigma < |1+\Gamma|$.

Theorem 7.8. *If* $\Gamma \in \mathbb{C} \sim (-1/2,0]$, $\mathrm{Re}\Gamma > -1/2$ *and*

$$\rho \geq 2|\Gamma(1+\Gamma)|/(|1+\Gamma|+|\Gamma|) = (\Sigma^2-\Delta^2)/2\Sigma,$$

then $E(\Gamma,\rho) \sim P_\alpha \neq \emptyset$ *if and only if*

$$v_1 = -(\rho-|\Gamma|)(|1+\Gamma|-\rho)e^{i2\alpha} \notin P_\alpha .$$

Remark. There exists $\rho \geq (\Sigma^2-\Delta^2)/2\Sigma$ such that $v_1 \notin P_\alpha$. This is, for example, the case if $\rho = \Sigma/2$. (See Theorem 7.6.)

Proof. We shall use the representations (4.11) and (4.9) of $E(\Gamma,\rho)$ and P_α respectively. Then $E(\Gamma,\rho) \sim P_\alpha \neq \emptyset$ if and only if

(7.15) $\quad \rho^2\Delta^2 - (r-|\Gamma(1+\Gamma)|+\rho^2)^2 > |\Gamma(1+\Gamma)| \cos^2\alpha$

for some r, $|v_1| \leq r \leq |v_2|$. Since

$$|v_1| = -|\Gamma(1+\Gamma)| - \rho^2 + \rho\Sigma \geq |\Gamma(1+\Gamma)| - \rho^2$$

for $\rho > (\Sigma^2-\Delta^2)/2\Sigma$, the left hand side of (7.15) attains its maximum for $r = |v_1|$. □

For the case $|\Gamma| < \rho \leq (\Sigma^2-\Delta^2)/2\Sigma$ we can still use (7.15).

Theorem 7.9. *If* $\Gamma \in ¢ \sim (-1/2,0]$, $\mathrm{Re}\Gamma > -1/2$ *and*

$$|\Gamma| = \frac{\Sigma-\Delta}{2} < \rho \leq \frac{\Sigma^2-\Delta^2}{2\Sigma} = \frac{2|\Gamma(1+\Gamma)|}{|1+\Gamma|+|\Gamma|},$$

then $E(\Gamma,\rho) \sim P_\alpha \neq \emptyset$ *only if* $|\tan\alpha| > 1$. *For these values of* α *there exist values of* $\delta = \arg(\Gamma/(1+\Gamma))/2$ *near* 0 *and values of* ρ *near* $(\Sigma^2-\Delta^2)/2\Sigma$ *such that* $E(\Gamma,\rho) \sim P_\alpha \neq \emptyset$.

Proof. From the proof of Theorem 7.8 follows that the left hand side of (7.15) attains its maximum for $r = |\Gamma(1+\Gamma)| - \rho^2$, since now

(7.16) $\qquad |v_1| \leq |\Gamma(1+\Gamma)| - \rho^2 < |v_2|,$

and thus $E(\Gamma,\rho) \sim P_\alpha \neq \emptyset$ if and only if

(7.17) $\qquad \rho^2\Delta^2 > |\Gamma(1+\Gamma)|\cos^2\alpha .$

This inequality has a solution if and only if it holds for $\rho = (\Sigma^2-\Delta^2)/2\Sigma$; that is

$$\Delta^2(\Sigma^2-\Delta^2)/\Sigma^2 > \cos^2\alpha\,.$$

This can clearly not be satisfied if $\alpha=0$; i.e., if $\Gamma \in \mathbb{R}$. Assume that $\Gamma \notin \mathbb{R}$, so that $\delta \neq 0$. Using Lemma 7.4 the last inequality can then be written

$$(1+\tan^2\delta)(1-\frac{\tan^2\delta}{\tan^2\alpha}) > 1$$

from which $\tan^2\alpha > 1+\tan^2\delta$ follows. This only can be satisfied if $|\tan\alpha| > 1$, and clearly is satisfied if $|\tan\delta|$ is small enough. □

The third case to consider is the case where $0 \notin \operatorname{Int} E(\Gamma,\rho)$.

Theorem 7.10. *If* $\Gamma \in \mathbb{C} \sim (-1/2,0]$, $\operatorname{Re}\Gamma > -1/2$ *and* $\rho \leq |\Gamma|$, *then* $E(\Gamma,\rho) \sim P_\alpha \neq \emptyset$ *for some* ρ *if and only if* $E(\Gamma,|\Gamma|) \sim P_\alpha \neq \emptyset$, *which only occurs if* $|\tan\alpha| > 2\sqrt{2}$. *For these values of* α *there exists values of* $\delta = \arg(\Gamma/(1+\Gamma))/2$ *near* $(\operatorname{sgn}\alpha)\tan^{-1}\sqrt{2}$ *and values of* ρ *near* $|\Gamma|$ *such that* $E(\Gamma,\rho) \sim P_\alpha \neq \emptyset$.

Proof. For $\rho \leq |\Gamma|$ (7.16) still holds. Hence, $E(\Gamma,\rho) \sim P_\alpha \neq \emptyset$ if and only if (7.17) holds. Clearly (7.17) holds for some $\rho \leq |\Gamma|$ if and only if it holds for $\rho = |\Gamma|$.

For $\Gamma \in \mathbb{R}^+$ we have $\cos^2\alpha = \Delta^2 = 1$ and (7.17) cannot hold. Assume that $\Gamma \notin \mathbb{R}$. Using Lemma 7.4 we see that (7.17) with $\rho = |\Gamma|$ is equivalent to

(7.18) $\quad (1+\tan^2\delta)(|\tan\alpha|-|\tan\delta|) > |\tan\alpha|+|\tan\delta|\,.$

From this

$$|\tan\alpha| > \frac{2+\tan^2\delta}{|\tan\delta|}$$

follows easily. The maximum of the right hand side is attained for $|\tan\delta| = \sqrt{2}$ and is $2\sqrt{2}$. We have $|\tan\delta| < |\tan\alpha|$, $(\operatorname{sgn}\alpha = \operatorname{sgn}\delta)$ so that $|\tan\delta| = \sqrt{2}$ is in the permissible range. □

As a consequence of the last three theorems we also have the following.

Corollary 7.11. *If* $\Gamma > -1/2$, *then* $E(\Gamma,\rho) \subseteq P_0$.

In view of Theorem 7.5 $(E(\Gamma,\rho) \cap (-\infty,-1/4) = \emptyset)$ and the monotonicity of $r(\theta)$, $(r(\theta)\,e^{i(\theta+2\alpha)} \in \partial E(\Gamma,\rho))$, it follows from (7.15) that $E(\Gamma,\rho) \sim P_\alpha$ is contained in one of the two half planes we get when we remove the line through -1/4, perpendicular to the axis of symmetry for $E(\Gamma,\rho)$, from $\mathbb{C}$. The following theorem shows that $E(\Gamma,\rho) \sim P_\alpha$ is always contained in the half plane which contains the vertex of P_α.

Theorem 7.12. *If* $\Gamma \in \mathbb{C} \sim (-1/2,0]$, $\mathrm{Re}\Gamma > -1/2$ *and* $\rho < |1+\Gamma|$, *then* $E(\Gamma,\rho) \sim P_\alpha$ *is contained in the sector*

$$S_\alpha := \begin{cases} [re^{i\theta}\colon \pi < \theta < \pi+4\alpha,\ r > 0] & \text{if } \alpha > 0, \\ [re^{i\theta}\colon \pi+4\alpha < \theta < \pi,\ r > 0] & \text{if } \alpha < 0. \end{cases}$$

Proof. If $\rho \geq (\Sigma^2-\Delta^2)/2\Sigma$, the result follows trivially from Theorem 7.8.

If $0 < \rho < (\Sigma^2-\Delta^2)/2\Sigma$, it suffices to prove that the two points

$$r(\theta_1)\,e^{i(\pm\theta_1+2\alpha)} \in \partial E(\Gamma,\rho), \text{ where } r(\theta_1) = |\Gamma(1+\Gamma)| - \rho^2,$$

are contained in S_α if $E(\Gamma,\rho) \sim P_\alpha \neq \emptyset$. Assume that $E(\Gamma,\rho) \sim P_\alpha \neq \emptyset$. Then $\Gamma \notin \mathbb{R}$ and by Lemma 7.4 and (7.17)

$$\rho^2 > (\Sigma^2-1)/4\,. \tag{7.19}$$

Assume first that $\rho > |\Gamma|$. Then $r(\theta_1) < r(-2\alpha\pm\pi)$ since

$$|\Gamma(1+\Gamma)| - \rho^2 < \frac{\Sigma^2+\Delta^2}{4} - \frac{\Sigma^2\Delta^2}{2} - \rho^2 + \Sigma\Delta\sqrt{\rho^2-(\Sigma^2-1)(1-\Delta^2)/4}$$

if and only if

$$\Sigma^4 - \Sigma^2 - \Sigma^2\Delta^2 + \Delta^2 < 4\Sigma^2\rho^2$$

which always holds if (7.19) holds. Hence $r(\theta_1)e^{i(\pm\theta_1+2\alpha)} \in S_\alpha$ and thus $E(\Gamma,\rho) \sim P_\alpha \subseteq S_\alpha$ by Theorem 7.5 and the monotonicity of $r(\theta)$.

If $\rho \le |\Gamma|$, the result follows by the same argument as above if we can prove that $\theta_0 > \pi - |2\alpha|$ when $E(\Gamma,\rho) \sim P_\alpha \neq \emptyset$, where θ_0 is given by (4.5)(a). But that is indeed so by (7.19) since then

$$\rho^2 - (\Sigma^2-1)(1-\Delta^2)/4 > 0. \qquad \square$$

8. Special classes of ovals. In the previous section we have encountered choices of ρ which lead to ovals of particular interest, either because of their simplicity or because of certain optimal properties that these ovals possess. We arrive at the following table.

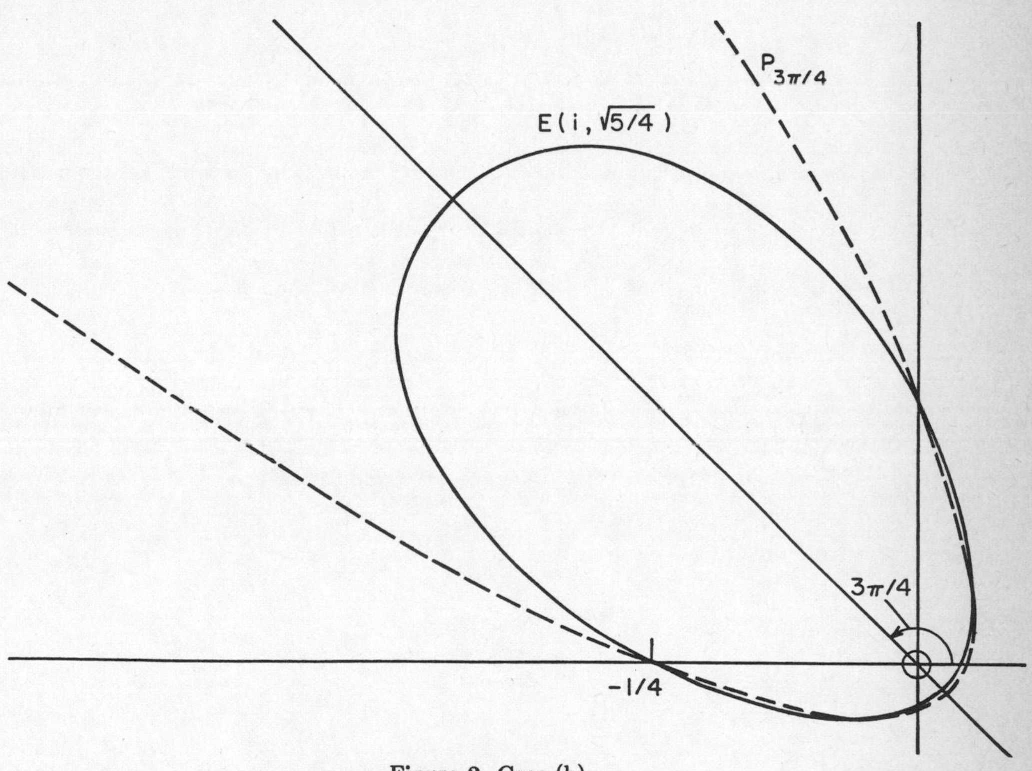

Figure 2 Case (b)

(a) $\rho^2 = |\Gamma|^2 = \dfrac{(\Sigma-\Delta)^2}{4}$ $\qquad$ $0 \in \partial E(\Gamma,\rho)$, longest diameter along axis of symmetry for the class $\rho \leq |\Gamma|$.

(b) $\rho^2 = |\Gamma+\frac{1}{2}|^2 = \dfrac{\Sigma^2+\Delta^2-1}{4}$ $\qquad$ $-\frac{1}{4} \in \partial E(\Gamma,\rho)$, cuts off largest possible part of negative real axis.

(c) $\rho^2 = \dfrac{\Sigma^2}{4}$ $\qquad$ largest distance of lower vertex to origin if $0 \in E(\Gamma,\rho)$.

(d) $\rho^2 = |\Gamma|^2 = |\Gamma+\frac{1}{2}|^2$ $\qquad$ $0, -\frac{1}{4} \in \partial E(\Gamma,\rho)$.

(e) $\rho^2 = |\Gamma(1+\Gamma)| = \dfrac{\Sigma^2-\Delta^2}{4}$ $\qquad$ $|a| = |\Gamma|\Delta$, very simple for $\Gamma > 0$.

All ovals initially involve three real parameters, ρ, α, δ, where $2\alpha = \arg(\Gamma(1+\Gamma))$, $2\delta = \arg(\Gamma/(1+\Gamma))$ and

$$0 < \rho < |1+\Gamma|, \ |\alpha| < \pi/2, \ 0 < |\delta| < |\alpha|, \ \alpha\delta \geq 0.$$

After ρ has been chosen, as indicated above, the two parameters α and δ remain. It is desirable to keep α fixed in order to study all ovals with the same axis of symmetry. It is for this reason that the choice of Σ and Δ as parameters, though advantageous in many respects, is not as good a choice as α and δ, if $\Gamma \notin \mathbb{R}$.

We begin with a closer study of case (a). Using the polar representation (4.l) we have

$$Y(\theta) = |\Gamma(1+\Gamma)|\cos\theta - |\Gamma|^2 \geq 0, \ D = 0$$

for $\cos\theta \geq \cos\theta_0$, and thus

$$r_-(\theta) = 0,$$

$$r_+(\theta) = 2Y(\theta) = \frac{\Sigma-\Delta}{2}((\Sigma+\Delta)\cos\theta - (\Sigma-\Delta)).$$

This means that $\partial E(\Gamma,|\Gamma|)$ is not differentiable at 0. Indeed, it has the one-sided tangents

$$\arg(\pm z) = \pm\theta_0 = \pm\cos^{-1}(|\Gamma|/|1+\Gamma|).$$

If $\Gamma > 0$, then $\Delta = 1$, $\Sigma = 2\Gamma + 1$ and

$$r_+(\theta) = 2\Gamma((\Gamma+1)\cos\theta - \Gamma) \quad \text{for} \quad \cos\theta \geq \frac{\Gamma}{1+\Gamma},$$

and the diameter of the oval is

$$|v_2| = r_+(0) = (\Sigma-\Delta)\Delta = 2\Gamma .$$

If $\Gamma \notin \mathbb{R}$, we can use Lemma 7.4:

$$r_+(\theta) = \frac{\cos^2\alpha(1+\tan^2\delta)}{2\tan^2\delta}(\tan\alpha-\tan\delta)((\tan\alpha+\tan\delta)\cos\theta-(\tan\alpha-\tan\delta))$$

for

$$\cos\theta \geq \cos\theta_0 = \frac{|\Gamma|}{|1+\Gamma|} = \frac{\tan\alpha-\tan\delta}{\tan\alpha+\tan\delta},$$

and the diameter of $E(\Gamma,\rho)$ is

$$|v_2| = (\Sigma-\Delta)\Delta = \frac{\cos^2\alpha(1+\tan^2\delta)}{\tan\delta}(\tan\alpha-\tan\delta).$$

In general we have that the diameter of $E(\Gamma,\rho)$ is given by

$$|v_2|-|v_1| = 2\rho\Delta \quad \text{if} \quad \rho \leq |\Gamma|.$$

This means that for a given Γ (or fixed α, δ) the maximum diameter along the axis of symmetry is assumed for $\rho = |\Gamma|$, if $\rho \leq |\Gamma|$. As $\Gamma \to \infty$ (i.e., $\delta \to 0$), this diameter $\to \infty$.

Case (b) is important (see Theorem 7.5) as a maximal oval, since it contains as much of the negative real axis as is permissible. Unfortunately, the expression for $r(\theta)$, with α, δ as parameters, is fairly complicated. Before deriving it we note that

$$a = \Gamma(1+\Gamma)\left(1-\frac{(\frac{1}{2}+\Gamma)(\frac{1}{2}+\bar{\Gamma})}{(1+\Gamma)(1+\bar{\Gamma})}\right)$$

$$= \frac{\Gamma}{1+\overline{\Gamma}} \left(\frac{3}{4}+\mathrm{Re}\Gamma\right)$$

so that $u \in E(\Gamma, |\Gamma+\frac{1}{2}|)$ can be written (see (3.2)) as

$$(8.1) \quad \left|u-\frac{\Gamma}{1+\overline{\Gamma}}\left(\frac{3}{4}+\mathrm{Re}\Gamma\right)\right| + \left|\frac{\frac{1}{2}+\Gamma}{1+\Gamma}\right||u| \le \frac{|\frac{1}{2}+\Gamma|}{|1+\Gamma|}\left(\frac{3}{4}+\mathrm{Re}\Gamma\right).$$

One can verify directly that $-\frac{1}{4} \in \partial E(\Gamma, |\Gamma+\frac{1}{2}|)$.

If $\Gamma \ge 0$, the ovals represent a connecting link between the Worpitzky disk and the Scott and Wall parabola:

Theorem 8.1. *If* $\Gamma \ge 0$, *then the following hold:*

A. $E(\Gamma,\Gamma+\frac{1}{2}) \subseteq P_0$, $\partial E(\Gamma,\Gamma+\frac{1}{2}) \cap \partial P_0 = [-1/4] = [v_1]$.

B. *The diameter of* $E(\Gamma,\Gamma+\frac{1}{2})$ *along the axis of symmetry is given by*

$$\frac{1}{4} + |v_2| = \Gamma + \frac{1}{2}, \quad \textit{since} \quad v_2 = \Gamma + \frac{1}{4}.$$

C. *For* $\Gamma = 0$ *we have* $E(\Gamma,\Gamma+\frac{1}{2}) = [u: |u| \le \frac{1}{4}] = W$.

For $\Gamma \to \infty$ *we have* $E(\Gamma,\Gamma+\frac{1}{2}) \to P_0$.

Proof. Parts A, B and C for $\Gamma = 0$ follow directly from previous results. To see what happens if $\Gamma \to \infty$, we use (8.1). From Part A we know that $E(\Gamma,\Gamma+\frac{1}{2}) \subseteq P_0$ for all Γ. It remains to prove that $E(\Gamma,\Gamma+\frac{1}{2})$ asymptotically fills P_0. (8.1) can be written as

$$(8.1)' \quad \left|u\,\frac{1+\Gamma}{3/4+\Gamma}-\Gamma\right| + \frac{1/2+\Gamma}{3/4+\Gamma}\,|u| \le 1/2 + \Gamma .$$

Set $k_n = (1+\Gamma)/(3/4+\Gamma)$, then

$$|uk_n-\Gamma| - \Gamma = \frac{-2\Gamma k_n \mathrm{Re}\, u + |u|^2 k_n^2}{|uk_n-\Gamma| + \Gamma}.$$

Hence (8.1)′ is equivalent to $|u|(k_n - \frac{2}{3+4\Gamma}) \leq \frac{1}{2} + \frac{2k_n\Gamma - |u|^2 k_n^2}{|uk_n-\Gamma| + \Gamma} \mathrm{Re}\, u$. As $\Gamma \to \infty$ this approaches the Scott and Wall parabola $[z\colon |z| - \mathrm{Re}\, z \leq 1/2]$. □

If $\Gamma \notin \mathbb{R}$, we have, using Lemma 7.4

$$|\Gamma(1+\Gamma)| = \frac{\Sigma^2-\Delta^2}{4} = \frac{1}{4\epsilon}(\tan^2\alpha - \tan^2\delta),$$

where

$$\epsilon = \sin^2\delta / \cos^2\alpha,$$

and

$$\rho^2 = \frac{\Sigma^2+\Delta^2-1}{4} = \frac{1}{4\epsilon}(\tan^2\alpha + \tan^2\delta - \epsilon),$$

$$D = \frac{\Sigma^2\Delta^2}{4} - \frac{1}{16} = \frac{1}{16\epsilon^2}(4\tan^2\alpha\tan^2\delta - \epsilon^2),$$

$$Y(\theta) = \frac{1}{4\epsilon}((\tan^2\alpha - \tan^2\delta)\cos\theta - \tan^2\alpha - \tan^2\delta + \epsilon).$$

Since $\rho > |\Gamma|$, we get the polar representation (4.1)

$$(8.2)\quad r(\theta) = Y(\theta) + \sqrt{Y^2(\theta)+D} = \frac{D}{\sqrt{Y^2(\theta)+D} - Y(\theta)}$$

$$= (4\tan^2\alpha\tan^2\delta - \epsilon^2)/$$

$$4\epsilon\Big(\sqrt{((\tan^2\alpha - \tan^2\delta)\cos\theta - \tan^2\alpha - \tan^2\delta + \epsilon)^2 + 4\tan^2\alpha\tan^2\delta - \epsilon^2}$$

$$- (\tan^2\alpha - \tan^2\delta)\cos\theta + \tan^2\alpha + \tan^2\delta - \epsilon\Big).$$

Using this expression we again get a bridge between the Worpitzky disk and the parabola P_α in the case $\Gamma \notin \mathbb{R}$:

Theorem 8.2. *If* $\Gamma \notin \mathbb{R}$, *then the following hold.*

A. $\partial E(\Gamma, |\Gamma+1/2|) \cap \partial P_\alpha \supseteq [-1/4]$.

B. $|\Gamma+1/2| > |\Gamma|$ *if and only if* $2\Sigma\Delta > 1$. *The diameter of* $E(\Gamma,|\Gamma+1/2|)$ *along the axis of symmetry is given by*

$$\begin{cases} |v_1|+|v_2| = 2\rho(|1+\Gamma|-\rho) = \dfrac{\sqrt{\Sigma^2+\Delta^2-1}\,(2\Sigma\Delta+1)}{2(\Sigma+\Delta+\sqrt{\Sigma^2+\Delta^2-1})} & \text{if } |\Gamma+\frac{1}{2}| > |\Gamma|, \\ |v_2|-|v_1| = 2\rho\Delta = \Delta\sqrt{\Sigma^2+\Delta^2-1} & \text{if } |\Gamma+\frac{1}{2}| \le |\Gamma| \end{cases}$$

C. *For* $\Gamma \to 0$ *we have* $E(\Gamma,|\Gamma+1/2|) \to [u\colon |u| \le 1/4] = W$.

For $\Gamma \to \infty$, α *fixed, we have* $E(\Gamma,|\Gamma+1/2|) \to P_\alpha$.

Remark. We no longer have that $E(\Gamma,|\Gamma+1/2|)$ is always $\subseteq P_\alpha$. For instance if $0 < |\alpha| < \pi/2$, then $E(0,1/2) \sim P_\alpha \neq \emptyset$. We do not even have that $v_1 \in P_\alpha$ for all choices of Γ. (See Theorem 7.6.)

Proof. Parts A, B and C for $\Gamma \to 0$ follow directly from previous results. To see what happens if $\Gamma \to \infty$, we use (8.2) with $\delta \to 0$. That gives

$$r(\theta) \to \frac{\cos^2\alpha}{2(1-\cos\theta)}$$

which gives the representation (4.8) of P_α. □

For case (c) we note that $|v_1|$ attains its maximum for $\rho = \Sigma/2$ if $\rho > |\Gamma|$. This gives

$$|v_1|_{\max} = (\frac{\Sigma}{2}-|\Gamma|)(|1+\Gamma|-\frac{\Sigma}{2}) = \frac{\Delta^2}{4}\,.$$

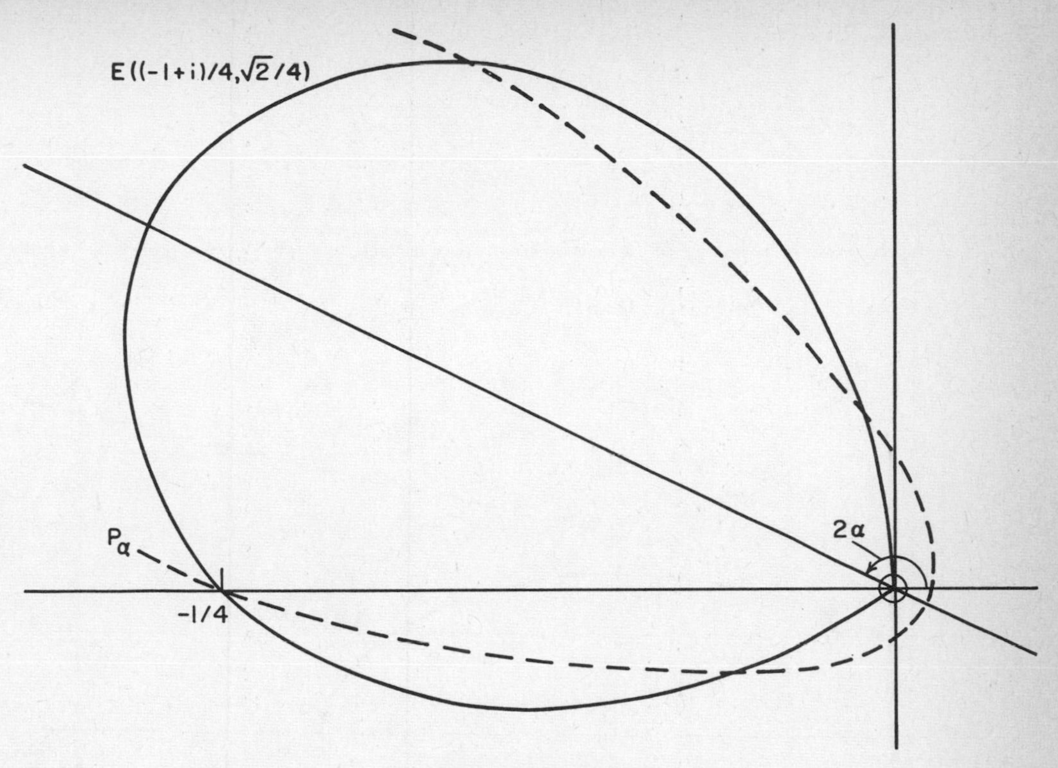

Figure 3 Case (d)

For case (d) it is convenient to use the parameter β obtained from $4\Gamma = -1 + i\beta$.

These are the only Γ for which $|\Gamma| = |\Gamma + \frac{1}{2}|$. We then have

$$16\Gamma(1+\Gamma) = -3 - \beta^2 + 2i\beta$$

$$16|\Gamma(1+\Gamma)| = \sqrt{(\beta^2+5)^2-16}$$

$$\cos 2\alpha = \frac{-(3+\beta^2)}{\sqrt{(\beta^2+5)^2-16}} = \frac{-x+2}{\sqrt{x^2-16}}, \quad x = 5 + \beta^2 .$$

For $r(\theta)$, in terms of β, one obtains

$$r(\theta) = \frac{1}{8}\left(\sqrt{\beta^4+10\beta^2+9}\cos\theta - 1 - \beta^2\right).$$

It is of interest to know for what values of α one can have an oval passing through both $z = 0$ and $z = -1/4$. The minimum for 2α is attained for $x = 8$. Thus

$$\cos 2\alpha_m = -6/\sqrt{64-16} = -\sqrt{3}/2$$

and

$$2\,\alpha_m = 5\pi/6\ .$$

For $x = 8$, $\beta^2 = 3$. The range of θ in the rotated oval is thus $|\theta| \le \cos^{-1}(1/\sqrt{3})$ which is, as expected, substantially larger than $\pi - 2\,\alpha_m = \pi/6$.

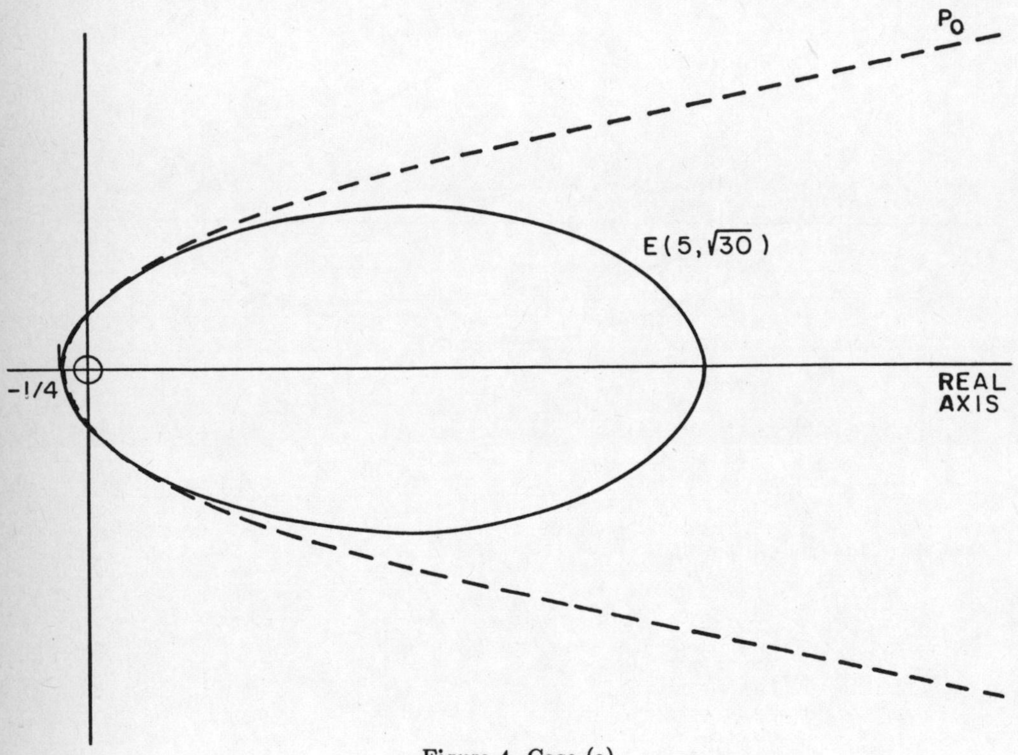

Figure 4 Case (e)

Finally, we turn to case (e). We note that

$$|a| = |\Gamma(1+\Gamma)|(1-\frac{\rho^2}{|1+\Gamma|^2}) = |\Gamma|\Delta\ .$$

If in particular $\Gamma > 0$, then $\Delta = 1$ and $a = |a| = \Gamma$. For $\Gamma > 0$ one thus has $z \in E(\Gamma, \sqrt{\Gamma(1+\Gamma)})$ if

$$|z-\Gamma| + \sqrt{\frac{\Gamma}{1+\Gamma}}\,|z| = \sqrt{\Gamma(1+\Gamma)}.$$

To compute $r(\theta)$ for $\Gamma \notin \mathbb{R}$ we proceed as in case (b). We have

$$D = \frac{\Delta^2(\Sigma^2-\Delta^2)}{4} = \frac{\cos^4\alpha}{4\sin^4\delta}\tan^2\delta(\tan^2\alpha-\tan^2\delta)$$

$$Y = \frac{\Sigma^2-\Delta^2}{4}(\cos\theta-1) = \frac{\cos^2\alpha}{4\sin^2\delta}(\tan^2\alpha-\tan^2\delta)(\cos\theta-1)$$

and hence

$$r(\theta) = \frac{D}{\sqrt{Y^2+D}-Y}$$

$$= \frac{\cos^2\alpha/\cos^2\delta}{1-\cos\theta+\sqrt{(1-\cos\theta)^2+4\tan^2\delta/(\tan^2\alpha-\tan^2\delta)}}.$$

Again P_α is obtained for the limiting value $\delta = 0$.

References

1. C. Baltus and W. B. Jones, Truncation error bounds for limit periodic continued fractions with $\lim a_n = 0$. *Numer. Math* *46* (1985), 541-569.

2. P. Henrici and P. Pfluger, Truncation error estimates for Stieltjes fractions. *Numer. Math* *9* (1966), 120-138.

3. K. L. Hillam and W. J. Thron, A general convergence criterion for continued fractions $K(a_n/b_n)$, *Proc. Amer. Math. Soc.* *16* (1965), 1256-1262.

4. L. Jacobsen, Modified approximants. Construction and applications. *Det Kgl. Norske Vit. Selsk. Skr.* (1983), No. 3, 1-46.

5. L. Jacobsen, General convergence of continued fractions. *Trans. Amer. Math. Soc.*, to appear.

6. L. Jacobsen, A theorem on simple convergence regions for continued fractions $K(a_n/1)$. These Lecture Notes.

7. L. Jacobsen, W. B. Jones and H. Waadeland, ____________________. These Lecture Notes.

8. W. B. Jones and W. J. Thron, Twin-convergence regions for continued fractions $K(a_n/1)$, *Trans. Amer. Math. Soc.* *150* (1970), 93-119.

9. W. B. Jones and W. J. Thron, Numerical stability in evaluating continued fractions. *Math. Comp.* *28* (1974), 795-810.

10. W. B. Jones and W. J. Thron, Truncation error analysis by means of approximant systems and inclusion regions. *Numer. Math.* *26* (1976), 117-154.

11. W. B. Jones and W. J. Thron, *Continued Fractions: Analytic Theory and Applications*, Encyclopedia of Mathematics and Its Applications, vol. 11, Addison-Wesley, Reading, Mass., 1980. Now available through Cambridge Univ. Press.

12. W. B. Jones, W. J. Thron, and H. Waadeland, Truncation error bounds for continued fractions $K(a_n/1)$ with parabolic element regions, *SIAM J. Numer. Anal.* *20* (1983), 1219-1230.

13. R. E. Lane, The value region problem for continued fractions, *Duke Math. J.* *12* (1945), 207-216.

14. W. Leighton and W. J. Thron, Continued fractions with complex elements. *Duke Math. J.* *9* (1942), 763-772.

15. G. Loria, *Spezielle algebraische und transzendente ebene Kurven, Theorie und Geschichte* I, 2. Aufl. B. G. Teubner, Leipzig and Berlin, 1910.

16. M. Overholt, The values of continued fractions with complex elements, *Det Kongelige Norske Vitenskabers Selskabs Skrifter,* (1983), No. 1, 109-116.

17. J. F. Paydon and H. S. Wall, The continued fraction as a sequence of linear transformations. *Duke Math. J.* *9* (1942), 360-372.

18. W. M. Reid, Parameterizations and factorizations of element regions for continued fractions $K(a_n/1)$, *Lecture Notes in Mathematics* No. 932, Springer-Verlag, Berlin, 1982.

19. W. M. Reid, Uniform convergence and truncation error estimates of continued fractions $K(a_n/1)$, Ph.D. thesis, University of Colorado, Boulder, 1978.

20. W. T. Scott and H. S. Wall, A convergence theorem for continued fractions, *Trans. Amer. Math. Soc.* *47* (1940), 155-172.

21. W. T. Scott and H. S. Wall, Value regions for continued fractions, *Bull. Amer. Math. Soc.* *47* (1941), 580-585.

22. W. J. Thron and H. Waadeland, Accelerating convergence of limit periodic continued fractions $K(a_n/1)$, *Numer. Math.* *34* (1980), 155-170.

23. J. Worpitzky, Untersuchungen über die Entwickelung der monodromen und monogenen Funktionen durch Kettenbrüche, *Jahresbericht, Friedrichs-Gymnasium und Realschule,* Berlin, 1865, 3-39.

Schur Fractions, Perron-Carathéodory Fractions and Szegö Polynomials, a Survey

W. B. Jones*
Department of Mathematics
Campus Box 426
University of Colorado
Boulder, CO 80309-0426 U.S.

O. Njåstad
Department of Mathematics
University of Trondheim, NTH
N-7034 Trondheim
Norway

W. J. Thron*
Department of Mathematics
Campus Box 426
University of Colorado
Boulder, CO 80309-0426, U.S.A.

CONTENTS

1. Introduction. When Carathéodory [3], in 1907, posed and solved the problem of characterizing, in terms of the coefficients c_n, those analytic functions $F(z) = \sum_{v=0}^{\infty} c_v z^v$ which map the unit disk into the right half plane he thought of it as an extension of a result of Landau, concerning functions analytic in the unit disk and not assuming the values 0,1. Later, in 1911, Toeplitz [19], Fischer [6] and Carathéodory and Fejer [4,5] gave different characterizations chiefly in terms of Toeplitz matrices. At that time they knew of connections to quadratic forms, quadrature formulae and moment problems.

The closely related problem of characterizing functions mapping the unit disk into itself was briefly considered by Carathéodory and Fejer [5]. In 1917/18 J. Schur [14] made very substantial progress on the problem by using a "continued fraction like" algorithm. Somewhat later, but independently, Hamel [8] employed modified approximants of the continued fraction

---

*The research of W.B.J. and W.J.T. was supported in part by the U.S. National Science Foundation under Grant No. DMS-8401717.

$$(1.1)\quad \frac{1}{b_0} - \frac{(1-|b_0|^2)b_1/b_0}{b_1 + zb_0} - \frac{(1-|b_1|^2)b_0b_1z}{b_2 + zb_1} - \cdots$$

as a tool in attacking the problem.

Szegö [16] became interested in Toeplitz matrices (1915,17) as well as in Hankel matrices (1918). In a paper on the latter topic he studied the "orthogonality system of polynomials" arising in this connection and pointed out these polynomials are the denominators of the approximants of a certain J-fraction. (The term "orthogonal polynomial" was, to the best of our knowledge, first used by Szegö in an article in 1921. The expression "orthogonal functions" was introduced by E. Schmidt in 1906.) In a series of papers (1918,20,21) Szegö showed that for Toeplitz matrices also there is an associated family of polynomials that can be considered as being orthogonal on the unit circle. He did not however (at that time) give the recurrence relations satisfied by these polynomials and did not relate them to continued fractions.

The recurrence relations involving the polynomials P_n as well as their reciprocals P_n^x appear first in Szegö's book [17] in 1939.

Starting in 1940 Geronimus [7] also became interested in these problems. In that context he introduced the continued fraction

$$(1.2)\quad 1 + \frac{2a_0z}{1-a_0z} - \frac{(1-|a_0|^2)a_1z/a_0}{1+a_1z/a_0} - \frac{(1-|a_1|^2)a_2z/a_1}{1+a_2z/a_1} - \cdots .$$

In 1944 Wall [20] replaced Schur's less familiar algorithm by the continued fraction

$$\gamma_0 + \frac{(1-|\gamma_0|^2)z}{\bar{\gamma}_0z} + \frac{1}{\gamma_1} + \frac{(1-|\gamma_1|^2)z}{\bar{\gamma}_1z} + \cdots ,$$

which Jones and Steinhardt [11] have called a <u>Schur fraction</u>.

The closely related continued fractions of Hamel and Geronimus have the disadvantage that they can be used only if all b_v (or a_v) do not vanish, which is an extraneous requirement. It is for this reason that we give preference to Schur fractions and hermitian PC-fractions where this assumption need not be made. These fractions embed the continued fraction (1.1) (or (1.2)) in the sense that the latter relate to the even approximants of the embedding fraction. Another way of looking at it is that between the P_n one interpolates the P_n^x in order to have a continued fraction which is always valid.

In this article we present the results of Carathéodory and Schur and their interrelations with trigonometric moment problems and polynomials orthogonal on the unit circle (Szegö polynomials) by making maximum use of Schur fractions and PC-fractions. The latter were recently introduced by the present authors [9] to study trigonometric and other strong moment problems.

Other, relatively new, advances treated in this paper include a study of the functions and continued fractions, under consideration, at ∞. That Schur fractions correspond to a power series at ∞ was first observed by Perron [13] and elaborated by Thron [18]. Here we are also able to give the explicit relation between the power series expression at 0 and that at ∞. In addition we trace a number of examples, originally investigated by Schur, starting at various places in the chain: Schur functions, Schur fractions, PC-fractions, Szegö polynomials, moments, moment distribution function, Szegö polynomials.

The material presented in this article is divided as follows: In Section 2 Schur functions, Schur fractions and the relationship between them are studied. In Section 3 we consider Carathéodory functions, Perron-Carathéodory fractions and their interrelationship. In Section 4 the connection between Schur functions and Schur fractions on the one hand and Carathéodory functions and Perron-Carathéodory fractions on the other hand is discussed. In Section 5 we point out the connection between Szegö polynomials and continued fractions. Section 6 contains examples illustrating some of the concepts taken up above. In Section 7 we briefly sketch how the trigonometric moment problem ties in with the topics treated above. We also give another example.

For basic information on continued fractions we refer to [12]. We give only a limited list of references. More extensive bibliographies can be found in some of the articles and books referred to here. Ammar and Gragg [22], [23] have recently given a fast algorithm for solving Toeplitz systems.

2. <u>Schur functions and Schur fractions</u>. A function f which is analytic in the open unit disk $[z: |z| < 1]$ and maps this disk into the closed disk $[z: |z| \leq 1]$ is called a <u>Schur function</u>. These functions were studied extensively by Schur in 1917/18 [14]. If $f(0) \in (-1,1)$ we shall call the function a <u>normalized</u> <u>Schur function</u>. We denote the class of all normalized Schur functions by $\mathcal{S}$, and define

$$\mathcal{S}_0 := [f \in \mathcal{S}: f(0) \neq 0]$$

We also introduce

$$\mathcal{D}_\infty := [g: g \text{ is analytic for } |z| > 1, |g(z)| > 1, \quad g(\infty) \in (-\infty,-1) \cup (1,\infty)] .$$

With every Schur function f we associate the <u>Schur reciprocal</u> $\hat{f}$

$$\hat{f}(z) := 1/\overline{f(1/\overline{z})} .$$

Note that

$$\hat{f}(z) \cdot \overline{f}(1/z) = 1,$$

where $\overline{h}(z) := \overline{h(\overline{z})}$; that is, only the coefficients (of the Taylor expansion) are conjugated, the argument z remains unchanged. Clearly, the mapping $f \to \hat{f}$ is a bijection from $\mathcal{S}_0$ onto $\mathcal{D}_\infty$.

Let the functions f and g have power series expansions

$$f(z) = a_0^{(0)} + \sum_{n=1}^{\infty} a_n z^n , \quad |z| < 1 , \tag{2.1}$$

$$g(z) = a_0^{(\infty)} + \sum_{n=1}^{\infty} a_{-n} z^{-n} , \quad |z| > 1 . \tag{2.2}$$

Then $g = \hat{f}$ iff

$$(\overline{a}_0^{(0)} + \sum_{n=1}^{\infty} \overline{a}_n z^n)(a_0^{(\infty)} + \sum_{n=1}^{\infty} a_{-n} z^{-n}) = 1 .$$

From this equation the coefficients of $\hat{f}$ can be found if those of f are known. In (2.1) set $a_0^{(0)} = a_0$; then H_n and H_n^* are the $(n+1)\times(n+1)$ matrices

$$H_n := \begin{pmatrix} a_0 & a_1 & \cdots & a_n \\ 0 & a_0 & \cdots & a_{n-1} \\ \vdots & & & \vdots \\ 0 & \cdots & 0 & a_0 \end{pmatrix} , \quad H_n^* := \begin{pmatrix} \overline{a}_0 & 0 & \cdots & 0 \\ \overline{a}_1 & \overline{a}_0 & \cdots & \\ & & & \\ \overline{a}_n & \overline{a}_{n-1} & \cdots & \overline{a}_0 \end{pmatrix} \tag{2.3}$$

Further let I_n be the $n\times n$ identity matrix and set

$$\Sigma_n := \begin{vmatrix} I_{n+1} & H_n \\ H_n^* & I_{n+1} \end{vmatrix} = |I_{n+1} - H_n^* H_n| . \tag{2.4}$$

We shall call the sequence $\{a_n\}$ positive S-definite if $a_0 \in \mathbb{R}$ and $\sum_n > 0$ for $n \geq 1$. The sequence will be called positive N-S-definite if $\bar{a}_0 \in \mathbb{R}$, $\sum_n > 0$ for $n = 1,\ldots,N-1$ and $\sum_n = 0$ for $n \geq N$. The following result is due to Schur [14, pp. 226-227].

Theorem 2.1. *A function $f \in \mathscr{S}_0$ iff the sequence $\{a_n\}$ (defined by (2.1) with $a_0^{(0)} = a_0$) is positive S-definite or positive N-S-definite for some N. The second case occurs iff f is a function of the form*

$$f(z) = \varepsilon \prod_{v=1}^{N} \frac{z+\omega_v}{1+\bar{\omega}_v z},$$

where $|\omega_v| < 1$ for $v = 1,\ldots,N$, $|\varepsilon| = 1$ *and* $\varepsilon \prod_{v=1}^{N} \omega_v \in \mathbb{R}$.

Another approach to the study of Schur functions, also due to Schur [14], is as follows.

Set $f_0(z) := f(z)$ and define inductively $\{f_n\}$ by the rule

$$f_{n+1}(z) := \frac{f_n(z)-\gamma_n}{z(1-\bar{\gamma}_n f_n(z))}, \quad \gamma_n := f_n(0), \quad n \geq 0 .$$

Then all the functions f_n are Schur functions and either $|\gamma_n| < 1$ for all $n \geq 0$, or the process terminates by some f_N being a constant with absolute value 1.

We now introduce the linear fractional transformations

(2.5) (a) $t_n(z,w) := \dfrac{\gamma_n+zw}{1+\bar{\gamma}_n zw}$, $n \geq 0$, $T_n(z,w) := T_{n-1}(z,t_n(z,w))$, $n \geq 1$,

(b) $T_0(z,w) := t_0(z,w)$

It is then clear that

$$f_0(z) = T_n(z,f_{n+1}(z))$$

provided $\gamma_v = f_v(0)$, $v = 0,\ldots,n$.

It is also true that for γ_n picked arbitarily subject only to the requirement $|\gamma_\nu| < 1$, $\nu = 1,\ldots,n$ the functions $T_n(z,w)$, $|w| < 1$ are Schur functions (considered as functions of z).

We may write

$$T_n(z,w) = \frac{C_n(z)zw + D_n(z)}{E_n(z)zw + F_n(z)}, \tag{2.6}$$

where the following recurrence relations (see [14, pp. 138-39], cf. also [18, p. 218]) hold

(2.7) (a) $$\begin{pmatrix} C_n \\ E_n \end{pmatrix} = z \begin{pmatrix} C_{n-1} \\ E_{n-1} \end{pmatrix} + \overline{\gamma}_n \begin{pmatrix} D_{n-1} \\ E_{n-1} \end{pmatrix}, \quad n \geq 1,$$

(2.7) (b) $$\begin{pmatrix} D_n \\ E_n \end{pmatrix} = \gamma_n z \begin{pmatrix} C_{n-1} \\ E_{n-1} \end{pmatrix} + \begin{pmatrix} D_{n-1} \\ E_{n-1} \end{pmatrix}, \quad n \geq 1,$$

(2.7) (c) $$C_0 = 1, \quad D_0 = \gamma_0, \quad E_0 = \overline{\gamma}_0, \quad F_0 = 1.$$

Thus C_n, D_n, E_n, F_n are polynomials in z of degree at most n. To express relationships between these polynomials we introduce the following concept. Let $\{H_n\}$ be a family of <u>indexed polynomials</u>, where $\deg H_n \leq n$. We define the <u>indexed reciprocal</u> H_n^x of H_n by

(2.8) $$H_n^x(z) := z^n \overline{H}_n(1/z).$$

In terms of this definition we have

(2.9) $$C_n = D_n^x, \quad D_n = C_n^x, \quad E_n = F_n^x, \quad F_n = E_n^x.$$

From this

(2.10) $$T_n(z,0) = [\overline{T_n(\tfrac{1}{\overline{z}}, \infty)}]^{-1}$$

follows. The "continued fraction like" algorithm can be transformed into a continued fraction by decomposing the t_n as follows

$$t_n(z,w) = \gamma_n + \frac{(1-|\gamma_n|^2)z}{\overline{\gamma}_n z + 1/w}.$$

We thus arrive at the continued fraction

(2.11) $$\gamma_0 + \frac{(1-|\gamma_0|^2)z}{\overline{\gamma}_0 z} + \frac{1}{\gamma_1} + \frac{(1-|\gamma_1|^2)z}{\overline{\gamma}_1} + \cdots, \quad |\gamma_n| \neq 1.$$

This derivation is due to Wall [20,21]. We shall call the continued fraction (2.11) the <u>Schur fraction</u> determined by the sequence $\{\gamma_n\}$ and denote it by $S\{\gamma_n\}$. We shall call a Schur fraction <u>positive</u> if $|\gamma_n| < 1$ for $n \geq 0$ and $\gamma_0 \in \mathbb{R}$. We write Σ (Σ^+) for the class of Schur fractions (positive Schur fractions) and denote by the subscript 0 the subclasses of those fractions for which $\gamma_0 \neq 0$. Schur fractions can be expressed in terms of linear fractional transformations

(2.12) $s_0(z,w) := \gamma_0 + w, \quad s_{2m}(z,w) := \frac{1}{\gamma_m + w}, \quad m \geq 1,$

$$s_{2m+1}(z,) := \frac{(1-|\gamma_m|^2)z}{\overline{\gamma}_m z + w}, \quad m \geq 0 .$$

Then $S_n(z,w)$ is, inductively, determined by

(2.13) $S_0\,(z,w) := s_0(z,w), \quad S_n(z,w) := S_{n-1}(z, s_n(z,w)), \quad n \geq 1 ,$

where the approximants of (2.11) are the quantities $S_n(z,0)$. We may write

(2.14) $$S_n(z,w) = \frac{A_n(z) + wA_{n-1}(z)}{B_n(z) + wB_{n-1}(z)} .$$

The polynomials A_n, B_n satisfy the following recurrence relations (see for example [6]):

(2.15) (a) $$\begin{pmatrix} A_{2m} \\ B_{2m} \end{pmatrix} = \gamma_m \begin{pmatrix} A_{2m-1} \\ B_{2m-1} \end{pmatrix} + \begin{pmatrix} A_{2m-2} \\ B_{2m-2} \end{pmatrix}, \quad m \geq 1 ,$$

(2.15) (b) $$\begin{pmatrix} A_{2m+1} \\ B_{2m+1} \end{pmatrix} = \overline{\gamma}_m z \begin{pmatrix} A_{2m} \\ B_{2m} \end{pmatrix} + (1-|\gamma_m|^2)z \begin{pmatrix} A_{2m-1} \\ B_{2m-1} \end{pmatrix}, \quad m \geq 1 ,$$

(2.15) (c) $A_0 = \gamma_0, \quad A_1 = z, \quad B_0 = 1, \quad B_1 = \overline{\gamma}_0 z .$

Thus we may write

(2.16)
$$\begin{aligned} A_{2m}(z) &= \gamma_0 + \dots + \gamma_m z^m , \\ B_{2m}(z) &= 1 + \dots + \overline{\gamma}_0 \gamma_m z^m , \\ A_{2m+1}(z) &= \gamma_0 \overline{\gamma}_m z + \dots + z^{m+1} , \\ B_{2m+1}(z) &= \overline{\gamma}_m z + \dots + \overline{\gamma}_0 z^{m+1} , \end{aligned}$$

The relation between A_n, B_n, S_n and C_n, D_n, E_n, F_n, T_n is easily seen to be

(2.17) $A_{2m+1} = zC_m, \quad A_{2m} = D_m, \quad B_{2m+1} = zE_m, \quad B_{2m} = F_m .$

Hence $T_m(z,w) = S_{2m+1}(z,1/w)$ and

(2.18)
$$\begin{aligned} \frac{A_{2m}}{B_{2m}} &= S_{2m+1}(z,\infty) = T_m(z,0) = \frac{D_m}{F_m} , \\ \frac{A_{2m+1}}{B_{2m+1}} &= S_{2m+1}(z,0) = T_m(z,\infty) = \frac{C_m}{E_m} . \end{aligned}$$

Our next result deals with the relationship between Schur fractions and Schur functions. For convenience we shall here and subsequently say that a terminating continued fraction corresponds to the function given by the value of the fraction.

Theorem 2.2. *Let $\{\gamma_n\}$ be a given finite or infinite sequence of complex numbers.*

(A) *Assume that $S\{\gamma_n\} \in \sum$. Then the sequence $\{A_{2m}/B_{2m}\}$ corresponds to a series $a_0^{(0)} + \sum_{n=1}^{\infty} a_n z^n$ at $z = 0$ with order of correspondence $(m+1)$. If $S\{\gamma_n\} \in \sum_0$, then the sequence $\{A_{2m+1}/B_{2m+1}\}$ corresponds to a series $a_0^{(\infty)} + \sum_{n=1}^{\infty} a_{-n} z^{-n}$ at $z = \infty$ with order of corespondence $(m+1)$. The two series satisfy*

$$(2.19) \qquad \overline{(a_0^{(0)}} + \sum_{n=1}^{\infty} \bar{a}_n z^n)(a_0^{(\infty)} + \sum_{n=1}^{\infty} a_{-n} z^n) = 1 .$$

(B) *Assume that $S\{\gamma_n\} \in \sum$, $\gamma_n \neq 0$ for all n. Then the sequence $\{A_{2m}/B_{2m}\}$ corresponds to the series $a_0^{(\infty)} + \sum_{n=1}^{\infty} a_{-n} z^{-n}$ at $z = \infty$ with order of correspondence m. If $S\{\gamma_n\} \in \sum_0$, $\gamma_n \neq 0$ for all n, then the sequence $\{A_{2m+1}/B_{2m+1}\}$ corresponds to the series $a_0^{(0)} + \sum_{n=1}^{\infty} a_n z^n$ at $z = 0$ with order of correspondence m.*

(C) *Assume that $S\{\gamma_n\} \in \sum^+$. Then the sequence $\{A_{2m}/B_{2m}\}$ converges to the sum $f(z)$ of the series $a_0^{(0)} + \sum_{n=1}^{\infty} a_n z^n$ for $|z| < 1$. If $S\{\gamma_n\} \in \sum_0^+$, then the sequence $\{A_{2m+1}/B_{2m+1}\}$ converges to the sum $g(z)$ of the series $a_0^{(0)} + \sum_{n=1}^{\infty} a_{-n} z^{-n}$ for $|z| > 1$. The function f is a Schur function and $g = \hat{f}$.*

Proof: We prove the theorem for the case that $\{\gamma_n\}$ is an infinite sequence. If $\{\gamma_n\}$ is finite and hence $S\{\gamma_n\}$ terminating, the results follow by similar, but simpler arguments. We set

$$(2.20) \qquad \Gamma_n := \prod_{v=0}^{n} (1-|\gamma_v|^2) , \quad n \geq 0 .$$

(A) Using the recurrence formulas (2.15) and the determinant formula for linear fractional transformations (see for example [12]) we get

$$\frac{A_{2m}}{B_{2m}} - \frac{A_{2m-2}}{B_{2m-2}} = \frac{\gamma_m \Gamma_{m-1} z^m}{B_{2m} B_{2m-2}} = \frac{\gamma_m \Gamma_{m-1} z^m}{1+\ldots+\bar{\gamma}_0^2 \gamma_{m-1} \gamma_m z^{2m-1}} ,$$

(2.21)

$$\frac{A_{2m+1}}{B_{2m+1}} - \frac{A_{2m-1}}{B_{2m-1}} = \frac{-\gamma_m \Gamma_{m-1} z^m}{z^{-2} B_{2m+1} B_{2m-1}} = \frac{-\gamma_m \Gamma_{m-1} z^{m-1}}{\bar{\gamma}_m \bar{\gamma}_{m-1} + \ldots + \gamma_0^2 z^{2m-1}} .$$

From these formulas the correspondence results under A follow (see [18, 219-220]). Formula (2.19) follows from the fact that A_{2m+1}/B_{2m+1} is the Schur reciprocal of A_{2m}/B_{2m}, see (2.17) and (2.10).

(B) Using the determinant formula we obtain

$$\text{(2.22)} \qquad \frac{A_{2m+1}}{B_{2m+1}} - \frac{A_{2m}}{B_{2m}} = \frac{A_{2m+1}B_{2m} - A_{2m}B_{2m+1}}{B_{2m}B_{2m+1}}$$

$$= \frac{\Gamma_m z^m}{\bar{\gamma}_m \bar{\gamma}_{m+1} + \ldots + \bar{\gamma}_m \gamma_0 z^{2m}}$$

It follows that $\{A_{2m+1}/B_{2m+1}\}$ corresponds to the same series as $\{A_{2m}/B_{2m}\}$ at $z = 0$ with order of correspondence m, and $\{A_{2m}/B_{2m}\}$ corresponds to the same series as $\{A_{2m+1}/B_{2m+1}\}$ at $z = \infty$ with order of correspondence m.

(C) The fact that the sequence $\{A_{2m}/B_{2m}\}$ converges to a Schur function follows from [14, p. 211]. Since the sequence corresponds to the series $a_0^{(0)} + \sum_{n=1}^{\infty} a_n z^n$ this series converges and its sum is the limit of the sequence. The analogous result for the sequence $\{A_{2m+1}/B_{2m+1}\}$ (making allowance for poles) follows from the fact that A_{2m+1}/B_{2m+1} is the Schur reciprocal of A_{2m}/B_{2m}. It follows from (A) that the sum of the series $a_0^{(\infty)} + \sum_{n=1}^{\infty} a_{-n} z^{-n}$ is the Schur reciprocal of the sum of the series $a_0^{(0)} + \sum_{n=1}^{\infty} a_n z^n$. □

Theorem 2.2 allows us to associate with every positive Schur fraction $S\{\gamma_n\}$ a unique normalized Schur function $f(z) = a_0^{(0)} + \sum_{n=1}^{\infty} a_n z^n$. We shall denote this mapping by Ψ_Σ, thus

$$S\{\gamma_n\} \to \Psi_\Sigma(S\{\gamma_n\}) .$$

For the problem of expressing the coefficients a_n in terms of the "Schur parameters" γ_n, see [14, p. 210].

Now let f be a given Schur function. We saw that we can write $f_m(z) = T_m(z, f_{m+1}(z))$ provided we set $\gamma_n = f_n(0)$. The sequence $\{\gamma_n\}$ so chosen satisfies $|\gamma_n| < 1$ (or the process terminates with $|\gamma_N| = 1$) and thus determines a $S\{\gamma_n\} \in \Sigma^+$. In this way we may associate with every Schur function a unique Schur fraction. We shall denote this mapping by $\Phi_{\mathscr{S}}$. Thus

$$f \to \Phi_{\mathscr{S}}(f) .$$

It follows from [14, p. 210-11] that the sequence $\{A_{2m}/B_{2m}\}$ of even approximants of the Schur fraction $\Phi_{\mathscr{S}}(f)$ converges to $f(z)$ for $|z| < 1$. Hence $\Psi_\Sigma\Phi_{\mathscr{S}}(f) = f$. We also have $\Phi_{\mathscr{S}}\Psi_\Sigma(S\{\gamma_n\}) = S\{\gamma_n\}$ since $S\{\gamma_n\}$ can correspond to at most one power series at $z = 0$. The problem of expressing the Schur parameters γ_n in terms of the coefficients a_n in the series expansion $a_0^{(0)} + \sum_{n=1}^{\infty} a_n z^n$ is taken up in [14, p. 209-10].

We may sum up these results in the following theorem.

Theorem 2.3. *The mapping $\Phi_{\mathscr{S}}$ is a bijection of the class $\mathscr{S}$ onto the class Σ^+. The class $\mathscr{S}_0$ is mapped onto the class Σ_0^+. The inverse mapping is Ψ_Σ.*

3. Carathéodory functions and Perron-Carathéodory fractions. A function F which is analytic in the open unit disk $[z: |z| < 1]$ and maps this disk into the closed right half-plane $[z: \text{Re } z \geq 0]$ is called a *Carathéodory function* (see for example [1]). We shall call the function *normalized* if $F(0) \in (0,\infty)$. We denote the class of all normalized Carathéodory functions by $\mathcal{C}$, and write $\mathcal{C}_0 = [F \in \mathcal{C}: F(0) \neq 1]$. We also set $\tilde{\mathcal{C}} = [G: G$ is analytic for $|z| > 1$, $\text{Re } G(z) \leq 0$, $G(\infty) \in (-\infty,0)]$ and write $\tilde{\mathcal{C}}_\infty = [G \in \tilde{\mathcal{C}}: G(\infty) \neq -1]$. With every Carathéodory function F we associate the *Carathéodory reciprocal*

$$\tilde{F}(z) = -[\overline{F(1/\overline{z})}] .$$

Note that

$$\tilde{F}(z) + \overline{F}(1/z) = 0 ,$$

and that $(\tilde{\mathcal{C}}_0) = \tilde{\mathcal{C}}_\infty$. The correspondence $F \to \tilde{F}$ is a bijection of $\mathcal{C}_0$ onto $\tilde{\mathcal{C}}_\infty$.

Let the functions F and G have power series expansions

$$F(z) = \mu_0^{(0)} + 2 \sum_{n=1}^{\infty} \mu_n z^n, \quad \text{for} \quad |z| < 1 , \tag{3.1}$$

$$G(z) = - \mu_0^{(\infty)} - 2 \sum_{n=1}^{\infty} \mu_{-n} z^{-n}, \quad \text{for} \quad |z| > 1 . \tag{3.2}$$

Then $G = \tilde{F}$ iff

$$\mu_0^{(\infty)} = \overline{\mu_0^{(0)}} , \quad \mu_{-n} = \bar{\mu}_n , \quad n \geq 1 .$$

In (3.1) set $\mu_0^{(0)} = \mu_0$. We introduce the matrix

$$M_n = \begin{pmatrix} \mu_0 & \mu_1 & \cdots & \mu_{n-1} & \mu_n \\ \bar{\mu}_1 & \ddots & & \vdots & \vdots \\ \vdots & & & \mu_0 & \\ \bar{\mu}_n & & \cdots & \bar{\mu}_1 & \mu_0 \end{pmatrix} \tag{3.3}$$

and let Δ_n be the (Toeplitz) determinant $\Delta_n = |M_n|$. The sequence $\{\mu_n\}$ will be called <u>positive definite</u> if $\Delta_n > 0$ for $n \geq 0$, positive N-definite if $\Delta_n > 0$ for $n = 0,\ldots N-1$, $\Delta_n = 0$ for $n \geq N$.

<u>Theorem</u> 3.1. <u>The function F with power series expansion</u> (3.1), <u>with $\mu_0^{(0)} = \mu_0$ is a normalized Carathéodory function iff the sequence $\{\mu_n\}$ is positive definite or positive N-definite, for some N. The second case occurs iff F is a function of the form</u>

$$F(z) = \sum_{v=1}^{N} \lambda_v \frac{t_v+z}{t_v-z} ,$$

<u>where $t_1,\ldots,t_N$ are distinct</u>, $|t_v| = 1$, $\lambda_v > 0$, $v = 1,\ldots,N$.

<u>Proof</u>: The result is contained in [14, pp. 229-30].

Let $F \in C$, set $F = F_0$, define

$$F_1 := \frac{\delta_0-F_0}{\delta_0+F_0} , \quad \delta_0 := F_0(0) \in \mathbb{R}_+$$

Note that $F_1 \in C$ and $F_1(0) = 0$. We then proceed inductively to define

$$F_{n+1} := \frac{\bar{\delta}_n z-F_n}{\delta_n F_n-z} , \quad \bar{\delta}_n := F_n'(0)$$

so that $F_n(0) = 0$ for all n and all $F_n \in C$. It is easy to see that this process is very similar to the one introduced by Schur. Again we have $|\delta_n| < 1$ for all $n \geq 1$ or the process terminates with $|\delta_N| = 1$. Following the analogy we define

$$(3.4) \qquad r_0(z,w) := \delta_0 \frac{1-w}{1+w}, \quad r_n(z,w) := z\,\frac{w+\overline{\delta}_n}{1+\delta_n w}$$

and, inductively,

$$(3.5) \qquad R_0(z,w) := r_0(z,w), \quad R_n(z,w) := R_{n-1}(z, r_n(z,w)), \quad n \geq 1 .$$

We may write

$$(3.6) \qquad R_n(z,w) = \frac{\pi_n(z)w + \omega_n(z)}{\rho_n(z)w + \tau_n(z)} .$$

The choice of $\pi_n, \omega_n, \rho_n, \tau_n$ is historically determined (we have not yet run out of the Roman alphabet). The following recurrence relations are valid (see [9]):

$$(3.7)\ (a) \qquad \binom{\pi_n}{\rho_n} = z\binom{\pi_{n-1}}{\rho_{n-1}} + \delta_n \binom{\omega_{n-1}}{\tau_{n-1}}, \quad n \geq 1 ,$$

$$(3.7)\ (b) \qquad \binom{\omega_n}{\tau_n} = \overline{\delta}_n z\binom{\pi_{n-1}}{\rho_{n-1}} + \binom{\omega_{n-1}}{\tau_{n-1}}, \quad n \geq 1 ,$$

$$(3.7)\ (c) \qquad \pi_0 = -\delta_0, \ \omega_0 = \delta_0, \ \rho_0 = 1, \ \tau_0 = 1 .$$

It follows that $\pi_n, \rho_n, \omega_n, \tau_n$ are polynomials in z of degree at most n.

We can break up the linear fractional transformation r_n into transformations that yield continued fractions as follows

$$k_0(z,w) := \delta_0 + w, \quad k_1(z,w) := -2\delta_0/(1+w) ,$$

$$k_{2m}(z,w) := \frac{1}{\overline{\delta}_m z + w}, \quad k_{2m+1}(z,w) := \frac{(1-|\delta_m|^2)z}{\delta_m + w}, \quad m \geq 1$$

so that

$$r_0(z,w) = k_0 \circ k_1(z, 1/w) ,$$

$$r_m(z,w) = 1/k_{2m} \circ k_{2m+1}(z, 1/w), \quad m \geq 1 .$$

Introduce $\tau(w) = 1/w$, then the above relations can be written as

$$(3.8) \qquad r_0 = k_0 \circ k_1 \circ \tau , \quad r_m = \tau \circ k_{2m} \circ k_{2m+1} \circ \tau , \quad m > 1 .$$

Define, inductively

$$K_0(z,w) := k_0(z,w), \quad K_m(z,w) := K_{m-1}(z,k_m(z,w)), \quad m \geq 1 ,$$

and set

$$(3.9) \qquad K_m(z,w) = \frac{P_m(z) + wP_{m-1}(z)}{Q_m(z) + wQ_{m-1}(z)} ,$$

which defines P_m, Q_m up to a factor of proportionality which is determined in (3.11). The continued fraction, $\delta_0 \in \mathbb{R}$, $|\delta_n| \neq 1$, $n \geq 1$,

$$(3.10) \qquad \delta_0 - \frac{2\delta_0}{1} + \frac{1}{\bar{\delta}_1 z} + \frac{(1-|\delta_1|^2)z}{\delta_1} + \frac{1}{\bar{\delta}_2 z} + \frac{(1-|\delta_2|^2)z}{\delta_2} + \cdots$$

shall be called the <u>hermitian Perron-Carathéodory fraction</u> (or hermitian PC-fraction) determined by the sequence $\{\delta_n\}$. It will be denoted by $HPC\{\delta_n\}$. We call a hermitian PC-fraction <u>positive</u> if $\delta_0 > 0$, $|\delta_n| < 1$, $n \geq 1$. The quantities $K_m(z,0) = P_m(z)/Q_m(z)$ are the mth approximants of (3.10) and the following recurrence relations are valid (see [9], [12]).

$$(3.11)\ (a) \qquad \binom{P_{2m}}{Q_{2m}} = \bar{\delta}_m z \binom{P_{2m-1}}{Q_{2m-1}} + \binom{P_{2m-2}}{Q_{2m-2}} , \quad m \geq 1 ,$$

$$(3.11)\ (b) \qquad \binom{P_{2m+1}}{Q_{2m+1}} = \delta_m \binom{P_{2m}}{Q_{2m}} + (1 - |\delta_m|^2) z \binom{P_{2m-1}}{Q_{2m-1}} , \quad m \geq 1 ,$$

$$(3.11)\ (c) \qquad P_0 = \delta_0, \quad P_1 = -\delta_0, \quad Q_0 = 1, \quad Q_1 = 1.$$

The P_m, Q_m are thus polynomials in z.

From (3.8) the following (see [9]) relationships between π_n, ω_n, ρ_n, τ_n, R_n and P_n, Q_n, K_n can be derived

$$(3.12) \qquad R_m(z,w) = K_{2m+1}(z,1/w) ,$$

$$(3.13) \qquad \begin{aligned} P_{2m}/Q_{2m} &= K_{2m}(z,\infty) = R_m(z,0) = \omega_m/\tau_m , \\ P_{2m+1}/Q_{2m+1} &= K_{2m+1}(z,\infty) = R_m(z,\infty) = \pi_m/\rho_m , \end{aligned}$$

$$(3.14) \qquad P_{2m+1} = \pi_m, \quad P_{2m} = \omega_m, \quad Q_{2m+1} = \rho_m, \quad Q_{2m} = \tau_m .$$

Furthermore

$$\rho_m^x = \tau_m, \quad \tau_m^x = \rho_m, \quad \pi_m^x = -\omega_m, \quad \omega_m^x = -\pi_m ,$$

so that

$$(3.15) \qquad R_m(z,0) = -\overline{[R_m(1/\bar{z},\infty)]} .$$

Additional results concerning positive PC-fractions are given in Section 5. We shall write Γ (Γ^+) for the class of hermitian PC-fractions (positive PC-fractions) and denote by the subscript zero the subclasses of those fractions for which $\delta_0 \neq 1$.

The next result deals with the relationship between PC-fractions and Carathéodory functions.

Theorem 3.2. *Let* $\{\delta_n\}$ *be a given finite or infinite sequence of complex numbers.*

(A) *Assume that* $HPC\{\delta_n\} \in \Gamma$. *Then the sequence* $\{P_{2m}/Q_{2m}\}$ *corresponds to a series* $\mu_0^{(0)} + 2\sum_{n=1}^{\infty} \mu_n z^n$ *at* $z = 0$ *with order of correspondence* $(m+1)$. *The sequence* $\{P_{2m+1}/Q_{2m+1}\}$ *corresponds to a series* $-\mu_0^{(\infty)} - 2\sum_{n=1}^{\infty} \mu_{-n} z^{-n}$ *at* $z = \infty$ *with order of correspondence* $(m+1)$. *The two series satisfy the condition*

$$(3.16) \qquad \mu_0^{(\infty)} = \overline{\mu_0^{(0)}} , \quad \mu_{-n} = \bar{\mu}_n , \quad n \geq 1 .$$

(B) *Assume that* $HPC\{\delta_n\} \in \Gamma$, $\delta_n \neq 0$ *for all* n. *Then the sequence* $\{P_{2m}/Q_{2m}\}$ *corresponds to the series* $-\mu_0^{(\infty)} - 2\sum_{n=1}^{\infty} \mu_{-n} z^{-n}$ *at* $z = \infty$ *with order of correspondence* m. *The sequence* $\{P_{2m+1}/Q_{2m+1}\}$ *corresponds to the series* $\mu_0^{(0)} + 2\sum_{n=1}^{\infty} \mu_n z^n$ *at* $z = 0$ *with order of correspondence* m.

(C) *Assume that* $HPC\{\delta_n\} \in \Gamma^+$. *Then the sequence* $\{P_{2m}/Q_{2m}\}$ *converges to the sum* $F(z)$ *of the series* $\mu_0^{(0)} + 2\sum_{n=1}^{\infty} \mu_n z^n$ *for* $|z| < 1$. *The sequence* $\{P_{2m+1}/Q_{2m+1}\}$ *converges to the sum* $G(z)$ *of the series* $-\mu_0^{(\infty)} - 2\sum_{n=1}^{\infty} \mu_{-n} z^{-n}$ *for* $|z| > 1$. *The function* $F \in C$ *and* $G = \tilde{F}$.

Proof: First assume that $\{\delta_n\}$ is an infinite sequence

(A) The correspondence result is contained in [9, Th. 2.1]. The relationship between the coefficients of the series follows from the fact that P_{2m+1}/Q_{2m+1} is the Carathéodory reciprocal of P_{2m}/P_{2m} see (3.13) and (3.15).

(B) This result is contained in [9, Th. 4.1].

(C) This result is essentially [9, Th. 3.2].

If the sequence $\{\delta_n\}$ is finite, and hence $\text{HPC}\{\delta_n\}$ is terminating the result follows by similar, but simpler, arguments. □

Theorem 3.2(C) allows us to associate with every positive PC-fraction $\text{HPC}\{\delta_n\}$ a unique normalized Carathéodory function $F(z) = \mu_0 + 2 \sum_{n=1}^{\infty} \mu_n z^n$. We shall denote this mapping by Ψ_Γ, thus

$$\text{HPC}\{\delta_n\} \rightarrow \Psi_\Gamma(\text{HPC}\{\delta_n\}) , \quad \text{if} \quad \text{HPC}\{\delta_n\} \in \Gamma^+ .$$

Let F be a given normalized Carathéodory function with power series expansion $F(z) = \mu_0 + 2 \sum_{n=1}^{\infty} \mu_n z^n$ and set $\mu_{-n} = \overline{\mu}_n$, $n \geq 1$. Note that $\mu_0 = F(0) > 0$. Let U_n denote the determinant

$$\Theta_n = \begin{vmatrix} \mu_{-1} & \mu_{-2} & \cdots & \mu_{-n} \\ \mu_0 & \mu_{-1} & \cdots & \mu_{-n+1} \\ \vdots & \vdots & \ddots & \vdots \\ \mu_{n-2} & \mu_{n-3} & \cdots & \mu_{-1} \end{vmatrix} , \quad n > 1 .$$

For every positive Toeplitz determinant Δ_n we define

$$\delta_n := (-1)^n \frac{\Theta_n}{\Delta_n} , \quad n \geq 1, \quad \delta_0 := \mu_0 .$$

Then $|\delta_n| < 1$ and the sequence $\{\delta_n\}$ gives rise to a (possibly terminating) positive PC-fraction $\text{HPC}\{\delta_n\}$. The sequence $\{P_{2m}/Q_{2m}\}$ of even approximants converges to the function F for $|z| < 1$ (see [9, Th. 2.2, 3.1 and 3.2]). In this way we may associate with every normalized Carathéodory function a unique positive PC-fraction. (The correspondence can also be established in the manner sketched earlier in this section). We shall denote this mapping by Φ_C. Thus

$$F \rightarrow \Phi_C(F) .$$

It follows from the above remarks that

$$\Psi_\Gamma \Phi_C(F) = F .$$

Also

$$\Phi_C \Psi_\Gamma(\text{HPC}\{\delta_n\}) = \text{HPC}\{\delta_n\} , \quad \text{for} \quad \text{HPC}\{\delta_n\} \in \Gamma^+ .$$

since $\text{HPC}\{\delta_n\}$ can correspond to at most one power series at $z = 0$.

We may sum up these results in the following theorem.

Theorem 3.3. *The mapping* Φ_C *is a bijection of the class* C *onto the class* Γ^+. *The class* C_0 *is mapped onto the class* Γ_0^+. *The inverse mapping is* Ψ_Γ.

4. Connection between Schur fractions and hermitian PC-fractions. Let $f \in \mathscr{S}$ and define F by

$$F(z) = \phi(f(z)) := \frac{1-f(z)}{1+f(z)} .$$

Then $F \in C$; and $F \in C_0$ iff $f \in \mathscr{S}_0$. In this way we associate with every Schur function a Carathéodory function and vice versa. The mapping ϕ is a bijection of $\mathscr{S}$ onto C and $\mathscr{S}_0$ is mapped exactly onto C_0.

Let $S\{\gamma_n\}$ be a positive Schur fraction and define

$$(4.1) \qquad \delta_0 := \frac{1-\gamma_0}{1+\gamma_0}, \quad \delta_n := \overline{\gamma}_n, \quad n \geq 1 .$$

Note that $\gamma_0 \in (-1,1)$ iff $\delta_0 > 0$ and $\gamma_0 \neq 0$ iff $\delta_0 \neq 1$. Thus the sequence $\{\delta_n\}$ determines a positive PC-fraction $HPC\{\delta_n\}$. We denote by $\Omega(S\{\gamma_n\})$ the hermitian PC-fraction whose coefficients δ_n are determined by (4.1). The mapping Ω is then a bijection from $\sum^+$ onto Γ^+ and $\sum_0^+$ is mapped onto Γ_0^+.

Now let $f \in \mathscr{S}$ and set $S\{\gamma_n\} = \Phi_\Gamma(f)$, $HPC\{\delta_n\} = \Omega(S\{\gamma_n\})$, $H := \Psi_\Gamma(HPC\{\delta_n\})$. Let A_n/B_n be the nth approximant of the Schur fraction $S\{\gamma_n\}$ and let P_n/Q_n be the nth approximant of the PC-fraction $HPC\{\delta_n\}$. With these agreements we have the following two theorems.

Theorem 4.1. *Set* $\alpha = 1/(1+\gamma_0)$. *Then for* $m \geq 0$

$$(4.2) \qquad \begin{aligned} &(a) \quad P_{2m+1} = \frac{\alpha}{z}(B_{2m+1} - A_{2m+1}), \\ &(b) \quad P_{2m} = \alpha(B_{2m} - A_{2m}), \\ &(c) \quad Q_{2m+1} = \frac{\alpha}{z}(B_{2m+1} + A_{2m+1}) \\ &(d) \quad Q_{2m} = \alpha(A_{2m} + B_{2m}) . \end{aligned}$$

Proof: By using the relations (4.1) and the recurrence relation (2.15) we find that the expressions on the right side in the equations (4.2) satisfy the recurrence relations (3.11), including the initial conditions. Since the solution of this system of equations is unique, the result follows. □

Using the mapping ϕ in conjunction with Theorem 4.1 we obtain

$$\phi\left(\frac{A_{2m}}{B_{2m}}\right) = \frac{P_{2m}}{Q_{2m}} , \quad \phi\left(\frac{\hat{A}_{2m+1}}{\hat{B}_{2m+1}}\right) = \frac{P_{2m+1}}{Q_{2m+1}} .$$

Since $\{A_{2m}/B_{2m}\}$ converges to f for $|z| < 1$ (see Theorem 2.2(C)) it follows that $\{P_{2m}/Q_{2m}\}$ converges to $\phi(f) = F$ for $|z| < 1$. Similarly $\{\hat{A}_{2m+1}/\hat{B}_{2m+1}\}$ converges to $\hat{f}$ and the sequence $\{P_{2m+1}/Q_{2m+1}\}$ converges to $(1-\hat{f})/(1+\hat{f}) = G$, for $|z| > 1$. One verifies easily that $G = \tilde{F}$. Note also that P_{2m}/Q_{2m} and P_{2m+1}/Q_{2m+1} are weak two-point Padé approximants for $(F,\tilde{F})$ of order $(m+1,m)$ and $(m,m+1)$, respectively.

We showed that $\{P_{2m}/Q_{2m}\}$ converges to F for $|z| < 1$. On the other hand $\{P_{2m}/Q_{2m}\}$ converges to $H = \Psi_\Gamma(\mathrm{HPC}\{\delta_n\})$. Thus $H = F = \phi(f)$. These results are summed up in the following theorem:

<u>Theorem</u> 4.2. <u>The following relations are valid</u>

$$\Phi_C(\phi(f)) = \Omega(\Phi_{\mathscr{S}}(f)) ,$$

$$\Psi_\Gamma(\Omega(S\{\gamma_n\}) = \phi(\Psi_\Sigma(S\{\gamma_n\})) ,$$

<u>In addition</u> $\{\phi(A_n/B_n)\}$ <u>is the sequence of approximants of the hermitian</u> PC-<u>fraction</u> $\Omega(S\{\gamma_n\})$.

5. <u>Szegö polynomials</u>. For any pair <p,q> of integers where $p \leq q$ we denote by $\Lambda_{p,q}$ the linear space of all functions of a complex variable z of the form

$$L(z) = \sum_{v=p}^{q} c_v z^v , \quad c_v \in \mathbb{C} . \tag{5.1}$$

We denote by Λ the linear space which is the union of all the spaces $\Lambda_{p,q}$. A function L belonging to Λ is called a <u>Laurent polynomial</u> or L-<u>polynomial</u>. Let $\{\mu_n\}_{-\infty}^{\infty}$ be a double sequence of complex numbers satisfying

$$\mu_{-n} = \overline{\mu}_n , \quad n \geq 0 ,$$

and assume that the series

$$\mu_0 + 2 \sum_{n=1}^{\infty} \mu_n z^n , \tag{5.2}$$

converges, for $|z| < 1$, to a normalized Carathéodory function $F(z)$. Then the sequence $\{\mu_n\}$ is either positive definite or positive N-definite for some N. To the sequence $\{\mu_n\}$ there corresponds a linear functional $\mathfrak{M}$ on Λ defined by

$$(5.3) \qquad \mathfrak{M}\left(\sum_{v=p}^{q} c_v z^v\right) = \sum_{v=p}^{q} c_v \mu_{-v} .$$

In terms of $\mathfrak{M}$ a functional $\langle \cdot , \cdot \rangle$ is defined on $\Lambda \times \Lambda$ by

$$(5.4) \qquad \langle P,Q\rangle := \mathfrak{M}(P(z)\cdot\overline{Q}(1/z)) .$$

If $\{\mu_n\}$ is positive N-definite, then $\langle \cdot , \cdot \rangle$ is an inner product on $\Lambda_{-N,N}$. If $\{\mu_n\}$ is positive definite, then $\langle \cdot , \cdot \rangle$ is an inner product on Λ. In the first case there exist monic polynomials $\rho_0(z),\ldots,\rho_N(z)$ such that $\langle \rho_n, z^v\rangle = 0$ for $v = 0,\ldots,n-1$, $n \leq N$. In the second case there exists an orthogonal sequence $\{\rho_n\}$ of monic polynomials, for which

$$\langle \rho_n, z^v\rangle = 0 \quad \text{for} \quad v = 0,\ldots,n-1$$

and all n. In both cases degree $\rho_n = n$. These polynomials are called the monic <u>Szegö polynomials</u> associated with the positive definite sequence $\{\mu_n\}$ (or with the normalized Carathéodory function F). The polynomials ρ_n and ρ_n^x may be expressed by the following determinant formulas

$$(5.5) \quad \rho_n(z) = \frac{1}{\Delta_{n-1}} \begin{vmatrix} \mu_0 & \cdots & \mu_{-n} \\ \vdots & & \vdots \\ \mu_{n-1} & & \mu_{-1} \\ 1 \; z & \cdots & z^n \end{vmatrix} , \quad \rho_n^x(z) = \frac{1}{\Delta_{n-1}} \begin{vmatrix} \mu_0 & \cdots & \mu_n \\ \vdots & & \vdots \\ \mu_{-(n-1)} & & \mu_1 \\ z^n \; z^{n-1} & \cdots & 1 \end{vmatrix}$$

$n \geq 1$

$\rho_0(z) = 1, \quad \rho_0^x(z) = 1.$

Note that

$$\rho_n(z) = \rho_n(0) + \ldots + z^n ,$$

$$\rho_n^x(z) = 1 + \ldots + \overline{\rho_n(0)}\; z^n .$$

It can be shown that

$$|\rho_n(0)| < 1 , \quad n > 1 .$$

Let $L(t,z)$ be an L-polynomial in t that is

$$L(t,z) = \sum_{v=p}^{q} q_v(z) t^v .$$

To emphasize that the functional $\mathcal{M}$ is operating on the variable t in $L(t,z)$ we write $\mathcal{M}_t$. Thus

$$\mathcal{M}_t(L(t,z)) = \sum_{v=p}^{q} q_v(z)\mu_{-v} .$$

We define the polynomial AP associated with the polynomial P by the formula

$$(5.6) \qquad AP(z) := \mathcal{M}_t\left(\frac{z+t}{z-t}\left(\frac{z}{t} P(t) - P(z)\right)\right) .$$

In particular we set

$$\pi_n := A\rho_n \qquad \omega_n := A\rho_n^x .$$

The relations between the Szegö polynomials, determined by the sequence $\{\mu_n\}$, and their associated polynomials on the one hand and the denominators and numerators of the positive PC-fraction, determined by $\{\mu_n\}$, on the other hand are given in the theorem below. For proofs of the results asserted in this section as well as for further information we refer to [9] and [10].

Theorem 5.1. Let $\{\mu_n\}$ be a positive definite sequence and let $\{\rho_n\}$ be the corresponding sequence of Szegö polynomials. Set

$$\delta_0 = \mu_0, \quad \delta_n = \rho_n(0), \quad n \geq 1 ,$$

and let P_n/Q_n be the nth approximant of the positive PC-fraction $HPC\{\delta_n\}$. Then the following equalities hold

$$(5.7) \quad Q_{2m+1} = \rho_m, \; Q_{2m} = \rho_m^x, \; P_{2m+1} = AQ_{2m+1} = \pi_m, \; P_{2m} = AQ_{2m} = \omega_m = \pi_m^x$$

If $\{\mu_n\}$ is positive N-definite then, these equalities hold for $m = 0,\ldots,N$.

It follows from (5.7) and (3.14) that the Szegö polynomials, their associated polynomials and their indexed reciprocals are the ρ_m, π_m, ω_m, τ_m introduced in (3.6) and satisfying the recurrence relations (3.7).

6. Examples. We shall work out a few concrete examples illustrating the relationship between positive Schur fractions, positive PC-fractions, Szegö polynomials, normalized Schur functions and normalized Carathéodory functions.

In the first example we start with a Schur function f, its associated Carathéodory function F and the series expansions of these functions. We determine the corresponding Schur fraction $S\{\gamma_n\}$ and PC-fraction $HPC\{\delta_n\}$ as well as the Szegö polynomials ρ_n.

In the last two examples we start with a Schur fraction $S\{\gamma_n\}$ or its associated PC-fraction $HPC\{\delta_n\}$. We then determine the numerators and denominators of the approximants of these fractions and in particular the Szegö polynomials ρ_n. From the approximants we then obtain the corresponding Schur functions and Carathéodory functions, together with their series expansions. Thus, in particular, we get the double sequence $\{\mu_n\}$ with respect to which $\{\rho_n\}$ is orthogonal.

<u>Example</u> 6.1. Set $f(z) = az$, $|a| \leq 1$. Then $\hat{f}(z) = z/\bar{a}$. We note that $f \in \mathscr{S} \sim \mathscr{S}_0$, $\hat{f} \notin \hat{\mathscr{S}}_\infty$. We have

$$F(z) = \frac{1-az}{1+az} = 1 + 2 \sum_{n=1}^{\infty} (-1)^n a^n z^n, \quad \text{for} \quad |z| < 1 ,$$

and

$$\tilde{F}(z) = \frac{\bar{a}-z}{\bar{a}+z} = -1 - 2 \sum_{n=1}^{\infty} (-1)^n (\bar{a})^n z^{-n} \quad \text{for} \quad |z| > 1 .$$

Thus

$$\mu_0 = 1, \quad \mu_n = (-1)^n a^n, \quad \mu_{-n} = (-1)^n (\bar{a})^n, \quad n \geq 1 .$$

We need to distinguish two cases.

(i) $|a| = 1$. Then the Schur algorithm yields $\gamma_0 = 0$, $\gamma_1 = a$, γ_n undefined for $n \geq 2$. The associated Schur fraction is therefore terminating and is as follows

$$(6.1) \qquad 0 + \frac{z}{0\cdot z\, +}\,\frac{1}{\bar{a}} = az .$$

The associated PC-fraction also terminates and is

$$(6.2) \qquad 1 - \frac{2}{1\, +}\,\frac{1}{az\, +}\,\frac{(1-|a|^2)z}{\bar{a}} = \frac{1-az}{1+az}$$

(ii) $|a| < 1$. Then the Schur algorithm gives $\gamma_0 = 0$, $\gamma_1 = a$, $\gamma_n = 0$ for $n \geq 2$. Thus

$$(6.3) \qquad S\{\gamma_n\} = 0 + \frac{z}{0\cdot z\, +}\,\frac{1}{\alpha\, +}\,\frac{(1-|a|^2)z}{\bar{a}z}\,\frac{1}{+\,0}\,\frac{1}{+\,0z\,+}\,\cdots$$

and

(6.4) $$\mathrm{HPC}\{\delta_n\} = 1 - \frac{2}{1} + \frac{1}{az} + \frac{(1-|a|^2)z}{\bar{a}} + \frac{1}{0\cdot z} + \frac{z}{0} + \cdots .$$

From the recurrence relations (2.15) one obtains

(6.5) $$A_{2m} = az, \quad A_{2m=1} = z^{m+1}, \quad B_{2m} = 1, \quad B_{2m+1} = \bar{a}\, z^m .$$

and hence, using (4.2)

(6.6) $$P_{2m} = 1 - az, \; P_{2m+1} = z^{m-1}(a-z), \; Q_{2m} = 1 + az, \; Q_{2m+1} = z^{m-1}(a+z)$$

In particular we have

(6.7) $$\rho_m = z^m + \bar{a}\, z^{m-1} .$$

<u>Example</u> 6.2. Let $\gamma_n = 1/(n+2)$, then $\alpha = 1/(1+\gamma_0) = 2/3$, $\delta_0 = (1-\gamma_0)/(1+\gamma_0) = 1/3$ and $\delta_n = \bar{\gamma}_n = 1/(n+2)$. Using (2.15) we find by induction

$$A_0 = 1/2, \quad A_1 = z, \quad B_0 = 1, \quad B_1 = z/2 .$$

(6.8) (a) $$A_{2m}(z) = \frac{1}{2m+2} \sum_{k=0}^{m} (k+2) z^{m-k}, \quad m \geq 1 ,$$

(6.8) (b) $$A_{2m+1}(z) = \frac{1}{2(m+2)} \left(2(m+2) z^{m+1} + \sum_{k=1}^{m} (m+1-k) z^{m-k+1}\right), \quad m \geq 1 ,$$

(6.8) (c) $$B_{2m}(a) = \frac{1}{2(m+2)} \left(\sum_{k=0}^{m-1} (k+1) z^{m-k} + 2(m+2)\right), \quad m \geq 1 .$$

(6.8) (d) $$B_{2m+1}(z) = \frac{1}{2(m+2)} \sum_{k=0}^{m} (m+2-k) z^{m-k+1}, \quad m \geq 1.$$

Using formulas (4.2) leads to

(6.9) (a) $$P_{2m}(z) = \frac{1}{3(m+2)} \left(- \sum_{k=1}^{m-1} z^{m-k} + m+ 2\right), \quad m > 1 ,$$

(6.9) (b) $$P_{2m+1}(z) = \frac{1}{3(m+2)} \left(-(m+2) z^m + \sum_{k=1}^{m} z^{m-k}\right), \quad m > 1,$$

(6.9) (c) $$Q_{2m}(z) = \frac{1}{3(m+2)} \left(3z^m + \sum_{k=1}^{m-1} (2k+3) z^{m-k} + 3(m+2)\right), \quad m > 1,$$

(6.9) (d) $$Q_{2m+1}(z) = \frac{1}{3(m+2)} \left(3(m+2) z^m + \sum_{k=1}^{m-1} (2m+3-2k) z^{m-k} + 3\right), \quad m > 1$$

Evaluating the sums (6.8) by standard methods we arrive at

$$(6.10)\ (a)\quad A_{2m}(z) = \frac{2z^{m+2} - z^{m+1} - (m+3)z + (m+2)}{2(m+2)(z-1)^2},$$

$$(6.10)\ (b)\quad B_{2m}(z) = \frac{z^{m+2} - (m+1)z^2 + mz}{2(m+2)(z-1)^2} + 1 .$$

It follows that

$$\lim_{m\to\infty} \frac{A_{2m}(z)}{B_{2m}(z)} = \lim_{m\to\infty} \frac{\frac{2}{m}\, z^{m+2} - \frac{1}{m}\, z^{m+1} - (1+\frac{3}{m})z + (1+\frac{2}{m})}{2(1+\frac{2}{m})(z-1)^2 + \frac{1}{m}\, z^{m+2} - (1+\frac{1}{m})z^2 + z},$$

$$= \frac{-z+1}{2(z-1)^2 - z^2 + z} = \frac{1}{2-z} \quad \text{for} \quad |z| < 1.$$

Thus by Theorem 2.2 $f(z) = 1/(2-z)$. (See Schur [14, II, p. 144].) Similarly

$$(6.11)\ (a)\quad A_{2m+1}(z) = \frac{mz^{m+2} - (m+1)z^{m+1} + z}{2(m+2)(z-1)^2} + z^{m+1}$$

$$(6.11)\ (b)\quad B_{2m+1}(z) = \frac{(m+2)z^{m+3} - (m+3)z^{m+2} - z^2 + 2z}{2(m+2)(z-1)^2}$$

and consequently

$$\lim_{m\to\infty} \frac{A_{2m+1}(z)}{B_{2m+1}(z)} = \lim_{m\to\infty} \frac{\frac{1}{m}\, z^{-m} + z - (1-\frac{1}{m}) + 2(1+\frac{2}{m})(z-1)^2}{-\frac{1}{m}\, z^{-m+1} + \frac{2}{m}\, z^{-m} - (1+\frac{3}{m})z + (1+\frac{2}{m})z^2}$$

$$= \frac{z - 1 + 2(z-1)^2}{-z + z^2} = \frac{2z-1}{z} \quad \text{for} \quad |z| > 1 .$$

The function $\frac{2z-1}{z}$ is indeed $\hat{f}$.
We also get

$$\lim_{m\to\infty} \frac{P_{2m}}{Q_{2m}} = \frac{1 - \lim \frac{A_{2m}}{B_{2m}}}{1 + \lim \frac{A_{2m}}{B_{2m}}} = \frac{z-1}{z-3} \quad \text{for} \quad |z| < 1 .$$

Thus by Theorem 3.2

$$(6.12)\qquad F(z) = \frac{z-1}{z-3} \quad \text{for} \quad |z| < 1.$$

Similarly

$$\lim_{m\to\infty} \frac{P_{2m+1}}{Q_{2m+1}} = \frac{1-z}{3z-1} \quad \text{for} \quad |z| > 1 .$$

As expected $(1-z)/(3z-1) = \tilde{F}$.
The functions F and $\tilde{F}$ have the series expansions

$$\text{(6.13) (a)} \quad F(z) = \frac{1}{3} + \sum_{n=1}^{\infty} \frac{-2}{3^{n+1}} z^n \quad \text{for} \quad |z| < 1 ,$$

$$\text{(6.13) (b)} \quad \tilde{F}(z) = -\frac{1}{3} + \sum_{n=1}^{\infty} \frac{2}{3^{n+1}} z^{-n} \quad \text{for} \quad |z| > 1 .$$

Thus

$$\text{(6.14)} \qquad \mu_0 = \frac{1}{3}, \quad \mu_n = \frac{-1}{3^{n+1}}, \quad \mu_{-n} = \frac{-1}{3^{n+1}}, \quad n \geq 1 .$$

The Szegö polynomials $\{\rho_n\}$ associated with the double sequence $\{\mu_n\}$ given by (6.14) are the polynomials Q_{2m+1}. Thus

$$\rho_n(z) = z^n + \sum_{k=1}^{n-1} (2n+3-2k) z^{n-k} + 1/(n+2) .$$

Example 6.3. Let $\gamma_0 = 1/2$, $\gamma_n = 2/(2n+1)$, $n \geq 1$. Then $\alpha = 1/(1+\gamma_0) = 2/3$, $\delta_0 = (1-\gamma_0)/(1+\gamma_0) = 1/3$, $\delta_n = \overline{\gamma}_n = 2/(2n+1)$, $n \geq 1$.

Starting with

$$A_0 = 1/2, \quad A_1 = z, \quad B_0 = 1, \quad B_1 = z/2$$

the recurrence relations (2.15) yield

$$\text{(6.15) (a)} \quad A_{2m}(z) = \frac{1}{2} + \sum_{k=1}^{m} \frac{2m-2k+2}{2m+1} z^k, \quad m \geq 1 ,$$

$$\text{(6.15) (b)} \quad A_{2m+1}(z) = \sum_{k=0}^{m} \frac{2k+1}{2m+1} z^{k+1}, \quad m \geq 1 ,$$

$$\text{(6.15) (c)} \quad B_{2m}(z) = \sum_{k=0}^{m} \frac{2m-2k+1}{2m+1} z^k, \quad m \geq 1 ,$$

$$\text{(6.15) (d)} \quad B_{2m+1}(z) = \sum_{k=0}^{m-1} \frac{2k+2}{2m+1} z^{k+1} + z^{m+1}/2, \quad m \geq 1 .$$

From formulas (4.2) we obtain

$$\text{(6.16) (a)} \quad P_{2m}(z) = \frac{1}{3} - \frac{2}{3(2m+1)} \sum_{k=1}^{m} z^k, \quad m \geq 1$$

(6.16) (b) $P_{2m+1}(z) = \frac{2}{3(2m+1)} \sum_{k=0}^{m-1} (4k+3)z^k + z^m, \quad m \geq 1$,

(6.16) (c) $Q_{2m}(z) = 1 + \frac{2}{3(2m+1)} \sum_{k=1}^{m} (4m-4k+3)z^k, \quad m \geq 1$,

(6.16) (d) $Q_{2m+1}(z) = \frac{2}{3(2m+1)} \sum_{k=0}^{m-1} (4k+3)z^k + z^m, \quad m \geq 1$.

As before A_{2m} and B_{2m} can be rewritten as

(6.17) (a) $A_{2m}(z) = \frac{1}{2} + \frac{2z^{m+2} - 2(m+1)z^2 + 2mz}{(2m+1)(z-1)^2}$

(6.17) (b) $B_{2m}(z) = 1 + \frac{z^{m+2} + z^{m+1} - (2m+1)z^2 + (2m-1)z}{(2m+1)(z-1)^2}$

It follows that

$$\lim_{m\to\infty} \frac{A_{2m}(z)}{B_{2m}(z)} = \frac{\frac{1}{2}(z-1)^2 - z^2 + z}{(z-1)^2 - z^2 + z} = \frac{1+z}{2} \quad \text{for} \quad |z| < 1$$

Thus (see Schur [14, II, p. 144])

(6.18) $f(z) = (1+z)/2$.

Similarly

(6.19) (a) $A_{2m+1}(z) = \frac{z + z^2 - (2m+3)z^{m+2} + (2m+1)z^{m+3}}{(2m+1)(z-1)^2}$,

(6.19) (b) $B_{2m+1}(z) = \frac{z^{m+1}}{2} + 2\,\frac{z - (m+1)z^{m+1} + mz^{m+2}}{(2m+1)(z-1)^2}$,

and hence

$$\lim_{m\to\infty} \frac{A_{2m+1}(z)}{B_{2m+1}(z)} = \frac{-2z + 2z^2}{z^2 + 1 - 2 + 2z} = \frac{2z}{z+1} .$$

This function is $\hat{f}$.

We also get

$$\lim \frac{P_{2m}}{Q_{2m}} = \frac{1-z}{3+z} \quad \text{and} \quad \lim \frac{P_{2m+1}}{Q_{2m+1}} = \frac{1-z}{1+3z} .$$

These functions are F and $\tilde{F}$, respectively. The functions F and $\tilde{F}$ have the series expansions

(6.20) (a) $F(z) = \frac{1}{3} + 2 \sum_{n=1}^{\infty} (-1)^n \frac{2}{3^{n+1}} z^n$, for $|z| < 1$,

(6.20) (b) $\tilde{F}(z) = -\frac{1}{3} - 2 \sum_{n=1}^{\infty} (-1)^n \frac{2}{3^{n+1}} z^{-n}$, for $|z| > 1$.

Thus

(6.21) $\mu_0 - 1/3, \quad \mu_n = 2(-1)^n/3^{n+1}, \quad \mu_{-n} = 2(-1)^n/3^{n+1}, \quad n \geq 1.$

The Szegö polynomials $\{\rho_n\}$ associated with the double sequence $\{\mu_n\}$ are the polynomials Q_{2n+1}, that is

(6.22) $$\rho_n(z) = \frac{2}{3(2n+1)} \sum_{k=0}^{n-1} (4k+3) z^k + z^n .$$

7. The trigonometric moment problem. The trigonometric moment problem (TMP) is the following. Given a double sequence $\{\mu_n\}_{-\infty}^{\infty}$ of complex numbers find necessary and sufficient conditions for the existence of a distribution function ψ on $[-\pi,\pi]$ such that

(7.1) $$\mu_n = \frac{1}{2\pi} \int_{-\pi}^{\pi} e^{-in\theta} \, d\psi(\theta), \quad n = 0, \pm 1, \pm 2, \ldots$$

(By a distribution function we mean a real valued, bounded, non-decreasing function). In the following we shall assume that $\mu_{-n} = \overline{\mu}_n$ for all n since this clearly is a necessary condition for the existence of a solution.

Let ψ be a solution of the problem and set

$$F(z) = \frac{1}{2\pi} \int_{-\pi}^{\pi} \frac{e^{i\theta} + z}{e^{i\theta} - z} \, d\psi(\theta) .$$

Then F is a normalized Carathéodory function and

(7.2) $$F(z) = \mu_0 + 2 \sum_{n=1}^{\infty} \mu_n z^n$$

for $|z| < 1$. It follows from Theorem 3.1 that the sequence $\{\mu_n\}$ is either positive definite or positive N-definite for some N. It is well known (see for example [1], [15]) that this condition is also sufficient for a solution to exist. The solution ψ is always unique. It follows from Theorem 3.1 that ψ has exactly N points of increase iff $\{\mu_n\}$ is positive N-definite and has an infinite number of points of increase iff $\{\mu_n\}$ is positive definite. These comments, together with theorems proved earlier in this article, lead to the following results.

Theorem 7.1. *The following conditions are equivalent:*

(A) *The* TMP *for* $\{\mu_n\}$ *has a solution with an infinite number of points of increase.*

(B) *The sequence* $\{\mu_n\}$ *is positive definite.*

(C) *There exists a non-terminating positive* PC-*fraction* HPC$\{\delta_n\}$ *such that the sequence* $\{P_{2m}(z)/Q_{2m}(z)\}$ *converges for* $|z| < 1$ *to the limit* (7.2).

(D) *There exists a non-terminating positive Schur fraction* $S\{\gamma_n\}$ *such that the sequence*

$$\left\{\frac{1 - A_{2m}(z)/B_{2m}(z)}{1 + A_{2m}(z)/B_{2m}(z)}\right\}$$

converges for $|z| < 1$ *to the limit* (7.2).

Theorem 7.2. *The following conditions are equivalent:*

(A) *The* TMP *for* $\{\mu_n\}$ *has a solution with* N *points of increase.*

(B) *The sequence* $\{\mu_n\}$ *is positive* N-*definite.*

(C) *There exists a terminating* HPC-*fraction*

$$\delta_0 - \frac{2\delta_0}{1} + \frac{1}{\overline{\delta}_1 z} + \frac{(1-|\delta_1|^2)z}{\delta_1} + \cdots + \frac{(1-|\delta_N|^2)z}{\delta_N}, \quad |\delta_n| < 1,\ 1 \le n \le N$$

whose value for $z < 1$ *is the function* F(z) *defined by* (7.2).

(D) *There exists a terminating Schur fraction*

$$\gamma_0 + \frac{(1-|\gamma_0|^2)z}{\overline{\gamma}_0 z} + \frac{1}{\gamma_1} + \cdots + \frac{1}{\gamma_N} + \frac{(1-|\gamma_N|^2)z}{\overline{\gamma}_N n}, \quad |\gamma_n| < 1,\ 1 \le n \le N$$

whose value for $|z| < 1$ *is*

$$f(z) = \frac{1 - F(z)}{1 + F(z)} \tag{7.3}$$

where F(z) *is given by* (7.2).

(E) *The series* (7.2) *converges for* $|z| < 1$ *and its sum* F *has the form*

$$F(z) = \sum_{v=1}^{N} \lambda_v \frac{t_v+z}{t_v-z},$$

$t_1,\ldots,t_N$ <u>distinct</u>, $|t_v| = 1$, $\lambda_v > 0$, $v = 1,\ldots,N$.

(F) <u>The series</u> (7.2) <u>converges for</u> $|z| < 1$ <u>and the function</u> t, <u>given by</u> (7.3), <u>has the form</u>

$$f(z) = \varepsilon \prod_{v=1}^{N} \frac{z+\omega_v}{1+\overline{\omega}_v z},$$

<u>where</u> $|\varepsilon| = 1$, $|\omega_v| < 1$, $v = 1,\ldots,N$, $\varepsilon \prod_{v=1}^{N} \omega_v \in \mathbb{R}$.

We conclude the paper with an example of a distribution function Ψ with its corresponding moment sequence, Schur fraction and PC-fraction.

<u>Example</u> 7.1. Set $\Psi(\theta) = \theta + \sin\theta$, then $\Psi'(\theta) = 1 + \cos\theta > 0$ and

$$\mu_n = \frac{1}{2\pi} \int_{-\pi}^{\pi} e^{-in\theta} \Psi'(\theta) d\theta .$$

We find

(7.4) $\mu_0 = 1$, $\mu_1 = 1/2$, $\mu_n = 0$ for $n \geq 2$.

The Carathéodory function F is thus given by $F(z) = 1 + z$ and the associated Schur function is

$$f(z) = \frac{1-(1+z)}{1+(1+z)} = \frac{-z}{2+z}.$$

The Schur algorithm yields

$$f_n(z) = \frac{(-1)^{n+1}}{(n+1)+nz}, \quad \gamma = \frac{(-1)^{n+1}}{n+1}, \quad n \geq 1, \quad \gamma_0 = 0 .$$

The Corresponding Schur fractions and PC-fractions are then

$$S\{\gamma_n\} = \frac{z}{0\cdot z} + \frac{1}{\frac{1}{2}} + \frac{(1-(\frac{1}{2})^2)z}{\frac{1}{2}z} + \cdots + \frac{1}{\frac{(-1)^{n+1}}{n+1}} + \frac{(1-(\frac{1}{2})^2)z}{\frac{(-1)^{n+1}}{n+1}z} + \cdots$$

$$HPC\{\delta_n\} = 1 - \frac{2}{1} + \frac{1}{z/2} + \frac{(1-(\frac{1}{2})^2)z}{1/2} + \cdots + \frac{1}{(-1)^{n+1}z/(n+1)} + \frac{(1-(\frac{1}{2})^2)z}{(-1)^{n+1}/(n+1)} + \cdots .$$

Summary of Notation

Schur	Carathéodory
$\mathscr{S} := [\text{normalized Schur functions}]$	$\mathbb{C} := [\text{normalized Carathéodory functions}]$
$= [f: f \text{ analytic for } \lvert z\rvert < 1, f(0) \in \mathbb{R}]$	$= [F: F(z) \text{ analytic for } \lvert z\rvert < 1, \operatorname{Re} F(z) \geq 0 \text{ for } \lvert z\rvert < 1, F(0) \in \mathbb{R}]$
$\mathscr{S}_0 := [f \in \mathscr{S}: f(0) \neq 0]$	$\mathbb{C}_0 := [F \in \mathbb{C}: F(0) \neq 1]$
$\hat{\mathscr{S}} := [g: g \text{ analytic for } 1\ \lvert z\rvert > 1, f(\infty) \in \mathbb{R} \quad [\infty]$	$\hat{\mathbb{C}} := [G: G \text{ analytic for } \lvert z\rvert > 1, \operatorname{Re} G(z) \leq 0 \text{ for } \lvert z\rvert > 1, G(\infty) \in \mathbb{R}]$
$\hat{\mathscr{S}}_\infty := [g \in \hat{\mathscr{S}}, g(\infty) \neq \infty]$	$\hat{\mathbb{C}}_\infty := [G \in \hat{\mathbb{C}}: G(\infty) \neq -1]$
$\hat{f}(z) := 1/\overline{f(1/\overline{z})}$ Schur reciprocal	$\tilde{F}(z) := -[\overline{F(1/\overline{z})}]$ Carathéodory reciprocal
$f_0(z) := f(z) \in \mathscr{S}$	$F_0(z) := F(z) \in \mathbb{C}$
$f_{n+1}(z) := \dfrac{f_n(z)-\gamma_n}{z(1-\overline{\gamma}_n f_n(z))}$, $\gamma_n = f_n(0),\ n \geq 0$	$F_{n+1}(z) := \dfrac{\overline{\delta}_n z - F_n(z)}{\delta_n F_n(z) - z}$, $\delta_n := F'(0)$
$t_n(z,w) := \dfrac{\gamma_n + zw}{1+\overline{\gamma}_n zw}$, $n \geq 0$	$r_0(z,w) := \delta_0 \dfrac{1-w}{1+w}$, $r_n(z,w) := \dfrac{w+\overline{\delta}_n}{1+\delta_n w}$, $n \geq 1$
$T_0 := t_0$, $T_n(z,w) = T_{n-1}(z,t_n(z,w))$, $n > 1$	$R_0 := r_0$, $R_n(z,w) = R_{n-1}(z,r_n(z,w))$, $n \geq 1$
$f_0 = T_n(z,f_{n+1}(z))$	$F_0 = R_n(z,F_{n+1}(z))$
$T_n(z,w) = \dfrac{C_n(z)zw+D_n(z)}{E_n(z)zw+F_b(z)}$	$R_n(z,w) = \dfrac{\pi_n(z)w+\omega_n(z)}{\rho_n(z)w+\tau_n(z)}$
$C_0 := 1,\ D_0 := \gamma_0,\ E_0 := \overline{\gamma}_0,\ F_0 := 1$	$\pi_0 = -\delta_0,\ \omega_0 := \delta_0,\ \rho_0 := 1,\ \tau_0 := 1$
$\binom{C_n}{E_n} = z\binom{C_{n-1}}{E_{n-1}} + \gamma \binom{D_{n-1}}{F_{n-1}},\ n \geq 1$	$\binom{\pi_n}{\rho_n} = z\binom{\pi_{n-1}}{\rho_{n-1}} + \delta \binom{\omega_{n-1}}{\tau_{n-1}},\ n \geq 1$
$\binom{D_n}{F_n} = \gamma\, z\binom{C_{n-1}}{E_{n-1}} + \binom{D_{n-1}}{F_{n-1}},\ n \geq 1$	$\binom{\omega_n}{\tau_n} = \delta\, z\binom{\pi_{n-1}}{\rho_{n-1}} + \binom{\omega_{n-1}}{\tau_{n-1}},\ n \geq 1$

Indexed Reciprocal

$$H_n^x(z) := z^n \overline{H}_n(1/z)$$

$C_n = D_n^x$, $D_n = C_n^x$, $E_n = F_n^x$, $F_n = E_n^x$

$T(z,0) = [\overline{T(1/z, \infty)}]^{-1}$

$$t_n(z,w) = \gamma_n + \frac{(1-|\gamma_n|^2)z}{\overline{\gamma}_n z + 1/w}$$

$\Sigma :=$ [Schur fractions $S\{\gamma_n\}$]

$$S\{\gamma_n\} := \gamma_0 + \frac{(1-|\gamma_0|^2)z}{\overline{\gamma}_0 z} \; \frac{1}{+ \gamma_1} \; \frac{(1-|\gamma_1|^2)z}{+ \overline{\gamma}_1} \; \frac{\cdots}{+}$$

$\Sigma^+ := [S\{\gamma_n\} \in \Sigma,\ \gamma_0 \in R,\ |\gamma_n| < 1,\ n \geq 1]$

$\Sigma_0 := [S\{\gamma_n\} \in \Sigma,\ \gamma_0 \neq 0]$

$\Sigma^+ := [S\{\gamma_n\} \in \Sigma^+,\ \gamma_0 \neq 0]$

$\tau_0(z,w) := \gamma_0 + w$

$$\tau_{2m}(z,w) := \frac{1}{\gamma_m + w}, \quad m \geq 1$$

$S_0 := s_0$, $S_n(z,w) = S_{n-1}(z, s_n w))$, $n \geq 1$

$$S_n(z,w) = \frac{A_n(z) + wA_{n-1}(z)}{B_n(z) + wB_{n-1}(z)}, \quad n \geq 0$$

$A_0 := \gamma_0$, $A_1 := z$, $B_0 := 1$, $B_1 := \overline{\gamma}_0 z$

$$\binom{A_{2m}}{B_{2m}} = \gamma_n \binom{A_{2m-1}}{B_{2m-1}} + \binom{A_{2m-2}}{B_{2m-2}}, \quad m > 1$$

$\omega_n^x = -\pi_n$, $\pi_n^x = -\omega_n$, $\tau_n^x = \rho_n$, $\rho_n^x = \tau$

$R(z,0) = -[\overline{R(1/z, \infty)}]$

$\tau_n = \tau_0 k_{2m} \circ j_{2m+1} \circ \tau$, $n \geq 1$,
$r_0 = k_0 \circ k_1 \circ \tau$
($\tau(z) = 1/z$, k_n defined below)

$\Gamma :=$ [hermitian PC-fraction $HPC\{\delta_n\}$]
$\delta_0 \quad R$, $|\delta_n| \neq 1$, $n \geq 1$

$$HPC\{\delta_n\} := \delta_0 - \frac{2\delta_0}{1} \; \frac{1}{+ \overline{\delta} z} \; \frac{(1-|\delta_1|^2)z}{+ \delta_1} \; \frac{1}{+ \overline{\delta}_2} \; \frac{(1-|\delta_n|^2)z}{+ \delta_2} \; \frac{\cdots}{+}$$

$\Gamma^+ := [HPC\{\delta_n\} \in \Gamma,\ \delta_0 > 0,\ |\delta_n| < 1,\ n \geq 1]$

$\Gamma_0 := [HPC\{\delta_n\} \in \Gamma,\ \delta_0 \neq 1]$

$\Gamma_0^+ := [HPC\{\delta_n\} \in \Gamma^+,\ \delta_0 \neq 1]$

$$k_0(z,w) = \delta_0 + w, \quad k_1(z,w) := \frac{-s\delta_0}{1+w},$$

$$k_{2m}(z,w) := \frac{1}{\overline{\delta}_m z + w}, \quad m \geq 1$$

$K_0 := k_0$, $K_n(z,w) := K_{n-1}(z, k_n(z,w))$, $n \geq 1$

$$K_n(z,w) = \frac{P_n(z) + wP_{n-1}(z)}{Q_n(z) + wQ_{n-1}(z)}, \quad n \geq 0$$

$P_0 := \delta_0$, $P_1 := -\delta_0$, $Q_0 := 1$, $Q_1 := 1$

$$\binom{P_{2m}}{Q_{2m}} = \delta_n \binom{P_{2m-1}}{Q_{2m-1}} + \binom{P_{2m-2}}{Q_{2m-2}}, \quad m > 1$$

$$\begin{pmatrix}A_{2m+1}\\B_{2m+1}\end{pmatrix} = \overline{\gamma}_m z\begin{pmatrix}A_{2m}\\B_{2m}\end{pmatrix} + (1 - |\gamma_n|^2) z\begin{pmatrix}A_{2m-1}\\B_{2m-1}\end{pmatrix},\ m \geq 1$$

$$\begin{pmatrix}P_{2m+1}\\Q_{2m+1}\end{pmatrix} = \delta_m\begin{pmatrix}P_{2m}\\Q_{2m}\end{pmatrix} + (1 - |\delta_n|^2) z\begin{pmatrix}P_{2m-1}\\Q_{2m-1}\end{pmatrix},\ m \geq 1$$

$$\Gamma_n = \prod_{n=1}^{m} (1 - |\delta_n|^2)$$

$$A_{2m}(z) = \gamma_0 + \cdots + \gamma_m z^m,\ m \geq 1$$

$$P_{2m}(z) = \delta_0 + \cdots - \delta_0 \overline{\delta}_m \Gamma_{m-1} z^m,\ m \geq 1$$

$$B_{2m}(z) = 1 + \cdots + \overline{\gamma}_0 \gamma_m z^m,\ m \geq 1$$

$$Q_{2m}(z) = 1 + \cdots + \overline{\delta}_m \Gamma_{m-1} z^m,\ m \geq 1$$

$$A_{2m+1}(z) = \gamma_0 \overline{\gamma}_m z + \cdots + z^{m+1},\ m \geq 1$$

$$P_{2m+1}(z) = \delta_0 \delta_m + \cdots \pm \delta_0 \Gamma_m z^m,\ m \geq 1$$

$$B_{2m+1}(z) = \overline{\gamma}_m z + \cdots + \overline{\gamma}_0 z^{m+1},\ m \geq 1$$

$$Q_{2m+1}(z) = \delta_m + \cdots + \Gamma_m z^m,\ m \geq 1$$

$$A_{2m+1} = zC_m,\ A_{2m} = D_m,\ m \geq 0$$

$$P_{2m+1} = \pi_m,\ P_{2m} = \omega_m,\ m \geq 0$$

$$B_{2m+1} = zE_m,\ B_{2m} = F_m$$

$$Q_{2m+1} = \rho_m,\ Q_{2m} = \tau_m$$

$$T_m(z,w) = S_{2m+1}(z, 1/2),\ m \geq 0$$

$$R_m(z,w) = K_{2m+1}(z, 1/w)$$

$$\frac{A_{2m}}{B_{2m}} = S_{2m+1}(z,\infty) = T_m(z,0) = \frac{D_m}{F_m}$$

$$\frac{P_{2m}}{Q_{2m}} = K_{2m+1}(z,\infty) = R_m(z,0) = \frac{\omega_m}{\tau_m}$$

$$\frac{A_{2m+1}}{B_{2m+1}} = S_{2m+1}(z,0) = T_m(z,\infty) = \frac{C_m}{E_m}$$

$$\frac{P_{2m+1}}{Q_{2m+1}} = K_{2m+1}(z,0) = R_m(z,\infty) = \frac{\pi_m}{\rho_m}$$

$$S\{\gamma_n\} \in \Sigma^+ \Rightarrow \psi_\Sigma(S\{\gamma_n\}) = f \in \mathscr{S}$$

$$HPC\{\delta_n\} \in \Gamma^+ \Rightarrow \psi_\Gamma(HPC\{\delta_n\}) = F \in \mathbb{C}$$

$$f \in \mathscr{S} \Rightarrow \Phi_{\mathscr{S}}(f) = S\{\gamma_n\} \in \Sigma^+$$

$$F \in \mathbb{C} \Rightarrow \Phi\ (F) = HPC\{\delta_n\} \in \Gamma^+$$

$$\psi_\Sigma \Phi_{\mathscr{S}}(f) = f,\ \Phi_{\mathscr{S}} \psi_\Sigma(S\{\gamma_n\}) = S\{\gamma_n\}$$

$$\psi_\Gamma \Phi_{\mathbb{C}}(F) = F,\ \Phi_{\mathbb{C}} \psi_\Gamma(HPC\{\delta_n\}) = HPC(\{\delta_n\})$$

REFERENCES

1. N. Akhiezer, The classical moment problem, Hafner, New York (1965).

2. G. Baker and P. Graves-Morris, "Padé Approximants. I, II," Encyclopedia of Mathematics and its Applications, 13,14, Addison-Wesley, Reading, Massachusetts (1980). Now available from Cambridge Univ. Press.

3. C. Carathéodory, "Über den Variabilitätsbereich der Koeffizienten von Potenzreihen die gegebene Werte nicht annehmen," Math. Ann. 64 (1907), 95-115.

4. C. Carathéodory, "Über den Variabilitätsbereich der Fourier'schen Konstanten von positiven harmonischen Funktionen," Rend. Circolo Math. Palermo 32 (1911), 193-217.

5. C. Carathéodory and L. Fejer, "Über den Zusammenhang der Extremen von harmonischen Funktionen mit ihren Koeffizienten und über den Picard-Landau'schen Satz," Rend. Circolo Math. Palermo 32 (1911), 218-239.

6. E. Fischer, "Über das Carathéodory'sche Problem, Potenzreihen mit positiv reellen Teil betreffend," Rend. Circolo Math. Palermo 32 (1911), 240-256.

7. Y. Geronimus, "On the trigonometric moment problem," Ann. of Math. (2) 47 (1946), 742-761.

8. G. Hamel, "Eine charakteristische Eigenschaft beschränkter analytischer Funktionen, Math. Ann. 78 (1917), 257-269.

9. W. B. Jones, O. Njåstad and W. J. Thron, "Continued Fractions associated with the trigonometric and other strong moment problems," Constructive Approximation, to appear.

10. W. B. Jones, O. Njåstad and W. J. Thron, "Szegö polynomials and Perron-Carathéodory fractions," in preparation.

11. W. B. Jones and A. Steinhardt, "Digital filters and continued fractions," Analytic Theory of Continued Fractions, (W. B. Jones, W. J. Thron and H. Waadeland, Eds.) Lecture Notes in Mathematics, No. 932, Springer, New York (1982), 129-151.

12. W. B. Jones and W. J. Thron, "Continued fractions: Analytic theory and applications," Encyclopedia of Mathematics and its Applications, 11, Addison-Wesley, Reading, Mass. (1980). Now available from Cambridge Univ. Press.

13. O. Perron, Lehre von den Kettenbrüchen, Band II, Teubner, Stuttgart (1957).

14. J. Schur, "Über Potenzreihen die im Inneren des Einheitskreises beschränkt sind," J. reine angewandte Math. 147 (1917), 205-232, 148 (1918/19), 122-145.

15. J. A. Shohat and J. D. Tamarkin, "The problem of moments," Math. Surveys, No. 1, Amer. Math. Society, Providence, R.I. (1943).

16. G. Szegö, Collected Papers, R. Askey Ed., vol. 1, Birhkäuser, Basel, Boston, Stuttgart (1982).

17. G. Szegö, Orthogonal Polynomials, Amer. Math. Society, Providence, R.I. (1939).

18. W. J. Thron, "Two-point Padé tables, T-fractions and sequences of Schur," Padé and Rational Approximation, (E. B. Saff and R. S. Varga, Eds.) Academic Press, New York (1977), 215-226.

19. O. Toeplitz, "Über die Fourier'sche Entwicklung positiver Funktionen," Rend. Circolo Math. Palermo 32 (1911), 191-192.

20. H. S. Wall, Analytic Theory of Continued Fractions, Van Nostrand, New York (1948).

21. H. S. Wall, "Continued fractions and bounded analytic functions," Bull. Amer. Math. Soc. 50 (1944), 110-119.

22. Gregory S. Ammar and William B. Gragg, "The generalized Schur algorithm for the superfast solution of Toeplitz systems," Proceedings of the Conference on Padé Approximation in Lancut, Poland (1985), to appear in the Springer Lecture Notes in Mathematics.

23. Gregory S. Ammar and William B. Gragg, "The implementation and use of the generalized Schur algorithm," Proceedings of the 7th International Symposium on the Mathematical Theory of Networks and Systems (MTNS-85), Stockholm (C. Byrnes and A. Lindquist, eds.), North Holland, to appear.

Equimodular Limit Periodic Continued Fractions*

N.J. Kalton and L.J. Lange

Department of Mathematics
University of Missouri
Columbia, Missouri 65211

1. Introduction.

Let

$$b_0 + \frac{a_1}{b_1} + \frac{a_2}{b_2} + \frac{a_3}{b_3} + \dots \tag{1.1}$$

be a continued fraction. Then the nth numerator A_n and denominator B_n satisfy $A_{-1} - 1$, $A_0 = b_0$, $B_{-1} = 0$, $B_0 = 1$ and for $n \geq 1$ the difference equation

$$x_n = b_n x_{n-1} + a_n x_{n-2}. \tag{1.2}$$

This equation can be written in the matrix form

$$\begin{pmatrix} x_n \\ x_{n-1} \end{pmatrix} = Q_n \begin{pmatrix} x_{n-1} \\ x_{n-2} \end{pmatrix}, \quad n - 1, 2, \dots, \tag{1.3}$$

where

$$Q_n = \begin{pmatrix} b_n & a_n \\ 1 & 0 \end{pmatrix}.$$

* The research of N.J. Kalton was supported in part by the National Science Foundation under grant number DMS 8301099. The research of L.J. Lange was supported in part by a grant from the Research Council of the Graduate School, University of Missouri - Columbia and a grant from the Nansen Foundation of Norway.

If $\lim_{n\to\infty} a_n = a \in \mathbb{C}$ and $\lim_{n\to\infty} b_n = b \in \mathbb{C}$, then $Q_n \to Q$ as $n \to \infty$, where

$$Q = \begin{pmatrix} b & a \\ 1 & 0 \end{pmatrix} .$$

The characteristic equation of Q is

$$\lambda^2 - b\lambda - a = 0. \tag{1.4}$$

In this simple limit periodic case, we say (1.1) is <u>equimodular</u> if the roots of (1.4) have equal absolute value.

More generally, suppose

$$\begin{aligned} &\lim a_{2n-1} = \alpha_1, \ \lim a_{2n} = \alpha_2 \\ &\lim b_{2n-1} = \beta_1, \ \lim b_{2n} = \beta_2, \end{aligned} \tag{1.5}$$

where α_i, $\beta_i \in C$, $i = 1,2$.
Using (1.3) we obtain

$$\begin{pmatrix} x_{n+1} \\ x_n \end{pmatrix} = Q_{n+1}Q_n \begin{pmatrix} x_{n-1} \\ x_{n-2} \end{pmatrix} = R_n \begin{pmatrix} x_{n-1} \\ x_{n-2} \end{pmatrix} , \ n \geq 1,$$

where

$$R_n = \begin{pmatrix} b_{n+1}b_n + a_{n+1} & b_{n+1}a_n \\ b_n & a_n \end{pmatrix} .$$

Under our hypothesis,

$$R_{2n-1} \to R = \begin{pmatrix} \beta_1\beta_2 + \alpha_2 & \beta_2\alpha_1 \\ \beta_1 & \alpha_1 \end{pmatrix}$$

and

$$R_{2n} \to S = \begin{pmatrix} \beta_1\beta_2 + \alpha_1 & \beta_1\alpha_2 \\ \beta_2 & \alpha_2 \end{pmatrix}$$

as $n\to\infty$. The matrices R and S have the common characteristic equation

$$\lambda^2 - \lambda(\alpha_1 + \alpha_2 + \beta_1\beta_2) + \alpha_1\alpha_2 = 0. \tag{1.6}$$

Under the conditions (1.5), we shall say (1.1) is <u>equimodular</u> if the roots of (1.6) have equal modulus.

Here we wish to point out what connection this definition has with the even and odd parts of (1.1) when they exist. If $b_{2k} \neq 0$, $k = 1, 2, \dots$, then by [6, Theorem 2.10] the partial numerators and denominators a_k^* and b_k^*, respectively, of the even part satisfy

$$\left.\begin{array}{l} a_k^* = -a_{2k-2}a_{2k-1}b_{2k-4}b_{2k} \\ b_k^* = a_{2k-1}b_{2k} + b_{2k-2}(a_{2k} + b_{2k-1}b_{2k}), \end{array}\right. \quad k = 3, 4, \dots$$

so

$$a_k^* \to a^* = -\alpha_1\alpha_2\beta_2^2$$

and

$$b_k^* \to b^* = \alpha_1\beta_2 + \beta_2(\alpha_1 + \beta_1\beta_2)$$

as $k \to \infty$. Applying (1.3) and (1.4) we have that the associated characteristic equation is

$$\lambda^2 - b^*\lambda - a^* = 0. \tag{1.7}$$

It is easy to see that, if the roots of (1.6) have equal modulus, then the roots of (1.7) have equal modulus. Similarly, if $b_{2k+1} \neq 0$, $k = 0, 1, 2, \dots$, then by [6, Theorem 2.11] the odd part of (1.1) exists and under conditions (1.5) its partial numerators and denominators tend to limits $\hat{a}$ and $\hat{b}$, respectively. The associated characteristic equation

$$\lambda^2 - \hat{b}\lambda - \hat{a} = 0$$

also has roots of equal modulus if this is the case for (1.6).

Our purpose in this paper is to make a rather extensive study of equimodular limit periodic continued fractions of the types $-K(-1/b_n)$ and $K(a_n/1)$. Most convergence theorems to date for limit periodic continued fractions have depended on the stipulation that the roots of (1.4) have unequal modulus. Under this hypothesis, a number of fruitful convergence results such as those given in paragraph 19 of Perron's book [10] were obtained. These results and Poincaré's theorem on finite differences (see [7, Theorem A]) served as useful tools in the work of Lange [7] on limit periodic δ-fractions. But, if the roots of (1.4) have the same modulus, we can no longer employ these results. In fact Louboutin [8] has recently shown that Poincaré's theorem cannot be extended when two characteristic roots have the same absolute value, even if some arithmetical constraints are imposed. This may help to explain why it appears from our investigations that the behavior of limit periodic continued fractions in equimodular cases is quite complicated.

Up to now, there appears to be only a modest number of scattered results in the literature relating directly to the work we are presenting here. One of the earliest important results for which the roots of (1.4) have equal absolute value is the well known Stern (1860) - Stolz (1886) Theorem (see [6, p. 79]). This theorem states that $K(1/b_n)$ diverges if $\Sigma|b_n|$ converges. In more recent times, Thron and Waadeland [11] have generated interest in this topic through their 1980 paper where, besides considering convergence acceleration questions, they studied various cases and examples concerned with continued fractions $K(a_n/1)$ where $\lim a_n = -1/4$. This paper and some work of Waadeland [12] led Jacobsen and Magnus [4] in 1983 to prove the result that $K(a_n/1)$ diverges if

$$a_n = -\frac{1}{4} - \frac{c}{16(n+\theta)(n+\theta+1)},$$

where $c > 1$, $\theta > -1$ are constants. In 1981 Heller and Roach [3] proved that $K(1/b_n)$ diverges if $b_{2n-1} \to u$, $b_{2n} \to v$, $-4 < uv < 0$, and both series $\Sigma \mid b_{2n-1} - u \mid$, $\Sigma \mid b_{2n} - v \mid$ converge. In 1984 Baltus [1] investigated continued fractions $K(a_n/1)$ with $a_n \to -1/4$ in section 6 of his doctoral thesis. Recently Jacobsen and Waadeland [5] have studied cases where $a_n \to -1/4$ or $a_n \to \infty$ and their connections.

In section 2 we study the convergence behavior of continued fractions $-K(-1/b_n)$, where $b_n \to b$, $-2 < b < 2$. The discriminant b^2-4 of (1.4) is negative in this case, so the roots of (1.4) not only have the same absolute value but they are unequal. Lemma 2.1 deals with the boundedness of solutions of certain second order difference equations. Besides possibly having applications in the study of difference equations, it has proved to be quite useful in our study of equimodular continued fractions. In Theorem 2.1 we show that $-K(-1/b_n)$ cannot converge finitely under the above conditions if the series $\Sigma \mid \bar{b}_n - b_{n+1} \mid$ converges. Theorem 2.2 shows how to construct continued fractions $-K(-1/b_n)$ with $b_n \to b$, $0 \le b < 2$, that diverge by oscillation.

In section 3 the main emphasis is on the study of continued fractions $-K(-1/b_n)$, where $b_n \to 2$ from the left at $n \to \infty$. In this case the roots of (1.4) are equal. The convergence behavior is considerably different under these conditions than for the case studied in section 2. Through a series of five theorems assuming various growth patterns of the b_n and using a variety of techniques in their proofs, we feel we have shed much light on the behavior of these continued fractions. Theorem 3.1 is a general convergence theorem. Theorem 3.2 states that $-K(-1/b_n)$ converges finitely if $b_n \ge 2 - 1/(4n^2)$ for $n \ge 1$. Theorems 3.3 and 3.4 are designed to show what effect the speed of growth of the b_n has on the convergence of $-K(-1/b_n)$, and as a by-product they show that the coefficient $-1/4$ of n^{-2} in Theorem 3.2 is sharp.

Though we learned a fair amount about continued fractions $-K(-1/b_n)$ when $b_n \to 0$ in section 2, we have chosen in section 4 to study this case in greater detail. In Theorem 4.1 we settle a question of Wall showing that the convergence of the series Σb_n is not sufficient for the divergence of $K(1/b_n)$. In Theorem 4.2 we show how to construct finitely convergent continued fractions $-K(-1/b_n)$, where $\lim b_n = 0$. Theorem 4.3 deals with a case where the b_n vary with a parameter, and it is designed to give information about S- and H-fractions on the "cut". In example 4.1 we demonstrate how this study has application to the question of whether or not a continued fraction corresponding to a function can actually converge to this function.

In section 5 we again study continued fractions $-K(-1/b_n)$, but here we do not necessarily demand convergence of the sequence $\{b_n\}$. Instead we assume $b_{2n-1} \to a$ and $b_{2n} \to b$, where in some cases we allow a and b to have nonzero imaginary parts but always $0 \leq ab < 4$. Under these conditions the discriminant $ab\,(ab - 4)$ of (1.6) is nonpositive. Therefore, the roots of (1.6) have equal absolute value. Lemma 5.1 is a statement about the boundedness of solutions of the difference equation

$$x_n = b_n x_{n-1} - x_{n-2}$$

that has proved to be quite useful in our study of equimodular continued fractions of this type. It may also prove to be useful to researchers in the field of difference equations. Theorem 5.1 essentially states that $-K(-1/b_n)$ cannot converge finitely if $a, b \in \mathbb{R}$, $0 < ab < 4$, $\Sigma\,|b_{2n+1} - b_{2n-1}| < \infty$, and $\Sigma\,|b_{2n+2} - b_{2n}| < \infty$. Theorem 5.2 deals with the construction of continued fractions with various types of convergence behavior for the two cases $a = b = 0$ and $a \neq 0$, $b = 0$. In our final result of this section, Theorem 5.3, we allow a and b to have nonzero imaginary parts. Theorem 5.3 states that, if $0 < ab < 4$ and a certain series converges, then $-K(-1/b_n)$ cannot converge finitely. If the case $a = b = 0$ is excluded, this theorem contains the main theorem in a paper by Heller and Roach [3] as a special case.

The following problem is considered in section 6: Given $0 < b < 2$, does there exist a finitely convergent continued fraction $-K(-1/b_n)$ such that $\lim b_n = b$? We, at this time, are able only to give a partial solution to this problem. In the proof of Theorem 6.1 we show how to construct a sequence $\{b_n\}$ such that $b_n \to b$ and $-K(-1/b_n)$ converges nonabsolutely to b, where $b = 2 \cos r\pi$ with $0 < r < 1/2$ a rational number. Theorem 6.2 says that for certain irrational numbers α, $0 < \alpha < 1/4$, there exists a sequence $\{b_n\}$ such that $b_n \to 2 \cos 2\alpha\pi$ and $-K(-1/b_n)$ converges absolutely to a finite limit.

After proving two needed lemmas in section 7, we give in Theorem 7.1 a matrix theoretic result whose conclusion involves the boundedness of products of matrices. This theorem has all the indications of being a powerful tool for investigating the convergence behavior of certain types of continued fractions. We point out through our Examples 7.1 how Theorem 7.1 might have been used to obtain some of our earlier results. It is our main force in the proofs of Theorems 7.2, 8.1, and 8.2. Theorem 7.2 is a divergence type theorem about continued fractions $-K(-1/b_n)$, where $b_n \to 0$ but $\Sigma \mid b_n - b_{n+1} \mid$ may diverge.

In our final section, section 8.1, we prove three theorems dealing with continued fractions $K(a_n/1)$ under various convergence conditions imposed on $\{a_n\}$. Theorem 8.1 says that, if $a_n < -1/4$, $a_n \to a < -1/4$, and $\Sigma \mid a_{n+1} - a_n \mid < \infty$, then $K(a_n/1)$ cannot converge finitely. Theorem 8.2 says that the same conclusion can be drawn about $K(a_n/1)$ if $a_{2n-1} \to a$, $a_{2n} \to b$, $K(a_n/1)$ is equimodular, and both series $\Sigma \mid a_{2n+1} - a_{2n-1} \mid$, $\Sigma \mid a_{2n+2} - a_{2n} \mid$ converge. Our concluding Theorem 8.3 deals with continued fractions $K(a_n/1)$, where $a_n \to -\infty$. It shows how to construct continued fractions of this type that diverge by oscillation and how to construct ones that converge finitely.

2. $-K(-1/b_n)$, <u>where</u> $\lim b_n = b(|b| < 2)$.

We begin this section with a fundamental lemma concerned with second order linear homogenous difference equations. The lemma plays an important role in the proofs of Theorems 2.1, 4.3 and 5.3.

<u>Lemma</u> 2.1: <u>If the sequence</u> $\{b_n\}$ <u>satisfies</u> $b_n \in \mathbb{C}$, $|b_n| < 2$, $b_n \to b \in \mathbb{R}$, $|b| < 2$, <u>and</u>

$$\sum_{n=1}^{\infty} |\bar{b}_n - b_{n+1}| < \infty, \tag{2.1}$$

<u>then all solutions of the difference equation</u>

$$x_n = b_n x_{n-1} - x_{n-2}, \quad n = 1, 2, \ldots \tag{2.2}$$

<u>are bounded</u>.

<u>Proof</u>: For n = 1, 2, ..., let

$$u_n = |x_n|^2 + |x_{n-1}|^2 - \mathrm{Re}(b_{n+1} x_n \bar{x}_{n-1})$$

and

$$u_n^* = |x_{n+1}|^2 + |x_n|^2 - \mathrm{Re}(\bar{b}_{n+1} x_{n+1} \bar{x}_n).$$

then, using (2.2), it is easily verified that

$$u_n^* = u_n, \quad n \geq 1.$$

Hence,

$$\begin{aligned} u_{n+1} - u_n &= u_{n+1} - u_n^* \\ &= \mathrm{Re}[(\bar{b}_{n+1} - b_{n+2})\, x_{n+1}\bar{x}_n] \\ &\leq |\bar{b}_{n+1} - b_{n+2}|\, |x_{n+1}|\, |x_n|. \end{aligned} \tag{2.3}$$

But,

$$u_n \geq |x_n|^2 + |x_{n-1}|^2 - |b_{n+1}|\ |x_n|\ |x_{n-1}|$$

$$(2.4) \qquad \geq (2 - |b_{n+1}|)\ \frac{|x_n|^2 + |x_{n-1}|^2}{2}$$

$$\geq (2 - |b_{n+1}|)\ |x_n|\ |x_{n-1}| .$$

Similarly, using the fact that $u_n = u_n^*$, we obtain

$$u_n \geq (2 - |b_{n+1}|)\ \frac{|x_{n+1}|^2 + |x_n|^2}{2}$$

$$(2.5) \qquad \geq (2 - |b_{n+1}|)\ |x_{n+1}|\ |x_n| .$$

It follows from inequalities (2.3) and (2.5) that

$$u_{n+1} \leq u_n \ (1 + \frac{|\bar{b}_{n+1} - b_{n+2}|}{2 - |b_{n+1}|}) ,$$

from which we obtain

$$u_{n+1} \leq u_1 \prod_{k=1}^{\infty} (1 + \frac{|\bar{b}_{k+1} - b_{k+2}|}{2 - |b_{n+1}|}) .$$

The last inequality shows that $\{u_n\}$ is bounded since $\sum_{n=1}^{\infty} |\bar{b}_n - b_{n+1}| < \infty$ and $|b_n| \to |b| < 2$ as $n \to \infty$. But from inequalities (2.4) we have

$$|x_n|^2 + |x_{n-1}|^2 \leq \frac{2\ u_n}{2 - |b_{n+1}|}$$

so that the x_n are bounded, and our proof is complete.

The following theorem tells us that, if a continued fraction $-K(-1/b_n)$ is to converge finitely when $\{b_n\}$ converges to a real limit $b(|b| < 2)$, restrictions must be put on the speed of convergence of $\{b_n\}$ and the sequence cannot be monotone when the b_n are real.

Theorem 2.1: *If the elements of the sequence* $\{b_n\}$ *satisfy* $b_n \in \mathbb{C}$, $|b_n| < 2$, $\lim b_n = b \in \mathbb{R}$, $|b| < 2$, *and*

$$\sum_{n=1}^{\infty} |\bar{b}_n - b_{n+1}| < \infty, \tag{2.6}$$

then the continued fraction

$$\frac{1}{b_1} - \frac{1}{b_2} - \frac{1}{b_3} - \cdots \tag{2.7}$$

cannot converge to a finite limit. If the further restriction $b_n \in \mathbb{R}$, $n \geq 1$, *is imposed, then condition* (2.6) *becomes*

$$\sum_{n=1}^{\infty} |b_n - b_{n+1}| < \infty, \tag{2.8}$$

which is satisfied, in particular, if $b_n \to b$ *monotonely as* $n \to \infty$.

Proof: The nth denominator B_n of (2.7) satisfies

$$B_{-1} = 0,\ B_1 = b_1;\ B_n = b_n B_{n-1} - B_{n-2},\ n \geq 1.$$

By Lemma 2.1, the sequence $\{B_n\}$ is bounded. In order for (2.7) to converge, at most a finite number of the B_n can be zero. So assume there exists a positive integer n_0 such that $B_n \neq 0$ for all $n \geq n_0$. But then the nth approximant f_n of (2.7) satisfies

$$f_n = f_{n_0} + \sum_{k=n_0+1}^{\infty} \frac{1}{B_k B_{k-1}},\ n > n_0 .$$

Since the B_k are bounded, the series

$$\sum_{k=n_0+1}^{\infty} \frac{1}{B_k B_{k-1}}$$

and therefore the sequence $\{f_n\}$ cannot converge to a finite limit. With this our proof is complete.

Our next theorem tells us how to create continued fractions $-K(-1/b_n)$ that diverge by oscillation when $\{b_n\}$ converges to a given limit b with the restriction $0 \leq b < 1$.

Theorem 2.2: *For* $n = 0, 1, 2, \ldots$ *let* $g_n = x_n + i\, y_n$, *where* $0 < y_n \leq 1$ *and* $x_n = (y_n(-y_n + 1/y_{n+1}))^{1/2}$. *Let* $b_n = g_n + 1/g_{n-1} = x_n + x_{n-1}y_n/y_{n-1}$. *If* $\lim y_n = L$, $0 < L \leq 1$, *then* $\lim b_n = 2(1-L^2)^{1/2}$ *and the continued fraction* (2.7) *diverges by oscillation.*

Proof: In the first part of our proof we let $b_n = g_n + 1/g_{n-1}$, where the only restriction on the g_n is that they are nonzero complex numbers. Let

$$f_n = \frac{1}{b_1} - \frac{1}{b_2} - \cdots - \frac{1}{b_n}$$

and set

$$F_n = \frac{1}{g_1} - \frac{1}{b_2} - \cdots - \frac{1}{b_n} .$$

Then

$$f_n = \frac{1}{(1/g_0) + (1/F_n)} . \tag{2.9}$$

Hence, if $F_n \to \infty$ as $n \to \infty$, then $f_n \to g_0$ as $n \to \infty$ and the tails of (2.7) are right tails. If we set $F_n = C_n/D_n$, where $C_0 = 0$, $C_1 = 1$, $D_0 = 1$, $D_1 = g_1$, and C_n, D_n each satisfy the difference equation

$$z_n = b_n z_{n-1} - z_{n-2}, \; n \geq 2,$$

then it can be established by induction that

$$D_n = g_1 g_2 \cdots g_n .$$

Now, since $F_n - F_{n-1} = 1/(D_n D_{n-1})$, we have that

$$F_n = F_0 + \sum_{k=1}^{n} 1/(D_k D_{k-1})$$

$$= \sum_{k=1}^{n} 1/(D_k D_{k-1}) \text{ (since } C_0 = 0)$$

$$(2.10) \qquad = g_0 \sum_{k=1}^{n} \prod_{j=1}^{k} 1/(g_j g_{j-1}) \ .$$

Finally, from (2.9) and (2.10) we have that

$$(2.11) \qquad f_n = \frac{g_0}{1 + 1/ \sum_{k=1}^{n} \prod_{j=1}^{k} 1/(g_j g_{j-1})} \ .$$

Now at this stage let $g_n = x_n + iy_n$, where the x_n and y_n satisfy the hypotheses of our theorem. Then

$$|g_n| = (x_n^2 + y_n^2)^{1/2} = (y_n / y_{n+1})^{1/2}$$

so

$$\sum_{k=1}^{n} \prod_{j=1}^{k} \frac{1}{|g_j g_{j-1}|} = \sum_{k=1}^{n} \prod_{j=1}^{k} \left(\frac{y_{j+1}}{y_j} \frac{y_j}{y_{j-1}}\right)^{1/2}$$

$$= \sum_{k=1}^{n} \left(\frac{y_k}{y_0} \frac{y_{k+1}}{y_1}\right)^{1/2} .$$

Since $\lim_{k\to\infty} y_k = L$, $0 < L \le 1$, it follows that the sequence $\{\sum_{k=1}^{n} (y_k y_{k+1}/(y_0 y_1))^{1/2}\}$ cannot converge to a finite limit. This same sequence cannot converge to ∞ either, for if so, $\{f_n\}$ and hence the continued fraction (2.7) would converge to g_0, a number whose imaginary part is not zero. This impossible because the b_n are real. Thus the conclusion must be that (2.7) diverges by oscillation.

3. $-K(-1/b_n)$, where $\lim b_n = 2$.

Our first theorem in this section is a general convergence type theorem for continued fractions $-K(-1/b_n)$ that does not demand that $\{b_n\}$ converge. It says that the continued fraction can converge finitely if the b_n are real or complex provided that the modulus of b_n satisfies a certain growth condition. The theorem leads us to a method for constructing finitely convergent continued fractions of this type, where $b_n \to 2$ from the left.

Theorem 3.1: *Let* $\{g_n\}_{n\geq 0}$ *be a sequence of positive real numbers, and let* $\{b_n\}$ *be a sequence of complex numbers satisfying*

(3.1) $$|b_n| \geq g_n + 1/g_{n-1}, \quad n = 1, 2, \ldots .$$

Then the continued fraction

(3.2) $$\frac{1}{b_1} - \frac{1}{b_2} - \frac{1}{b_3} - \ldots$$

converges to a finite value in the disk $|z| \leq g_0$. *If, in particular,*

(3.3) $$b_n \geq 2 - \frac{1}{4n(n+1/2)}, \quad n = 1, 2, \ldots,$$

then (3.2) *converges to a number in the interval* $[-3/2, 3/2]$.

Proof: If $\{\gamma_n\}_{n\geq 0}$ is an arbitrary sequence of positive real numbers, then the continued fraction

$$K = \frac{1}{\gamma_0}\left(\frac{\gamma_0\gamma_1}{\gamma_1 b_1} - \frac{\gamma_1\gamma_2}{\gamma_2 b_2} - \frac{\gamma_2\gamma_3}{\gamma_3 b_3} - \ldots\right)$$

is equivalent to the continued fraction (3.2). By the Pringsheim Convergence Theorem (see [10, Satz 2.19]) K converges to a finite limit contained in the disk $|z| \leq 1/\gamma_0$ if

(3.4) $$\gamma_n |b_n| \geq 1 + \gamma_n\gamma_{n-1}, \quad n = 1, 2, \ldots .$$

Thus to complete our proof it is sufficient to set $\gamma_n = 1/g_n$ in (3.4). Finally, in order to see that (3.2) converges if (3.3) holds, we have only to set

$$g_n = 1 + \frac{1}{2(n+1)}, \quad n = 0, 1, 2, \ldots .$$

We now show that condition (3.3) of Theorem 3.1 can be sharpened. The method of proof is somewhat novel in that it couples the theory of extensions for continued fractions with a 1905 convergence result of Pringsheim.

<u>Theorem</u> 3.2: <u>The continued fraction</u> (3.2) <u>converges to a finite limit if</u>

$$b_n \geq 2 - \frac{1}{4n^2}, \quad n = 1, 2, \ldots .$$

<u>Proof</u>: The continued fraction (3.2) can be extended to

$$(3.5) \qquad \frac{1}{b_1+1} - \frac{1}{1} - \frac{1}{b_2+2} - \frac{1}{1} - \frac{1}{b_3+2} - \frac{1}{1} - \cdots,$$

and (3.5) can be extended to

$$(3.6) \qquad -1 + \frac{1}{1} - \frac{1}{b_1+2} - \frac{1}{1} - \frac{1}{b_2+2} - \frac{1}{1} - \frac{1}{b_3+2} - \frac{1}{1} - \cdots .$$

The 2n+1 st approximant of (3.6) is the nth approximant of (3.2). The continued fraction (3.6) (and therefore (3.2)) will converge finitely if

$$(3.7) \qquad \frac{1}{1} - \frac{1}{b_1+2} - \frac{1}{1} - \frac{1}{b_2+2} - \frac{1}{1} - \frac{1}{b_3+2} - \frac{1}{1} - \cdots$$

converges to a finite limit. Now let

$$p_1 = \frac{5}{4} > 1 \; ; \; p_n = \frac{2n-1}{n}, \quad n = 2, 3, \ldots,$$

and let

$$c_{2n-1} = 1, \; c_{2n} = b_n + 2, \; n = 1, 2, \ldots .$$

Then, according to [10, Satz 2.21], (3.7) converges to a finite limit if

(3.8) $$\frac{1}{|c_n c_{n-1}|} \le \frac{-1+p_n}{p_n p_{n-1}}, \; n = 2, 3, \ldots .$$

Inequality (3.8) will be satisfied if

(3.9) $$|2 + b_n| \ge \frac{p_{2n} p_{2n-1}}{-1+p_{2n}}$$

and

(3.10) $$|2 + b_n| \ge \frac{p_{2n} p_{2n+1}}{-1+p_{2n+1}}$$

are both satisfied. Since

$$\frac{p_1 p_2}{-1+p_2} = \frac{p_2 p_3}{-1+p_3} = \frac{15}{4},$$

it follows that (3.9) and (3.10) are satisfied for $n = 1$ if $b_1 \ge 2 - 1/4$. Since

(3.11) $$\frac{p_n p_{n+1}}{-1+p_{n+1}} = 4 - \frac{1}{n^2}, \; n = 2, 3, \ldots,$$

it follows that the sequence $\{p_n p_{n+1}/(1 + p_{n+1})\}$ is increasing. Hence, for $n \ge 2$, (3.9) will be satisfied if (3.10) is satisfied. But, in view of (3.11), (3.10) is equivalent to

$$|2 + b_n| \geq 4 - \frac{1}{4n^2}, \quad n = 2, 3, \ldots,$$

which is clearly satisfied if $b_n \geq 2 - 1/(4n^2)$. This completes our proof of Theorem 3.2.

Theorem 3.3: _Let_ b _be a constant such that_ $0 \leq b \leq 1$, _and let the sequence_ $\{b_n\}_{n\geq 1}$ _be defined by_

$$b_n = \left(1 + \frac{1}{n+b}\right)^{1/2} + \left(1 - \frac{1}{n+b}\right)^{1/2} .$$

Then $0 < b_n < 2$, $\lim b_n = 2$, _and the continued fraction_ (3.2) _converges to_ $(1 + 1/b)^{1/2}$ _if_ $0 < b \leq 1$ _and to_ ∞ _if_ $b = 0$.

Remarks. If we choose b_n as in Theorem 3.3 and let $b_n^* = 2 - 1/(4n^2)$, then after a few calculations it can be seen that $\lim(2-b_n)/(2-b_n^*)=1$; so $2 - b_n \sim 2 - b_n^*$ as $n \to \infty$. This tells us that the coefficient $-1/4$ of n^{-2} in b_n^* is sharp with respect to having finite convergence of (3.2). For by Theorem 3.2, $-K(-1/b_n^*)$ converges to a finite limit, but as we shall show in the proof below of Theorem 3.3, $-K(-1/b_n)$ converges finitely or to ∞ according as $0 < b \leq 1$ or $b = 0$.

Proof: If we set $g_n = (1 + 1/(n + b))^{1/2}$ for $n \geq 0$ and $0 < b \leq 1$, then the fact that (3.2) converges finitely follows immediately from Theorem 3.1. But under these hypotheses we can give a more direct argument that tells us more. As in the proof of Theorem (2.2), the nth approximant f_n of (3.2) is given by

$$f_n = \frac{1}{\frac{1}{g_0} + \frac{1}{F_n}} ,$$

where

$$F_n = g_0 \sum_{k=1}^{n} \prod_{j=1}^{k} \frac{1}{g_j g_{j-1}} .$$

Simple calculations will show that

$$F_n = (1 + b) \sum_{k=1}^{n} \frac{1}{((k+1+b)(k+b))^{1/2}}, \quad 0 < b \leq 1,$$

so that $F_n \to \infty$ as $n \to \infty$. It follows that $\lim f_n = g_0 = (1 + 1/b)^{1/2}$.

If $b = 0$, it is easily verified by induction that the denominators B_n of (3.2) are given by $B_n = (n + 1)^{1/2}$, $n = 0, 1, 2, \ldots$. Hence,

$$f_n = \sum_{k=1}^{n} \frac{1}{B_k B_{k-1}} = \sum_{k=1}^{n} \frac{1}{((k+1)k)^{1/2}};$$

so $f_n \to \infty$ as $n \to \infty$, and our proof is complete.

<u>Theorem</u> 3.4: <u>For the continued fraction</u> (3.2) <u>let</u> B_n <u>be its</u> n<u>th denominator and let</u>

$$b_n \geq 2 - \frac{a(1-a)}{n(n-1+a)}, \quad n = 1, 2, \ldots,$$

<u>where</u> a <u>is a positive constant. Then</u>

$$h_n := B_n/B_{n-1} \geq 1 + a/n; \; B_n \geq \prod_{k=1}^{n}(1+a/k) = O(n^a).$$

<u>Furthermore</u>, (3.2) <u>converges absolutely to a finite value if</u> $a > 1/2$; <u>it converges</u> (<u>possibly to</u> ∞) <u>if</u> $a > 0$; <u>and, in particular, it converges to</u> ∞ <u>if</u>

$$b_n = 2 - \frac{a(1-a)}{n(n-1+a)}$$

<u>with</u> $0 < a \leq 1/2$.

<u>Remarks.</u> If we set $c_n = 2 - a(1-a)/(n(n-1+a))$ for $a > 0$ and $d_n = 2 - 1/(4n^2)$, then

$$2 - c_n \sim 4a(1 - a)(2 - d_n)$$

and, in particular,

$$2 - c_n \sim 2 - d_n$$

if $a = 1/2$. By Theorem 3.2 $-K(-1/b_n)$ converges finitely if $b_n \geq d_n$, and by Theorem 3.4 $-K(-1/b_n)$ converges to ∞ if $b_n = c_n$, where $0 < a \leq 1/2$. Thus these examples again tell us that the estimate for the b_n in Theorem 3.2 is sharp if we wish to have finite convergence.

<u>Proof</u>: Since the h_n satisfy the recurrence relation

$$h_{n+1} = b_{n+1} - 1/h_n,$$

it is easily established by induction that

$$h_n \geq 1 + a/n.$$

Because

$$B_n = \prod_{k=1}^{n} h_k,$$

it follows that

$$B_n \geq \prod_{k=1}^{n} (1 + a/k),$$

where it can be verified by induction that equality holds if

$$b_n = 2 - \frac{a(1-a)}{n(n-1+a)}, \quad a > 0.$$

Since the nth approximant f_n of (3.2) is given by

$$f_n = \sum_{k=1}^{n} \frac{1}{B_k B_{k-1}},$$

we can use Raabe's test and $\sum_{k=1}^{\infty} 1/(C_k C_{k-1})$ as a comparison series, where $C_n = \prod_{k=1}^{n}(1+a/k)$, to verify that (3.2) has the convergence behavior indicated in the statement of our Theorem.

Stirling's formula for the Gamma function Γ comes in handy for determining the growth behavior of $\prod_{k=1}^{n}(1+a/k)$. Using it and the functional relation $\Gamma(x+1) = x\Gamma(x)$ we obtain

$$\prod_{k=1}^{n}(1+a/k) = \frac{1}{\Gamma(1+a)} \frac{\Gamma(n+1+a)}{\Gamma(n+1)}$$

$$\sim \frac{e^{-a}}{\Gamma(1+a)} (n+a)^{a}(1+a/n)^{1/2}, \ n \to \infty.$$

Thus $\prod_{k=1}^{n}(1+a/k) = O(n^{a})$ as asserted, and we are finished with the proof.

__Theorem__ 3.5: __Let__ $\{y_n\}_{n\geq 0}$ __be any sequence satisfying__ $0 < y_n < 1$, $y_{n+1} \leq y_n$, $\lim y_n = 0$, $\lim y_n/y_{n+1} = 1$, __and__ $\sum_{n=0}^{\infty} y_n = \infty$. __For__ $n \geq 0$, __let__ $x_n = (-y_n^2 + y_n/y_{n+1})^{1/2}$ __and__ $g_n = x_n + iy_n$. __Finally, let__

$$b_n = g_n + 1/g_{n-1} = x_n + x_{n-1}y_n/y_{n-1}.$$

__Then__ $0 < b_n < 2$, $\lim b_n = 2$, __and the continued fraction__ (3.2) __diverges by oscillation__. __In particular__, (3.2) __diverges by oscillation if__ $\{y_n\}$ __is defined by__ $y_n = a/(n+1)$, __where__ $0 < a < \sqrt{2}$.

__Remarks__. If $y_n = a/(n+1)$ in Theorem 3.5, where $0 < a < \sqrt{2}$, then b_n becomes

$$b_n = (1 + 1/(n+1) - a^2/(n+1)^2)^{1/2} + (1 - 1/(n+1) - a^2/(n+1)^2)^{1/2}.$$

Here we also use this formula to define b_n if $a = 0$. Setting $c_n = 2 - 1/(4n^2)$, it can be established that

$$2 - b_n \sim (1 + 4a^2)(2 - c_n), \ n \to \infty.$$

By Theorem 3.1, taking $g_n = ((n+2)/(n+1))^{1/2}$, $-K(-1/b_n)$ converges finitely if $a = 0$. By Theorem 3.2, $-K(-1/c_n)$ converges finitely also. But, as we shall prove below, $-K(-1/b_n)$ diverges by oscillation if $0 < a < \sqrt{2}$. Thus our discussion here points out that the simple estimate in Theorem 3.2 for the elements of (3.2) provides a sharp dividing line between finite convergence and divergence by oscillation for these continued fractions whose elements approach 2 from the left.

<u>Proof:</u> We have that

$$|g_n| = (x_n^2 + y_n^2)^{1/2} = (y_n/y_{n+1})^{1/2}. \tag{3.12}$$

Let $\theta_n = \tan^{-1}(y_n/x_n)$, $n \geq 0$. Then

$$\theta_n = \tan^{-1}(y_n y_{n+1}/(1 - y_n y_{n+1}))^{1/2} . \tag{3.13}$$

It follows from the conditions on the y_n that $0 < \theta_n < \pi/2$, $\theta_{n+1} \leq \theta_n$, and $\theta_n \to 0$ as $n \to \infty$.

We now investigate the convergence behavior of the series

$$\sum_{n=1}^{\infty} \prod_{k=1}^{n} \frac{1}{g_k g_{k-1}} . \tag{3.14}$$

Using (3.12) and (3.13), (3.14) can be written in the form

$$\sum_{n=1}^{\infty} (y_{n+1}y_n/(y_1y_0))^{1/2}\exp(-i\sum_{j=1}^{n}(\theta_j + \theta_{j-1})). \tag{3.15}$$

Choose ε so that $0 < \varepsilon < \pi/2$. Then the hypotheses guarantee that there exists an integer N_0 such that for $N > N_0$ there exists an integer $m \geq 0$ for which

$$0 \leq \varepsilon/2 \leq \sum_{j=N+1}^{N+1+m} 2(y_j y_{j-1})^{1/2} \leq \sum_{j=N+1}^{N+1+m} 2\left(\frac{y_j y_{j-1}}{1-y_j y_{j-1}}\right)^{1/2} \leq \varepsilon .$$

Hence, for $N + 1 \le k \le N + 1 + m$, we have

$$0 \le \sum_{j=N+1}^{k} (\theta_j + \theta_{j-1}) \le 2\sum_{j=N+1}^{k} \theta_{j-1} \le \sum_{j=N+1}^{N+1+m} 2\left(\frac{y_j y_{j-1}}{1-y_j y_{j-1}}\right)^{1/2} \le \varepsilon,$$

so

$$\cos \sum_{j=N+1}^{k} (\theta_j \theta_{j-1}) \ge \cos \varepsilon > 0.$$

Let S_n denote the nth partial sum of (3.15). Then

$$|S_{N+m+1} - S_{N-1}| = |(y_{N+1}y_N/(y_1y_0))^{1/2} + S_{N+1+m} - S_N|$$

$$\ge (y_1y_0)^{-1/2}((y_{N+1}y_N)^{1/2} \sum_{k=N+1}^{N+1+m} (y_{k+1}y_k)^{1/2} \cos \varepsilon)$$

$$\ge (y_1y_0)^{-1/2}(\sum_{k=N+1}^{N+1+m} (y_{k+1}y_k)^{1/2}) \cos \varepsilon$$

$$\ge (y_1y_0)^{1/2}(\varepsilon/2) \cos \varepsilon > 0.$$

Hence the series (3.15) diverges and its nth partial sum S_n cannot converge to a finite limit. Suppose $S_n \to \infty$ as $n \to \infty$. Then the nth approximant f_n of (3.2) is given by

$$f_n = g_0/(1 + 1/S_n)$$

for n large enough. But now $f_n \to g_0$ as $n \to \infty$, which is impossible since $\mathrm{Im}(g_0) \ne 0$ and the elements of (3.2) are real. Hence the series (3.15), and therefore the continued fraction (3.2), must diverge by oscillation. This completes our proof of Theorem 3.5.

4. $-K(-1/b_n)$ <u>where</u> $b_n \to 0$ <u>and a problem of Wall.</u>

It is well known that the continued fraction $K(1/b_n)$ diverges if the series $\Sigma|b_n|$ converges. In 1948 Wall [13], on the bottom of page 33 of his book, raised the question of whether the simple convergence of the series Σb_n is sufficient for the divergence of $K(1/b_n)$. To our knowledge an answer to this question has not appeared in the literature to date. In our next theorem we give an example showing that convergence of Σb_n is not sufficient for the divergence of $K(1/b_n)$.

<u>Theorem</u> 4.1: <u>Let the sequence</u> $\{b_n\}$ <u>be defined by</u> $b_1 = 1$, $b_{2n+1} = 2(-1)^n/\sqrt{(n+1)}$, $b_{2n} = (-1)^{n-1}/(\sqrt{(n+1)} + \sqrt{n})$, $n \geq 1$. <u>Then both the continued fraction</u> $\overset{\infty}{\underset{n=1}{K}}(1/b_n)$ <u>and the series</u> $\sum_{n=1}^{\infty} b_n$ <u>converge to finite limits.</u>

<u>Proof:</u> We shall first establish that $\Sigma b_n < \infty$. If S_n denotes the nth partial sum of Σb_n, then

$$S_{2n+2} = 1 + \sum_{k=1}^{n+1}(-1)^{k-1}/(\sqrt{(k+1)} + \sqrt{k}) + \sum_{k=1}^{n} 2(-1)^k/\sqrt{(k+1)}.$$

Since the partial sums appearing on the right hand side of the last equation are partial sums of convergent alternating series, it follows that $\{S_{2n+2}\}$ converges. Because $S_{2n+1} = S_{2n+2} - b_{2n+2}$ and $b_{2n+2} \to 0$ as $n \to \infty$ it is therefore also true that $\{S_n\}$ converges. Thus Σb_n converges finitely as asserted.

We shall now establish the convergence of $K(1/b_n)$. It is easily verified that $K(1/b_n)$ is equivalent to $-K(-1/c_n)$, where

$$c_1 = b_1, \; c_{2n+1} = b_{2n+1}, \; c_{2n} = -b_{2n}, \; n \geq 1.$$

The denominators B_n of $-K(-1/c_n)$ satisfy

$$B_0 = B_1 = 1; \; B_n = c_n B_{n-1} - B_{n-2}, \; n \geq 2.$$

Using this difference equation for the B_n it follows by induction that

$$B_{2n} = (-1)^n\sqrt{(n+1)},\ B_{2n-1} = 1,\ n \geq 1.$$

The nth approximant f_n of $-K(-1/c_n)$ is given by

$$f_n = \sum_{k=1}^{n} 1/(B_k B_{k-1}).$$

Hence,

$$f_{2n+1} = \sum_{k=1}^{2n+1} 1/(B_k B_{k-1}) = 1 + \sum_{k=1}^{n} 2/B_{2k} = 1 + \sum_{k=1}^{n} 2(-1)^k/\sqrt{(k+1)}.$$

Since $\sum_{k=1}^{\infty} 2(-1)^k/\sqrt{(k+1)}$ is a convergent alternating series, it follows that $\{f_{2n+1}\}$ converges. But then $\{f_{2n}\}$ converges to the same limit since $f_{2n} = f_{2n+1} - 1/B_{2n} = f_{2n+1} + (-1)^{n-1}/\sqrt{(n+1)}$. Thus $-K(-1/c_n)$ and $K(1/b_n)$ converge to finite limits, and our proof of Theorem 4.1 is complete.

Our next theorem shows that there is a wealth of finitely convergent continued fractions $-K(-1/b_n)$ with the property that $b_n \to 0$ through real values:

Theorem 4.2: *Let $\{\alpha_k\}$ and $\{\beta_k\}$ be any sequences of real numbers satisfying*

$$0 < \beta_k < \alpha_k,\ k \geq 1;\ \lim_{k\to\infty} \alpha_k = 0$$

with the additional properties

$$\sum_{k=1}^{\infty} \beta_k < \infty,\ \sum_{k=1}^{\infty} \alpha_k = \infty.$$

For $n \geq 1$, *let*

$$b_{2n-1} = \alpha_n / \prod_{k=1}^{n-1} \left(1 - \frac{\beta_k}{\beta_k + \sum_{j=1}^{k}(\alpha_j - \beta_j)}\right)^2$$

$$b_{2n} = \beta_n \alpha_1^{-2} / \prod_{k=1}^{n-1} \left(1 + \frac{\alpha_{k+1}}{\sum_{j=1}^{k}(\alpha_j - \beta_j)}\right)^2 .$$

Then $\lim b_n = 0$ _and the continued fraction_

$$(4.1) \qquad \frac{1}{b_1} - \frac{1}{b_2} - \frac{1}{b_3} - \dots$$

converges to a finite limit.

Proof: After a modest amount of computation it can be verified that the sequence $\{B_n\}_{n\geq 0}$ of denominators of (4.1) is determined by the formulas

$$B_{2n} = (-1)^n \prod_{k=1}^{n} \left(1 - \frac{\beta_k}{\beta_k + \sum_{j=1}^{k}(\alpha_j - \beta_j)}\right)$$

$$(n \geq 0)$$

$$B_{2n+1} = (-1)^n \alpha_1 \prod_{k=1}^{n} \left(-1 + \frac{\alpha_{k+1}}{\sum_{j=1}^{k}(\alpha_j - \beta_j)}\right).$$

From these formulas it is readily seen that

$$b_{2n-1} = \alpha_n / (B_{2n-2})^2, \quad b_{2n} = \beta_n / (B_{2n-1})^2, \quad n \geq 1.$$

If we set $C_n = B_n B_{n-1}$, then

$$C_{2n} = \sum_{k=1}^{n} (\beta_k - \alpha_k) < 0, \quad C_{2n+1} = \alpha_{n+1} + \sum_{k=1}^{n} (\alpha_k - \beta_k) > 0.$$

Hence $C_{2n} \to -\infty$, $C_{2n+1} \to \infty$ as $n \to \infty$, and the C_n alternate in sign. Also

$$|C_{2n+1}| - |C_{2n}| = \alpha_{n+1} > 0, \quad |C_{2n}| - |C_{2n-1}| = \beta_n > 0,$$

so the sequence $\{|C_n|\}$ is increasing. Therefore, the series $\sum_{n=1}^{\infty} 1/C_n$ converges to a finite limit, so that (4.1) also converges finitely. This concludes our argument for Theorem 4.2.

Perron [10, p. 192] calls a continued fraction of the form

$$(4.2) \qquad \frac{1}{b_1 z} + \frac{1}{b_2} + \frac{1}{b_3 z} + \frac{1}{b_4} + \frac{1}{b_5 z} + \frac{1}{b_6} + \cdots$$

an H-fraction (after Hamburger) if $b_k \in \mathbb{R}$, $b_{2k-1} > 0$, and $b_{2k} \neq 0$, $k = 1, 2, \ldots$. If all the b_k are positive, he calls (4.2) an S-fraction (after Stieltjes). If we multiply (4.2) by -1, replace its parameter z by $-x^2$, and apply an equivalence transformation, we obtain the continued fraction

$$(4.3) \qquad \frac{1}{b_1 x^2} - \frac{1}{b_2} - \frac{1}{b_3 x^2} - \frac{1}{b_4} - \frac{1}{b_5 x^2} - \frac{1}{b_6} - \cdots .$$

The following result holds for these continued fractions:

<u>Theorem</u> 4.3: <u>Let</u> $x \in \mathbb{R}$ <u>and let</u> $\{b_n\}$ <u>be a sequence of real numbers such that</u> $\lim b_n = 0$ <u>and</u> $\Sigma\, |b_n - b_{n+1}| < 0$. <u>Then the continued fraction</u> (4.3) <u>cannot converge to a finite limit.</u>

<u>Proof</u>: Let B_n be the nth denominator of (4.3). Then, $B_0 = 1$, $B_1 = x^2 b_1$ and

$$B_{2n+1} = x^2 b_{2n+1} B_{2n} - B_{2n-1}; \quad B_{2n} = b_{2n} B_{2n-1} - B_{2n-2}, \quad n \geq 1.$$

Now define $\{C_n\}$ by

$$C_{2n+1} = B_{2n+1}, \quad C_{2n} = xB_{2n}, \quad n = 0, 1, 2, \ldots .$$

Then

$$C_n = xb_n C_{n-1} - C_{n-2}, \quad n \geq 2.$$

Let $N \geq 2$ be a fixed positive integer such that $|xb_n| < 2$ if $n \geq N$. Then

$$C_{n+N} = xb_{n+N}C_{n+N-1} - C_{n+N-2}, \quad n \geq 0.$$

That is, the C_{n+N} satisfy the difference equation

$$x_n = \hat{b}_n x_{n-1} - x_{n-2},$$

where $\hat{b}_n = xb_{n+N}$. Since

$$\Sigma\ |b_n - b_{n+1}| < \infty \Rightarrow \Sigma\ |x|\ |b_{n+N} - b_{n+1+N}| < \infty,$$

it follows from Lemma 2.1 that $\{C_{n+N}\}$ is bounded. Thus the sequence $\{C_n\}$ (hence also the sequence $\{b_n\}$) is bounded, so the continued fraction (4.3) cannot converge finitely.

<u>Example</u> 4.1. An immediate application of Theorem 4.3 and one of the type that helped motivate our study of equimodular limit periodic continued fractions is the following:

In 1982 Elbert [2] proved that the function

$$M(x) := \sum_{n=1}^{\infty} \frac{2n}{(n^2+x^2)^2}$$

is asymptotic to the series $\sum_{n=0}^{\infty} (-1)^n B_{2n} x^{-2n-2}$ as $x \to \infty$, where B_{2n} is the 2nth Bernoulli number. Thus it follows that

$$x^2 M(x) \sim \sum_{n=0}^{\infty} (-1)^n B_{2n} x^{-2n}.$$

We claim without supplying a proof here that the latter asymptotic series corresponds to the continued fraction

(4.4)
$$1 - \frac{x^{-2}/4}{b_1} - \frac{x^{-2}/4}{b_2} - \frac{x^{-2}/4}{b_3} - \cdots ,$$

where $b_n = 1/n + 1/(n+1)$, $n = 1, 2, \ldots$.
For $x \neq 0$ the convergence behavior of (4.4) is the same as that of

(4.5)
$$\frac{x^{-2}/4}{b_1} - \frac{x^{-2}/4}{b_2} - \frac{x^{-2}/4}{b_3} - \cdots .$$

But (4.5) is equivalent to

(4.6)
$$\frac{1}{4x^2 b_1} - \frac{1}{b_2} - \frac{1}{4x^2 b_3} - \frac{1}{b_4} - \frac{1}{4x^2 b_5} - \cdots .$$

Since
$$\sum_{n=1}^{\infty} |b_n - b_{n+1}| = 2 \sum_{n=1}^{\infty} (n(n+2))^{-1} < \infty,$$

it follows from Theorem 4.3 that (4.6) cannot converge to a finite limit if x is real. Thus (4.4) cannot converge to $x^2 M(x)$ for any real values of x, even though the continued fraction and the function are related through the indicated correspondence.

5. $-K(-1/b_n)$, <u>where</u> $b_{2n-1} \to a$, $b_{2n} \to b$, $0 \leq ab < 4$.

We begin this section with a few simple but interesting examples. The continued fraction

$$\frac{1}{0} - \frac{1}{b_2} - \frac{1}{0} - \frac{1}{b_4} - \frac{1}{0} - \frac{1}{b_6} - \cdots$$

is divergent because its odd denominators B_{2n-1} are zero. For the continued fraction

$$\frac{1}{1} - \frac{1}{0} - \frac{1}{1} - \frac{1}{0} - \frac{1}{1} - \frac{1}{0} - \cdots$$

the denominators are given by

$$B_{2n} = (-1)^n, \ n \geq 0; \ B_{2n-1} = (-1)^{n-1}n, \ n \geq 1.$$

Its nth approximant is $\sum_{k=1}^{n} 1/(B_k B_{k-1})$, so the continued fraction converges to 0 because

$$\sum_{k=1}^{\infty} 1/(B_k B_{k-1}) = 1 - 1 + 1/2 - 1/2 + 1/3 - 1/3 + 1/4 - \ldots = 0$$

More generally, the continued fraction

$$\frac{1}{b_1} - \frac{1}{0} - \frac{1}{b_3} - \frac{1}{0} - \frac{1}{b_5} - \frac{1}{0} - \ldots$$

has denominators B_n that satisfy

$$B_{2n} = (-1)^n, \ n \geq 0; \ B_{2n-1} = (-1)^{n-1} \sum_{k=1}^{n} b_{2k-1}, \ n \geq 1.$$

It follows that this continued fraction cannot converge to a finite limit if $\Sigma b_{2k-1} < \infty$. Clearly, if the b_{2k-1} are nonnegative and $\sum_{k=1}^{\infty} b_{2k-1} = \infty$, then the continued does converge to a finite limit.

We are now ready to make a more comprehensive study of how continued fractions $-K(-1/b_n)$ behave when the sequences $\{b_{2n}\}$ and $\{b_{2n-1}\}$ have limits that are not necessarily the same. We first prove a lemma that is concerned with boundedness of solutions of difference equations. It is the heart of our proof of Theorem 5.1. Hopefully, the lemma will also be of interest to researchers in the field of difference equations.

Lemma 5.1: _Let_ $\{b_n\}$ _be a sequence of positive real numbers whose elements satisfy the conditions_ $0 < b_n b_{n+1} < 4$, $\lim b_{2n-1} = a$, $\lim b_{2n} = b$, $0 < ab < 4$, _and_

$$\sum_{n=1}^{\infty} |b_{2n+1} - b_{2n-1}| < \infty, \ \sum_{n=1}^{\infty} |b_{2n+2} - b_{2n}| < \infty.$$

<u>Then all solutions of the difference equation</u>

(5.2) $$x_n = b_n x_{n-1} - x_{n-2}, \quad n = 1, 2, \ldots$$

<u>are bounded</u>.

<u>Proof</u>: Let

(5.3) $$u_n := b_{n+1}x_n^2 + b_{n+2}x_{n-1}^2 - b_{n+1}b_{n+2}x_n x_{n-1}, \quad n \geq 1,$$

where the x_n satisfy (5.2). Then it is easily verified that

(5.4) $$u_n = b_{n+1}x_{n+2}^2 + b_{n+2}x_{n+1}^2 - b_{n+1}b_{n+2}x_{n+2}x_{n+1}.$$

Using (5.3) and (5.4), we obtain

$$u_{n+2} = b_{n+3}x_{n+2}^2 + b_{n+4}x_{n+1}^2 - b_{n+3}b_{n+4}x_{n+2}x_{n+1}.$$

$$= u_n + b_{n+3}x_{n+2}^2(b_{n+3} - b_{n+1})/b_{n+3} + b_{n+4}x_{n+1}^2(b_{n+4} - b_{n+2})/b_{n+4}$$

$$+ (b_{n+1}b_{n+2} - b_{n+3}b_{n+4})x_{n+2}x_{n+1}$$

(5.5) $$\leq u_n + \gamma_n(b_{n+3}x_{n+2}^2 + b_{n+4}x_{n+1}^2)$$

$$+(b_{n+1}b_{n+2} - b_{n+3}b_{n+4})x_{n+2}x_{n+1},$$

where

$$\gamma_n = \max\{|b_{n+3} - b_{n+1}|/b_{n+3}, |b_{n+4} - b_{n+2}|/b_{n+4}\}.$$

With the aid of formula (5.3) we obtain from (5.5) that

$$u_{n+2} \leq u_n + \gamma_n u_{n+2} + (\gamma_n b_{n+3}b_{n+4} + b_{n+1}b_{n+2} - b_{n+3}b_{n+4})x_{n+2}x_{n+1}.$$

But,

$$u_n = b_{n+1}x_n^2 + b_{n+2}x_{n-1}^2 - b_{n+1}b_{n+2}x_n x_{n-1}$$

(5.6) $$\geq (\sqrt{(b_{n+1}b_{n+2})})(2 - \sqrt{(b_{n+1}b_{n+2})})|x_n x_{n-1}| \geq 0$$

Now let

$$\delta_n = \gamma_n + \frac{\gamma_n b_{n+3} b_{n+4} + b_{n+2}|b_{n+1} - b_{n+3}| + b_{n+3}|b_{n+2} - b_{n+4}|}{(\sqrt{(b_{n+3}b_{n+4})})(2 - \sqrt{(b_{n+3}b_{n+4})})}.$$

Then, using (5.6), it follows from (5.5) that

$$u_{n+2} \le u_n + \delta_n u_{n+2}.$$

Hence

(5.7) $$u_{n+2} \le u_n/(1 - \delta_n)$$

for n large enough, say $n \ge n_0$. The hypotheses of our Lemma guarantee that $\sum_{n=1}^{\infty} \delta_{2n-1}$ and $\sum_{n=1}^{\infty} \delta_{2n}$ converge. Therefore, the products $\pi(1-\delta_{2n})$ and $\pi(1 - \delta_{2n-1})$ converge, so it follows from (5.7) that for $n \ge n_1$

$$u_{2n+2} \le u_{2n_1}/\prod_{n=n_1}^{\infty}(1-\delta_{2n}),\quad u_{2n+1} \le u_{2n_1-1}/\prod_{n=n_1}^{\infty}(1 - \delta_{2n-1}),$$

where n_1 is a fixed positive integer such that $2n_1 - 1 \ge n_0$. Hence the sequence $\{u_n\}$ is bounded. From (5.3) we obtain

(5.8) $$u_n \ge (b_{n+1}x_n^2 + b_{n+2}x_{n-1}^2)(2 - \sqrt{(b_{n+1}b_{n+2})})/2.$$

Since $\{u_n\}$ is bounded, it now follows from (5.8) that $\{x_n\}$ is bounded and our proof is complete.

Theorem 5.1: Let $\{b_n\}$ be a sequence of positive real numbers whose elements satisfy the conditions $0 < b_n b_{n+1} < 4$, $\lim b_{2n-1} = a$, $\lim b_{2n} = b$, $0 < ab < 4$, and

(5.9) $$\sum_{n=1}^{\infty}|b_{2n+1} - b_{2n-1}| < \infty,\quad \sum_{n=1}^{\infty}|b_{2n+2} - b_{2n}| < \infty.$$

Then the continued fraction

$$(5.10)\qquad \frac{1}{b_1} - \frac{1}{b_2} - \frac{1}{b_3} - \dots$$

cannot converge to a finite limit. We note that condition (5.9) is satisfied, in particular, if $b_{2n-1} \to a$ and $b_{2n} \to b$ monotonely as $n\to\infty$.

Proof: By Lemma (5.1), the sequence of denominators $\{B_n\}$ of (5.10) is bounded. Hence, by an argument similar to the one given in the proof of Theorem 2.1, (5.10) cannot converge to a finite limit.

Our next theorem gives us some indication of what can happen with continued fractions $-K(-1/b_n)$ when at least one of the sequences $\{b_{2n-1}\}$, $\{b_{2n}\}$ converges to 0.

Theorem 5.2: Let $\{c_n\}_{n\geq 0}$ be any sequence satisfying $0 < c_n < 1$. Let the sequence $\{b_n\}_{n\geq 1}$ be defined by

$$b_{2n} = (c_{2n}/(1 - c_{2n+1}))\prod_{k=0}^{n}(1 - c_{2k+1})/(1 + c_{2k}),\ n \geq 1$$

$$b_{2n+1} = c_{2n+1} \prod_{k=0}^{n}(1 + c_{2k})/(1 - c_{2k+1}),\ n \geq 0.$$

(A) If $\sum_{n=0}^{\infty} c_n < \infty$, then $\lim b_n = 0$ and the continued fraction (5.10) diverges by oscillation.

(B) If $\sum_{n=0}^{\infty} c_n = \infty$, $\sum_{n=0}^{\infty} c_n^2 < \infty$, and $c_{n+1} \leq c_n$ for $n \geq 0$, then $\lim b_{2n} = 0$, $\lim b_{2n+1}$ may be finite or infinite, and (5.10) converges to $-1/(1 + c_0)$.

(C) In particular, if $a > 0$ is a constant and

$$c_{2n+1} = c_{2n} = 1/(2n+1 + 1/a),\ n \geq 0,$$

then

$$\lim b_{2n} = 0,\ \lim b_{2n+1} = a$$

<u>and</u> (5.10) <u>converges to</u> $-a/(2a+1)$.

<u>Proof</u>: After setting $c_{-1} = 0$, we define the sequence $\{g_n\}_{n\geq 0}$ by

(5.11) $\quad g_{2n} = -\prod_{k=0}^{n}(1-c_{2k-1})/(1+c_{2k}),\ g_{2n+1} = \prod_{k=0}^{n}(1+c_{2k})/(1-c_{2k+1}).$

Then it can be verified that the b_n in our theorem are given by

(5.12) $\quad b_n = g_n+1/g_{n-1},\ n \geq 1.$

Now let

(5.13) $\quad t_n = \prod_{k=1}^{n} 1/(g_k g_{k-1}) = (-1)^n \prod_{k=1}^{n}(1 + (-1)^k c_k),$

and set F_n equal to the nth partial sum of the series

(5.14) $\quad \sum_{k=1}^{\infty} t_k.$

Then the nth approximant f_n of (5.10) is given by

(5.15) $\quad f_n = g_0(1 + 1/F_n).$

If the series $\sum_{k=1}^{\infty} c_k$ converges, then the product $\prod_{k=1}^{\infty}(1 + (-1)^k c_k)$ in (5.13) converges, so that $t_n \not\to 0$ as $n \to \infty$. Hence, in this case (5.14) cannot converge to a finite limit. We have

(5.16) $\quad F_{2n} = \sum_{k=1}^{n} (t_{2k} + t_{2k-1}) = \sum_{k=1}^{n}(c_{2k}/(1+c_{2k}))\prod_{j=1}^{2k}(1+(-1)^j c_j).$

Therefore, $\{T_{2n}\}$ converges if $\Sigma c_{2n} < \infty$, which is the case since $\Sigma c_n < \infty$. But then (5.14) must diverge by oscillation. Thus (5.15) implies that (5.10) diverges by oscillation. This completes the proof of part (A).

If the c_n satisfy the stated conditions in part (B), then the product

$$\prod_{k=1}^{\infty}(1 + (-1)^k c_k)$$

converges. Hence from (5.16) we have that $F_{2n} \to \infty$ as $n \to \infty$. But then $F_{2n+1} \to \infty$ also as $n \to \infty$, since

$$F_{2n+1} = F_{2n} - \prod_{k=1}^{2n+1}(1 + (-1)^k c_k).$$

So from (5.15) we see that $f_n \to g_0$ as $n \to \infty$, and therefore (5.10) converges to $g_0 = -1/(1 + c_0)$ in this case.

It remains to justify part (C). If

$$c_{2n+1} = c_{2n} = 1/(2n+1 + 1/a),\ n \geq 0,\ a > 0,$$

then $0 < c_n < 1$, $c_{n+1} \leq c_n$, $\Sigma c_n = \infty$, and $\Sigma c_n^2 < \infty$; so that by part (B), (5.10) converges to $g_0 = -a/(2a + 1)$. Since the hypotheses guarantee that $c_n \to 0$ monotonely in this case, it follows that $b_{2n} \to 0$ as $n \to \infty$. But now

$$\begin{aligned} b_{2n+1} &= c_{2n+1} \prod_{k=0}^{n}(1+c_{2k})/(1-c_{2k+1}) \\ &= (2n+1+1/a)^{-1} \prod_{k=1}^{n}(2(k+1)+1/a)/(2k+1/a) \\ &= (2n+1+1/a)^{-1}(2n+2+1/a)a, \end{aligned}$$

so $b_{2n+1} \to a$ as $n \to \infty$. With this the proof of our theorem is complete.

The following theorem extends a result of Heller and Roach [3].

<u>Theorem</u> 5.3: <u>Let the elements of the sequence</u> $\{b_n\}$ <u>satisfy</u> $b_n \in \mathbb{C}$, $\lim b_{2n-1} = u \in \mathbb{C}$, $\lim b_{2n} = v \in \mathbb{C}$, <u>where</u> $uv \in \mathbb{R}$ <u>and</u> $0 < uv < 4$. <u>Let</u> $\{c_n\}$ <u>be defined by</u>

$$c_{2n-1} = (v/u)b_{2n-1},\ c_{2n} = b_{2n},\ n = 1, 2, \ldots .$$

Then, if

$$(5.17) \qquad \sum_{n=1}^{\infty} |\bar{c}_n / \bar{v} - c_{n+1} / v|$$

converges, the continued fraction (5.10) _cannot converge to a finite limit. In particular,_ (5.17) _converges if both series_

$$(5.18) \qquad \sum_{n=1}^{\infty} |b_{2n-1}-u|, \quad \sum_{n=1}^{\infty} |b_{2n}-v|$$

converge.

Proof: The denominators B_n of (5.10) satisfy

$$B_0 = 1, \; B_1 = b_1; \; B_n = b_n B_{n-1} - B_{n-2}, \; n \geq 2.$$

For n = 0, 1, 2, ... set

$$D_{2n+1} = B_{2n+1}, \; D_{2n} = B_{2n}\sqrt{(uv)}/v.$$

Then

$$D_n = D_{n-1} c_n \sqrt{(uv)}/v - D_{n-2}, \; n \geq 2.$$

There exists an integer N such that

$$x_n = d_n x_{n-1} - x_{n-2}$$

where $d_n = c_{n+N}\sqrt{(uv)}/v$ satisfies $|d_n| < 2$. Hence by Lemma 2.1, the sequence $\{D_{n+N}\}_{n\geq 0}$ is bounded. Therefore, $\{D_n\}$ and $\{B_n\}$ are also bounded. Thus $\{1/(B_n B_{n-1})\}$ cannot converge to 0 so (5.10) cannot converge finitely. Since it is easily seen that

$$|\bar{c}_{2n} / \bar{v} - c_{2n+1}/v| \leq |b_{2n}-v|/|v| + |b_{2n+1} - u|/|u|$$

and

$$|\overline{c}_{2n-1}/\overline{v} - c_{2n}/v| \le |b_{2n-1}-u|/|u| + |b_{2n} - v|/|v|,$$

it follows that the convergence of the series (5.18) imply the convergence of (5.17). We are now finished with the proof of Theorem 5.3.

6. Construction of $\{b_n\}$, where $\lim b_n = b(0<b<2)$ and $-K(-1/b_n) < \infty$.

In this section we consider the problem of finding for a given $b(0<b<2)$ a corresponding real sequence $\{b_n\}$ such that $b_n \to b$ and $-K(-1/b_n)$ converges finitely. We conjecture that there always exists such a sequence $\{b_n\}$, though in this section we are able only to verify this conjecture for certain b's in the interval (0,2). A stronger conjecture is that the demand of finite convergence of $-K(-1/b_n)$ can be replaced by the demand of absolute finite convergence.

Theorem 6.1: Let p and q be positive integers with no common factor such that $0 < p/q < 1/2$, and let $b = 2\cos(\pi p/q)$. Let $\{\varepsilon_n\}$ be any sequence satisfying $0 < \varepsilon_n < 1$, $\lim \varepsilon_n = 0$, and $\sum_{n=1}^{\infty}\varepsilon_n = \infty$. Let the sequence $\{b_n\}$ be defined by $b_n = (1 - \varepsilon_m)b$ if $n = mq - 1$, $m \ge 1$, and let $b_n = b$ otherwise. Then $\lim b_n = b$, $0 < b < 2$, and the continued fraction

$$\frac{1}{b_1} - \frac{1}{b_2} - \frac{1}{b_3} - \dots \tag{6.1}$$

converges nonabsolutely to b.

Proof: Let

$$D_n := (\sin(n+1)\alpha)/(\sin\alpha),\quad n = 0, 1, \dots, q-1, \tag{6.2}$$

where $\alpha := \pi p/q$. Then

$$\sum_{k=1}^{q-2}(D_k D_{k-1})^{-1} = \sin^2\alpha \sum_{k=1}^{q-2}(\sin k\alpha \ \sin(k+1)\alpha)^{-1}$$

$$= \sin\alpha \sum_{k=1}^{q-2}(\cot k\alpha - \cot(k+1)\alpha)$$

$$= (\sin\alpha)(\cot\alpha - \cot(q-1)\alpha)$$

$$(6.3) \qquad = 2\sin\alpha \cot\alpha = b.$$

If θ is any real number such that $\cos(\theta+k\alpha) \neq 0$, $k = 0, 1, \ldots, q$, then

$$\sum_{k=1}^{q}(\cos(\theta+(k-1)\alpha)\cos(\theta+k\alpha))^{-1} = (\sin\alpha)^{-1}\sum_{k=1}^{q}(\tan(\theta+k\alpha) - \tan(\theta+(k-1)\alpha))$$

$$(6.4) \qquad = (\sin\alpha)^{-1}(\tan(\theta+q\alpha) - \tan\theta) = 0.$$

Now that we have established (6.3) and (6.4) our goal is to determine the denominators B_n of (6.1). It is easily verified that

$$(6.5) \qquad B_n = D_n, \ n = 0, 1, \ldots, q-2.$$

We claim that the B_n for $n \geq q-1$ are given by the formulas

$$B_{(2m-1)q-1} = (-1)^p b \sum_{k=1}^{2m-1} \varepsilon_k$$

$$B_{2mq-1} = b \sum_{k=1}^{2m} \varepsilon_k$$

$$(6.6) \qquad B_{(2m-1)q-2} = (-1)^{p+1} \qquad (m \geq 1)$$

$$B_{2mq-2} = -1$$

$$B_{mq+r} = D_{r+1}B_{mq-1} - D_r B_{mq-2},$$

where $r = 0, 1, \ldots, q-2$. These formulas can be established by mathematical induction, using the fact that the B_n satisfy

$$(6.7)\qquad B_0 = 1,\ B_1 = b,\ B_n = b_n B_{n-1} - B_{n-2},\ n \geq 2.$$

It follows from formulas (6.2), (6.5), and (6.6) that $B_n \neq 0$ for all $n \geq 0$. Since the B_n are not zero, the nth approximant f_n of (6.1) is given by

$$(6.8)\qquad f_n = \sum_{k=1}^{n} 1/(B_k B_{k-1}).$$

We assert that

$$(6.9)\qquad f_{(n+1)q-2} = b,\ n \geq 0.$$

We shall now verify (6.9) by induction. Clearly (6.9) is true for $n = 0$ by formula (6.3). Assume (6.9) holds for some n. Then

$$f_{(n+2)q-2} = \sum_{k=1}^{(n+2)q-2} (B_k B_{k-1})^{-1}$$

$$= \sum_{k=1}^{(n+1)q-2} (B_k B_{k-1})^{-1} + \sum_{k=(n+1)q-1}^{(n+2)q-2} (B_k B_{k-1})^{-1}$$

$$= b + \sum_{k=(n+1)q-1}^{(n+2)q-2} (B_k B_{k-1})^{-1}.$$

But there exist real constants θ and λ, $\lambda \neq 0$, such that

$$B_{(n+1)q-2+k} = \lambda \cos(\theta + k\alpha),\ k = 0, 1, \ldots, q.$$

Hence, we have from above that

$$f_{(n+2)q-2} = b + \lambda^{-2} \sum_{k=1}^{q} (\cos(\theta + (k-1)\alpha)\cos(\theta + k\alpha))^{-1}.$$

By formula (6.4), the term on the right involving Σ is zero; hence $f_{(n+2)q-2} = b$ and (6.9) holds for all $n \geq 0$ by the induction principle.

Since $\sum_{k=1}^{\infty}\varepsilon_k = \infty$, it follows from formulas (6.2), (6.5), and (6.6) that $|B_nB_{n-1}| \to \infty$ as $n \to \infty$. Hence from (6.8) and (6.9) we have $f_n \to b$, so that (6.1) converges to b. However, the series $\sum_{k=1}^{\infty} 1/(B_kB_{k-1})$ cannot converge absolutely. This follows from the fact that, with the aid of formulas (6.6), we have

$$\sum_{m=1}^{\infty}(|B_{2mq-1}B_{2mq-2}|)^{-1} = b^{-1}\sum_{m=1}^{\infty}\left(\sum_{k=1}^{2m}\varepsilon_k\right)^{-1} = \infty.$$

The latter series diverges by a theorem of Abel and Dini which states that if S_n is the nth partial sum of a positive term series Σa_n and $\Sigma a_n = \infty$, then $\Sigma a_n/S_n = \infty$. This concludes our proof of Theorem 6.1.

Theorem 6.2: *Let* α *be an irrational number satisfying* $0 \quad \alpha < 1/4$ *such that, for any rational approximation* p/q,

$$(6.10) \qquad |\alpha - p/q| \geq 1/(16q^3).$$

Let $b = 2 \cos 2\alpha\pi$. *Then there exists a sequence* $\{b_n\}$ *such that* $0 < b_n < 2$, $\lim b_n = b$, *and the continued fraction* (6.1) *converges absolutely to a finite limit*.

Proof: First we establish that there exists an α as described in the Theorem. There must be an irrational number β in the interval (0,1) such that

$$|\beta - p/q| \geq 1/(4q^3)$$

for any rational approximation p/q. For if not, then every irrational number in (0,1) is in an open interval of radius $(4q^3)^{-1}$ at some p/q, where $q \geq 2$, $0 < p < q$. The Lebesgue measure of the union of these intervals is at most

$$(1/2)\sum_{q=1}^{\infty} q(q^{-3}) = \pi^2/12 < 1.$$

This says that the measure of the set of irrationals in $(0,1)$ is less than 1, which is a contradiction. Hence the desired β exists, and we get the desired α by setting $\alpha = \beta/4$.

By the Dirichlet Principle, for any integer $N > 1$ there exists integers m and n such that $1 \le n < N$ and

$$(6.11) \qquad |\alpha - m/n| \le 1/(nN).$$

Thus, in view of (6.10), it follows that $1/(16n^3) \le 1/(nN)$ and $n \ge \sqrt{N}/4$. Note that

$$(6.12) \qquad 1/(16n^2) \le |n\alpha - m| \le 1/N.$$

Now for any β, $0 < \beta < 1$, there exists k, $1 \le k \le 16n^2$, such that

$$|kn\alpha - km - \beta| \le 1/N.$$

Setting $\ell = kn$ we see that $\ell \le 16n^3 \le 16N^3$ and hence for any β, $0 < \beta < 1$, there exists $\ell \in \mathbb{N}$, $1 \le \ell \le 16N^3$, and $m \in \mathbb{Z}$ so that

$$(6.13) \qquad |\ell\alpha - m - \beta| \le 1/N.$$

Now let $\theta = 2\alpha\pi$. Since $\tan x$ is continuously differentiable in a neighborhood of $\pi/2 - \theta$, we can find $0 < \delta < \pi/2$ and a constant $M > 0$ such that, if

$$|\Phi - (\pi/2 - \theta)| \le \delta,$$

then

$$|\tan \Phi - \cot \theta| \le M|\Phi + \theta - \pi/2|.$$

Now suppose $0 < \gamma < 1$. We claim that there is a constant $C = C(\alpha)$ such that whenever $\{x_k\}$ is a sequence of the form

$$x_k = \cos(k\theta + \mu),$$

where $0 < \pi/2 - \mu < 2\pi$, then for some $k \le C\gamma^{-3}$ we have

$$|x_k| \le \gamma |x_{k-1}|. \tag{6.14}$$

In fact we may suppose $0 < \gamma < M\delta \sin\theta$, since if we establish it in this case, then by altering C we get the general case. Then, since $\gamma \le 2\pi M \sin\theta$, there is an integer N with

$$2\pi M\gamma^{-1}\sin\theta \le N \le 4\pi M\gamma^{-1}\sin\theta.$$

By (6.13) there exists $\ell \in \mathbb{N}$, $1 \le \ell \le 16N^3$, such that if $\beta = 1/4 - \mu/(2\pi)$ we have for some $m \in \mathbb{N}$ that

$$|\ell\alpha - m - \beta| \le 1/N.$$

Thus

$$|\ell\theta - 2m\pi - \pi/2 + \mu| \le 2\pi/N \le \gamma(M\sin\theta)^{-1} \le \delta,$$

and hence

$$|\tan((\ell-1)\theta + \mu) - \cot\theta| \le \gamma/(\sin\theta)$$

or

$$|\cos(\ell\theta + \mu| \le \gamma\ |\cos((\ell-1)\ \theta \pm \mu)|.$$

That is, $|x_\ell| \le \gamma|x_{\ell-1}|$. Now we note

$$\ell \le 16N^3 \le (16(4\pi)^3M^3\sin^3\theta)\gamma^{-3} = C\gamma^{-3}$$

and our claim (6.14) is established.

Now fix $0 < a < 1/3$. Let $b = 2\cos\theta$. We will select the sequence b_n by induction. We suppose $b_1 = b$, $B_0 = 1$, $B_1 = b$. Then the denominators B_n of (6.1) are given by the difference equation

$$B_n = b_n B_{n-1} - B_{n-2}.$$

In order to fix the sequence $\{b_n\}$ we shall fix a sequence of indices n_k with $0 = n_0 < n_1 < n_2 < \ldots$, and put

$$b_n = b \text{ if } n \notin \{n_k\}$$

$$= b(1 - (1/2)k^{-a}) \text{ if } n = n_k (k = 1, 2, \ldots).$$

The sequence $\{n_k\}$ is determined by the condition that n_k is the first integer $\ell > n_{k-1}$ such that

$$|b\, B_{\ell-1} - B_{\ell-2}| < (b/4)k^{-a}|B_{\ell-1}|.$$

For convenience let us put $n_k = \infty$ if this condition is not satisfied for any $\ell > n_{k-1}$. We shall use C in the proof to denote a constant depending only on α (or θ or b) which may vary from line to line. We prove by induction that $n_k < \infty$ and $B_{n_k} \neq 0$ for all k. In fact this is true for $k = 0$. Suppose $n_{k-1} < \infty$ and $B_{n_{k-1}} \neq 0$. Then the difference equation

$$x_p = b\, x_{p-1} - x_{p-2}$$

subject to the initial conditions

$$x_{-1} = B_{n_{k-1}-1},\ x_0 = B_{n_{k-1}}$$

has a solution of the form

$$x_p = \lambda \cos (p\theta + \mu),$$

where $0 < \pi/2 - \mu < 2\pi$ and $\lambda \neq 0$. Hence, if $(b/4)k^{-a} < M\delta \sin\theta$, there exists $p \leq C(b/4)^3 k^{-3a}$ such that

$$|x_p| \leq (b/4)k^{-a} \, |x_{p-1}|.$$

Let p_0 be the first such p. Then

$$n_k = n_{n-1} + p_0$$

and

$$B_{n_{k-1}+p} = x_p$$

for $0 \leq p < p_0$. We deduce that

$$n_k - n_{k-1} \leq C \; k^{-3a}, \tag{6.15}$$

where C depends only on α. In particular, $n_k < \infty$. Next note that, if $p < p_0$, then

$$|x_p| \geq (b/4)k^{-a} |x_{p-1}|; \tag{6.16}$$

so that $B_n \neq 0$ for $n_{k-1} \leq n < n_k$. Now

$$\begin{aligned} |B_{n_k}| &= |bB_{n_k-1} - B_{n_k-2} - (b/2)k^{-a}B_{n_k-1}| \\ &\geq (b/4)k^{-a} \, |B_{n_k-1}| \neq 0. \end{aligned} \tag{6.17}$$

Thus we see that $n_k < \infty$ for all k and that $n_k - n_{k-1} \leq C \; k^{-3a}$. Combining (6.16) and (6.17) we further have

$$|B_n| \geq (b/4)k^{-a} \, |B_{n-1}| \tag{6.18}$$

for $n_{k-1} < n \leq n_k$. Since, for all n,

$$|B_n| \le b|B_{n-1}| + |B_{n-2}|$$

we conclude for $n_{k-1} < n \le n_k$ that

$$|B_n| \le (b + 4k^a/b)\ |B_{n-1}|. \tag{6.19}$$

Hence, from (6.18) and (6.19) we have

$$|B_{n-1}|^2 \le (4k^a/b)\ |B_nB_{n-1}|$$

$$|B_n|^2 \le (b + 4k^a/b)\ |B_nB_{n-1}|;$$

so that

$$|B_n|^2 + |B_{n-1}|^2 \le C\ k^a\ |B_nB_{n-1}|, \tag{6.20}$$

where $C = C(\alpha)$.

Now let

$$E_n = B_n^2 + B_{n-1}^2 - bB_nB_{n-1}.$$

Then $E_n \ge 0$ and $E_n \le C\ k^a\ |B_nB_{n-1}|$. In fact, if $n_{k-1} < n < n_k$, $E_n = E_{n-1}$. If $n = n_k$,

$$B_n^2 + B_{n-1}^2 - b_nB_nB_{n-1} = B_{n-1}^2 + B_{n-2}^2 - b_nB_{n-1}B_{n-2};$$

so

$$E_n = E_{n-1} + (b_n - b)(B_n - B_{n-2})B_{n-1}.$$

Now

$$B_n - B_{n-2} = b_nB_{n-1} - 2\ B_{n-2}$$

$$= (b_n - 2b)B_{n-1} + 2(bB_{n-1} - B_{n-2}).$$

Hence,

$$(b_n - b)B_{n-1}(B_n - B_{n-2}) = (b_n - b)(b_n - 2b)B_{n-1}^2$$

$$+ 2(bB_{n-1} - B_{n-2})(b_n - b)B_{n-1}.$$

Now

$$|bB_{n-1} - B_{n-2}| \le (b/4)k^{-a}\ |B_{n-1}|.$$

So that

$$|2(bB_{n-1} - B_{n-2})B_{n-1}| \le (b/2)k^{-a}\ |B_{n-1}|^2$$

while

$$|(b_n - 2b)B_{n-1}^2| \ge b|B_{n-1}|^2.$$

We conclude that

$$(b_n - b)B_{n-1}(B_n - B_{n-2}) \ge (1/2)(b - b_n)(2b - b_n)B_{n-1}^2$$

$$\ge C^{-1}k^{-a}B_{n-1}^2,$$

where $C = C(\alpha)$. Thus

$$E_n - E_{n-1} \ge C^{-1}k^{-2a}\ |B_{n-1}B_{n-2}|$$

$$\ge C^{-1}k^{-3a}E_{n-1}$$

or

$$E_n \ge (1 + C^{-1}k^{-3a})E_{n-1}$$

if $n = n_k$. Therefore, if $e_k = E_{n_k}$, we have

$$e_k \ge \exp(C^{-1}k^{-3a})e_{k-1},\ k = 2, 3, \ldots,$$

where $C = C(\alpha)$. Hence

$$e_k \geq \exp(C^{-1}k^{1-3a}), \; k = 1, 2, \ldots$$

for some constant C. If $n_{k-1} < n \leq n_k$,

$$|B_n B_{n-1}| \geq C^{-1}k^{-a}E_n$$

$$\geq C^{-1}k^{-a}\exp(C^{-1}k^{1-3a}).$$

Hence,

$$|B_n B_{n-1}|^{-1} \leq C\, k^a \exp(-C^{-1}k^{1-3a}).$$

Thus

$$\sum_{n_{k-1}+1}^{n_k} |B_n B_{n-1}|^{-1} \leq C\, k^{4a} \exp(-C^{-1}k^{1-3a}).$$

Since $0 < a < 1/3$, we have $\sum_{n=1}^{\infty} |B_n B_{n-1}|^{-1} < \infty$. The nth approximant f_n of (6.1) is given by $f_n = \sum_{k=1}^{n} (B_k B_{k-1})^{-1}$, and hence (6.1) converges absolutely for our choice of $\{b_n\}$. This completes our proof of Theorem 6.2.

7. A Matrix Theoretic Result with Applications to Continued Fractions.

In this section we first state and sketch the proofs of two lemmas that are needed to prove Theorem 7.1. This theorem is a powerful matrix theoretic result that gives sufficient conditions for boundedness of products of matrices. In the latter part of this section and through the first two theorems in Section 8 we show how Theorem 7.1 can be used to obtain convergence information about equimodular limit periodic continued fractions. We have not attempted to obtain an exhaustive list of applications of Theorem 7.1, though we suspect there are many in the fields of continued fractions and difference equations.

Lemma 7.1: *Let* A_0 *be an* $n\times n$*-matrix. Suppose* λ_0 *is an eigenvalue of* A_0 *which is a simple root of the characteristic polynomial. Suppose* $A_0\underline{x}_0 = \lambda_0\underline{x}_0$ *where* $\underline{x}_0 \neq \underline{0}$. *Then there is an open neighborhood* U *of* A_0 *and* C^1*-maps* $\sigma: U \to \mathbb{C}^n$, $\lambda: U \to \mathbb{C}$ *such that*

$$\sigma(A_0) = \underline{x}_0$$

$$\lambda(A_0) = \lambda_0$$

$$A\sigma(A) = \lambda(A)\sigma(A).$$

Proof: Consider the $(n+1)$-equations

$$A\underline{x} = \lambda\underline{x}$$

$$\underline{x}^*\underline{x}_0 = \|\underline{x}_0\|^2 \qquad (\underline{x}^* = \bar{\underline{x}}^T),$$

where the norm is the Euclidean norm. The result follows from the Implicit Function Theorem once we check that the following matrix is non-singular.

$$B_0 = \left(\begin{array}{c|c} A_0 - \lambda_0 I & -\underline{x}_0 \\ \hline \underline{x}_0^* & 0 \end{array}\right).$$

Suppose

$$B\begin{pmatrix}\underline{y}\\ \mu\end{pmatrix} = 0.$$

Then

$$(A_0 - \lambda_0 I)\underline{y} - \mu \underline{x}_0 = 0$$

$$\bar{\underline{x}}_0^* \ \underline{y} = 0$$

So

$$(A_0 - \lambda_0 I)^2 \underline{y} = (A - \lambda_0 I)\mu \underline{x}_0 = 0.$$

Hence, (see Nering [9, Corollary 8.4]), since λ_0 is a simple root,

$$A_0 \underline{y} = \lambda_0 \underline{y}$$

and

$$\underline{y} = \alpha \underline{x}_0.$$

Now $\alpha \underline{x}_0^* \underline{x}_0 = 0 \Rightarrow \alpha = 0 \Rightarrow \underline{y} = 0$, and therefore $\mu = 0$. This completes our proof of Lemma 7.1.

Lemma 7.2: *Let* Q *be an* nxn-*matrix with* n *distinct eigenvalues. Suppose* $Q_m \to Q$ *as* $m \to \infty$ *and*

$$\Sigma \|Q_m - Q_{m+1}\| < \infty,$$

where the matrix norm is the operator norm. Then there exists an integer N *and invertible matrices* $P_m (m \geq N)$ *such that*

$$P^{-1} Q_m P_m = \Delta_m$$

is diagonal $(m \geq N)$ *and*

$$\Sigma \|P_m - P_{m+1}\| < \infty.$$

$$\Sigma \|P_m^{-1} - P_{m+1}^{-1}\| < \infty.$$

<u>Proof</u>: There is an open neighborhood U of Q and C^1-maps $\sigma_i : U \to \mathbb{C}^n (i \le n)$, $\lambda_i : U \to \mathbb{C}$ such that

$$\sigma_i(Q) = \underline{x}_i, \ \lambda_i(Q) = \lambda_i,$$

where $\lambda_1, \ldots, \lambda_n$ are the eigenvalues of Q and $\underline{x}_1, \ldots, \underline{x}_n$ are the eigenvectors. Let

$$P_m = [\sigma_1(Q_m), \ldots, \sigma_n(Q_m)]$$

as long as $Q_m \in U$. Since the irreducible matrices are open and $P \to P^{-1}$ is C^1, we can determine N so that the P_m for $m \ge N$ satisfy the conditions. This finishes our sketch of a proof for Lemma 7.2.

<u>Theorem</u> 7.1: <u>Let</u> $\{Q_m\}$ <u>be a sequence of</u> nxn-<u>matrices such that</u>

(i) $Q_m \to Q$ <u>as</u> $m \to \infty$, <u>where</u> Q <u>has distinct eigenvalues</u>.
(ii) $\Sigma \|Q_m - Q_{m+1}\| < \infty$.

<u>Let</u> $R_m = \mu_m I + \tau_m Q_m$ for $m \ge 1$, <u>where</u> $\{\mu_m\}$ <u>and</u> $\{\tau_m\}$ <u>are arbitrary sequences of real numbers</u>. <u>Let</u> r_m <u>be the spectral radius of</u> R_m. <u>Suppose</u>

$$\sup_n \left(\prod_{k=1}^{n} r_k\right) < \infty.$$

<u>Then there exists a constant</u> $K < \infty$ <u>such that</u>

$$\|R_m R_{m-1} \ldots R_1\| \le K, \ m \ge 1.$$

<u>Proof</u>: If $X \in \mathbb{C}^n$, let

$$X_m = R_m \ldots R_1 X.$$

Then

$$X_m = R_m X_{m-1},\ X_0 = X,\ m \geq 1.$$

So by Lemma 7.2, for each $m \geq N$ there exists P_m such that

$$\begin{aligned} P_m^{-1} X_m &= P_m^{-1} R_m P_m R_m^{-1} X_{m-1} \\ &= D_m P_m^{-1} X_{m-1}, \end{aligned}$$

where D_m is diagonal. In what follows the norm of a vector will be the Euclidean norm and the norm of a matrix will be the operator norm. From the above we obtain

$$\begin{aligned} \| P_m^{-1} X_m \| &\leq r_m \| P_m^{-1} X_{m-1} \| \\ &\leq r_m (\| P_{m-1}^{-1} X_{m-1} \| + \| (P_m^{-1} - P_{m-1}^{-1}) X_{m-1} \|) \\ &\leq r_m (\| P_{m-1}^{-1} X_{m-1} \| + \| (P_m^{-1} P_{m-1} - I) P_{m-1}^{-1} X_{m-1} \|) \\ &\leq r_m (1 + \| P_m^{-1} \| \| P_m - P_{m-1} \|) \| P_{m-1}^{-1} X_{m-1} \| \end{aligned}$$

Now, if

$$\sup_k \prod_{m=N}^{k} r_m (1 + \| P_m^{-1} \| \| P_m - P_{m-1} \|) < \infty,$$

we are done. In fact $\{ \| P_m^{-1} \| \}$ is bounded, $\Sigma \| P_m - P_{m-1} \| < \infty$, and hence

$$\| P_m^{-1} X_m \| \leq K$$

for some constant K independent of m. Since $\{P_m\}$ is bounded,

$$\| X_m \| \leq K',$$

and as this is true for all X, $\| R_m \ldots R_1 \|$ is bounded.

Examples 7.1.

Given $\{b_n\}$, $b_n \in \mathbb{R}$, suppose

$$x_n = b_n x_{n-1} - x_{n-2}, \quad n = 1, 2, \ldots .$$

Let

$$\underline{x}_n = \begin{pmatrix} x_n \\ x_{n-1} \end{pmatrix} .$$

Then $\underline{x}_n = Q_n \underline{x}_{n-1}$, where

$$Q_n = \begin{pmatrix} b_n & -1 \\ 1 & 0 \end{pmatrix} .$$

If $b_n \to b$, then $Q_n \to Q$, where

$$Q = \begin{pmatrix} b & -1 \\ 1 & 0 \end{pmatrix} .$$

Q has distinct eigenvalues if $|b| < 2$. Letting r_n denote the spectral radius of Q_n, we note that $r_n = 1$ if $|b_n| \le 2$. Hence, if $b_n \in \mathbb{R}$, $|b_n| \le 2$, $b_n \to b(|b| < 2)$ and $\Sigma|b_n - b_{n+1}| < \infty$, the requirements for Q_n in Theorem 7.1 are satisfied. By taking $\mu_n \equiv 0$ and $\tau_n \equiv 1$ so that $R_n = Q_n$, it follows from this theorem that $\{x_n\}$ is bounded. Thus Lemma 2.1 follows from Theorem 7.1 when the b_n are real.

If we change the definition of $\underline{x}_n$ above to

$$\underline{x}_n = \begin{pmatrix} x_{2n} \\ x_{2n-1} \end{pmatrix} ,$$

then $\underline{x}_n = Q_n \underline{x}_{n-1}$, where now

$$Q_n = \begin{pmatrix} b_{2n}b_{2n-1} & -b_{2n} \\ b_{2n-1} & -1 \end{pmatrix}.$$

It is not difficult to see that if the b_n meet the hypotheses of Lemma 5.1, then Q_n meets the requirements of the Q_n in Theorem 7.1. Hence again, if we set $\mu_n \equiv 0$ and $\tau_n \equiv 1$ in Theorem 7.1, it follows that Lemma 5.1 is implied by this theorem. We have now finished the discussion we intended under Examples 7.1.

We have seen earlier that if $b_n \in \mathbb{R}$, $|b_n| < 2$, $\lim b_n = 0$, and $\Sigma\ |b_n - b_{n+1}| < \infty$, then $-K(-1/b_n)$ cannot converge to a finite limit. In our next theorem we give some indication of what can be said if the condition $\Sigma|b_n - b_{n+1}| < \infty$ is not required, that is if $\Sigma|b_n - b_{n+1}| = \infty$ is allowed.

__Theorem__ 7.2: __Let__ $\{b_n\}$ __be a sequence whose elements satisfy__ $b_n \in \mathbb{R}$, $|b_n| < 2$, $0 < b_n b_{n+1} < 4$, $\lim b_n = 0$, __and__

(i) $\lim b_{2n-1}/b_{2n} = L \neq 0, \infty$

(ii) $\Sigma|b_{2n-1} - b_{2n+1}| < \infty$ __or__ $\Sigma|b_{2n} - b_{2n+2}| < \infty$

(iii) $\Sigma|b_{2n-1}/b_{2n} - b_{2n+1}/b_{2n+2}| < \infty$.

__Then__

$$\frac{1}{b_1} - \frac{1}{b_2} - \frac{1}{b_3} - \cdots \tag{7.1}$$

__cannot converge to a finite limit__. __In particular__, __if__ a, b, c, d __are constants satisfying__

$$a > 0,\ c > 0,\ a+b > 1/2,\ c+d > 1/2,\ a \neq c$$

<u>and</u>

$$b_{2n-1} = 1/(an+b), \; b_{2n} = 1/(cn+d),$$

<u>then</u> (7.1) <u>cannot converge finitely</u>.

<u>Proof</u>: If we again let B_n denote the nth denominator of (7.1), then

$$\begin{pmatrix} B_{2n+2} \\ B_{2n+1} \end{pmatrix} = M_n \begin{pmatrix} B_{2n} \\ B_{2n-1} \end{pmatrix},$$

where

$$M_n = \begin{pmatrix} b_{2n+2}b_{2n+1} - 1 & -b_{2n+2} \\ b_{2n+1} & -1 \end{pmatrix}.$$

Now

$$M_n = b_{2n+2}Q_n - I$$

and

$$M_n = b_{2n+1}R_n - I,$$

where

$$Q_n = \begin{pmatrix} b_{2n+1} & -1 \\ b_{2n+1}/b_{2n+2} & 0 \end{pmatrix}$$

and

$$R_n = \begin{pmatrix} b_{2n+2} & -b_{2n+2}/b_{2n+1} \\ 1 & 0 \end{pmatrix}.$$

We have that

$$Q_n \to Q = \begin{pmatrix} 0 & -1 \\ L & 0 \end{pmatrix}$$

and

$$R_n \to R = \begin{pmatrix} 0 & -1/L \\ 1 & 0 \end{pmatrix}.$$

The characteristic roots of Q are $\pm\, i\sqrt{L}$ and those of R are $\pm\, i/\sqrt{L}$. The series $\Sigma\|Q_n - Q_{n+1}\|$ converges if (i) and (iii) holds and $\Sigma|b_{2n-1} - b_{2n+1}| < \infty$, or the series $\Sigma\|R_n - R_{n+1}\|$ converges if (i) and (iii) holds and $\Sigma|b_{2n} - b_{2n+2}| < \infty$. Since $0 < b_n b_{n+1} < 4$, the characteristic roots of M_n are distinct with common modulus 1. Thus it follows from Theorem 7.1 that

$$\{\|M_n M_{n-1} \cdots M_1\|\}_{n\geq 1}$$

is bounded. Hence the sequences $\{B_{2n}\}$ and $\{B_{2n-1}\}$ are bounded so that $\{B_n\}$ is bounded. It follows that the nth term of the series $\Sigma\,(B_n B_{n-1})^{-1}$ cannot converge to 0, which implies that the continued fraction (7.1) cannot converge to a finite limit.

8. Equimodular Limit Periodic $K(a_n/1)$.

In this section we study continued fractions $K(a_n/1)$, where the a_n lie on the negative real axis and the convergence structure of the sequence $\{a_n\}$ is varied. Theorem 8.1 deals with the case $a_n \to a < -1/4$. Theorem 8.2 is concerned with the case $a_{2n-1} \to a < 0$, $a_{2n} \to b < 0$, where a and b are chosen so that $K(a_n/1)$ is equimodular. Finally, Theorem 8.3 gives information about what happens when $a_n \to -\infty$.

Theorem 8.1: *If $a_n < -1/4$, $n = 1, 2, \ldots$, and $\lim a_n = a$, where $a < -1/4$; then the continued fraction*

$$(8.1) \qquad \frac{a_1}{1} + \frac{a_2}{1} + \frac{a_3}{1} + \ldots$$

cannot converge to a finite limit if

$$\sum_{n=1}^{\infty} |a_{n+1} - a_n| < \infty.$$

Proof: The denominators B_n of (8.1) satisfy the difference equation

$$B_0 = B_1 = 1; \; B_n = B_{n-1} + a_n B_{n-2}, \; n \geq 2.$$

Let the sequence $\{C_n\}$ be defined by

$$B_n = C_n P_n^{1/2},$$

where

$$P_n = \prod_{k=1}^{n} (-a_k).$$

Then

$$C_{n+2} P_{n+2}^{1/2} = C_{n+1} P_{n+1}^{1/2} + a_{n+2} C_n P_n^{1/2}$$

from which we obtain

$$c_{n+2} = c_{n+1}/(-a_{n+2})^{1/2} - c_n(a_{n+2}/a_{n+1})^{1/2}.$$

Writing the last equation as a matrix equation, we have

$$\begin{pmatrix} c_{n+2} \\ c_{n+1} \end{pmatrix} = A_{n+1} \begin{pmatrix} c_{n+1} \\ c_n \end{pmatrix} ,$$

where

$$A_{n+1} = \begin{pmatrix} (-a_{n+2})^{-1/2} & -(a_{n+2}/a_{n+1})^{1/2} \\ 1 & 0 \end{pmatrix}$$

The eigenvalues λ_i, $i = 1, 2$, of A_{n+1} satisfy

$$\lambda^2 - \lambda(-a_{n+2})^{-1/2} + (a_{n+2}/a_{n+1})^{1/2} = 0.$$

It follows from this equation that

$$|\lambda_i| = (a_{n+2}/a_{n+1})^{1/4}, \ i = 1, 2.$$

Hence the spectral radius r_{n+1} of A_{n+1} is given by

$$r_{n+1} = (a_{n+2}/a_{n+1})^{1/4}.$$

Thus

$$\prod_{k=1}^{n} r_k = (a_{n+1}/a_1)^{1/4}.$$

So

$$\sup_n \left(\prod_{k=1}^{n} r_k\right) < \infty,$$

since $\{a_n\}$ is a convergent (hence bounded) sequence.

Clearly $A_n \to A$ as $n \to \infty$, where

$$A = \begin{pmatrix} (-a)^{-1/2} & -1 \\ 1 & 0 \end{pmatrix} .$$

The eigenvalues of A have modulus 1 and are unequal since $a < -1/4$. Since

(8.2)
$$\sum_{n=1}^{\infty} |(-a_{n+1})^{-1/2} - (-a_n)^{-1/2}| = \sum_{n=1}^{\infty} \frac{|a_{n+1} - a_n|}{(a_n a_{n+1})^{1/2}((-a_n)^{1/2} + (-a_{n+1})^{1/2})} ,$$

it is easy to see that (8.2) converges because of our assumptions that $a_n \to a < -1/4$ and the series

$$\sum_{n=1}^{\infty} |a_{n+1} - a_n| \tag{8.3}$$

converges. The hypotheses that $a_n \to a < -1/4$ and (8.3) converges also guarantee the convergence of the series

$$\sum_{n=1}^{\infty} |(a_{n+2}/a_{n+1})^{1/2} - (a_{n+1}/a_n)^{1/2}| . \tag{8.4}$$

This is easier seen once we observe the relation

$$\left|(a_{n+2}/a_{n+1})^{1/2} - (a_{n+1}/a_n)^{1/2}\right| = \frac{|a_{n+1} - a_{n+2}|}{(-a_{n+1})^{1/2}((-a_{n+2})^{1/2} + (-a_{n+1})^{1/2})}$$
$$+ \frac{|a_{n+1} - a_{n+2}|}{(-a_{n+1})^{1/2}((-a_{n+1})^{1/2} + (-a_n)^{1/2})}$$
$$+ \frac{|a_{n+1} - a_n|}{(a_{n+1}a_n)^{1/2}} .$$

Hence the convergence of (8.3) implies the convergence of

$$\sum_{n=1}^{\infty} \|A_n - A_{n+1}\| .$$

Therefore, by Theorem 7.1 the sequence $\{C_n\}$ is bounded.

Now (8.1) can converge to a finite limit only if the series

$$\text{(8.5)} \qquad \sum_{n=N}^{\infty}(-1)^n(B_nB_{n-1})^{-1}\prod_{k=1}^{n}a_k$$

converges for N large enough. But

$$(-1)^{n+1}(B_{n+1}B_n)^{-1}\prod_{k=1}^{n}a_k = (C_{n+1}C_n)^{-1}(-a_{n+1})^{1/2} \not\to 0$$

as $n \to \infty$ since $\{C_n\}$ is bounded and $a_{n+1} \to a < -1/4$. Therefore, (8.5) cannot converge to a finite limit, and we have completed our proof.

<u>Theorem</u> 8.2: <u>If</u> $a_n < 0$, $n = 1, 2, \ldots$, $\lim a_{2n-1} = a < 0$, <u>and</u> $\lim a_{2n} = b < 0$. <u>where</u>

$$\text{(8.6)} \qquad ((-a)^{1/2} - (-b)^{1/2})^2 < 1 < ((-a)^{1/2} + (-b)^{1/2})^2,$$

<u>then the continued fraction</u> (8.1) <u>cannot converge to a finite limit if both series</u>

$$\text{(8.6)} \qquad \sum_{n=1}^{\infty} |a_{2n-1} - a_{2n+1}|, \quad \sum_{n=1}^{\infty} |a_{2n} - a_{2n+2}|$$

<u>converge</u>.

<u>Proof</u>: We define $\{B_n\}$ and $\{C_n\}$ as in the proof of Theorem 8.1. Then

$$\begin{pmatrix} C_{2n+2} \\ C_{2n+1} \end{pmatrix} = Q_n \begin{pmatrix} C_{2n} \\ C_{2n-1} \end{pmatrix} \qquad , n = 1, 2, \ldots,$$

where

$$Q_n = \begin{pmatrix} (a_{2n+2}a_{2n+1})^{-1/2} - \left(\frac{a_{2n+2}}{a_{2n-1}}\right)^{1/2} & -\left(\frac{a_{2n+1}}{a_{2n}}\right)^{1/2}(-a_{2n+2})^{-1/2} \\ (-a_{2n+1})^{-1/2} & -\left(\frac{a_{2n+1}}{a_{2n}}\right)^{1/2} \end{pmatrix}.$$

It follows from our hypotheses that $\lim Q_n = Q$, where

$$Q = \begin{pmatrix} (ab)^{-1/2} - (b/a)^{1/2} & -(a/b)^{1/2}(-b)^{-1/2} \\ (-a)^{-1/2} & -(a/b)^{1/2} \end{pmatrix}.$$

The characteristic equation of Q is

$$\lambda^2 + \lambda((a/b)^{1/2} + (b/a)^{1/2} - (ab)^{-1/2}) + 1 = 0.$$

Condition (8.6) guarantees that the discriminant of this equation is negative so that its roots are unequal, though they have equal absolute value 1. The characteristic equation of Q_n will have a negative discriminant provided

$$\begin{aligned} &((a_{2n+1}^2 a_{2n+2}/a_{2n})^{1/4} + (a_{2n+2}^2)^{1/4})^2 > 1 \\ (8.8)\qquad &((a_{2n+1}^2 a_{2n+2}/a_{2n})^{1/4} - (a_{2n+2}^2)^{1/4})^2 < 1. \end{aligned}$$

Since $a_{2n-1} \to a < 0$, $a_{2n} \to b < 0$ and (8.6) holds, there exists a positive integer N such that (8.8) holds for $n \geq N$. Hence for $n \geq N$, it can be seen that the spectral radius r_n of Q_n is given by

$$r_n = (a_{2n+2}/a_{2n})^{1/4}.$$

Thus the product $\prod_{k=n}^{\infty} r_k$ converges to $(-b/a_{2N})^{1/4}$, so that $\{\prod_{k=1}^{n} r_k\}$ is bounded. After making some tedious estimates it can be seen that $\Sigma\|Q_n - Q_{n+1}\| < \infty$. Thus it follows from Theorem 7.1 that the sequence $\{C_n\}$ is bounded. As in the last part of the proof of Theorem 8.1, we can now come to the conclusion that (8.1) cannot converge to a finite limit under the given hypotheses.

<u>Theorem</u> 8.3: <u>Let</u> $\{h_n\}_{n\geq 0}$ <u>be a sequence of real numbers satisfying</u>

(8.9) $$h_n > 1,\ h_{n+1} \geq h_n,\ h_n \to \infty \text{ as } n \to \infty.$$

<u>Let</u> $\{a_n\}_{n\geq 1}$ <u>be defined by</u>

(8.10) $$a_{2n-1} = -h_{2n-2}(-1+h_{2n-1});\ a_{2n} = -h_{2n-1}(1+h_{2n}).$$

<u>Then the continued fraction</u> (8.1) <u>diverges by oscillation if</u> $\Sigma\ 1/h_k < \infty$. <u>If</u> $\Sigma\ 1/h_k = \infty$ <u>and</u> $\Sigma\ 1/h_k^2 < \infty$, <u>then</u> (8.1) <u>converges</u> to h_0.

<u>Proof</u>: Set

$$t_n = -(1 + (-1)^n/h_n),\ n \geq 1.$$

Then it can be established that

(8.11) $$f_n - h_0 = -h_0/(1 + t_1 + t_1t_2 + \dots + t_1t_2 \dots t_n),$$

where f_n is the nth approximant of (8.1). Set

$$C_n = \prod_{k=1}^{n} t_k = (-1)^n \prod_{k=1}^{n} (1 + (-1)^k/h_k).$$

Then $\{|C_n|\}$ is bounded if the infinite product

(8.12) $$\prod_{k=1}^{\infty} (1 + (-1)^k/h_k)$$

converges. Clearly (8.12) converges absolutely if $\Sigma\ 1/h_k < \infty$. If $\Sigma\ 1/h_k = \infty$ and the h_k satisfy (8.9), then $\Sigma\ (-1)^k/h_k < \infty$ and (8.12) converges if and only if $\Sigma\ 1/h_k^2 < \infty$. If (8.12) converges, then

$\lim |c_n| \neq 0$, so that the series Σc_n cannot converge to a finite limit. In this case, it follows from (8.11) that the only way (8.1) can converge is that $\lim |\sum_{k=1}^{n} c_k| = \infty$. Let

$$T_n = \sum_{k=1}^{n} c_k.$$

Then

$$T_{2n} = \sum_{k=1}^{2n} c_k = \sum_{k=1}^{n} (c_{2k-1} + c_{2k}).$$

But

$$c_{2n} + c_{2n-1} = \prod_{k=1}^{2n} (1 + (-1)^k/h_k) - \prod_{k=1}^{2n-1} (1 + (-1)^k/h_k)$$

$$= c_{2n-1}/h_{2n}$$

So

$$T_{2n} = \sum_{k=1}^{n} c_{2k-1}/h_{2k}. \tag{8.13}$$

Now if $\Sigma 1/h_k < \infty$, then c_{2n-1} tends to a negative limit and $\Sigma c_{2n-1}/h_{2n} < \infty$, so that $\{T_{2n}\}$ converges to a finite limit. Hence, in this case, the continued fraction (8.1) must diverge by oscillation.

If $\Sigma 1/h_k = \infty$ but $\Sigma 1/h_k^2 < \infty$, the product (8.12) is still convergent. In this case it follows (with the aid of (8.9) and (8.13)) that $T_{2n} \to -\infty$ as $n \to \infty$. Since $T_{2n+1} = c_{2n+1} + T_{2n}$, it is also true that $T_{2n+1} \to -\infty$. Hence, using (8.11), we have that (8.1) converges to h_0. This completes our proof of Theorem 8.3.

REFERENCES

1. Christopher Baltus, Limit-Periodic Continued Fractions: Value Regions and Truncation Error Bounds, Ph.D. Thesis, University of Colorado, Boulder, 1984.

2. A. Elbert, Asymptotic expansion and continued fraction for Mathieu's series, Periodica Mathematica Hungarica 13 (1982) 1-8.

3. Robert Hellar and F.A. Roach, A generalization of a classical necessary condition for convergence of continued fractions, Pac. J. Math. 95 (1981) 307-310.

4. Lisa Jacobsen and Arne Magnus, On the convergence of limit periodic continued fractions $K(a_n/1)$, where $a_n \to -1/4$, Lecture Notes in Math., Vol. 1105, Springer-Verlag, Berlin, Heidelberg, New York, and Tokyo (1984) 243-248.

5. Lisa Jacobsen and Haakon Waadeland, Even and odd parts of limit periodic continued fractions, to appear.

6. William B. Jones and W.J. Thron, Continued Fractions: Analytic Theory and Applications, Vol. II, Addison-Wesley, Reading, MA, 1980.

7. L.J. Lange, σ-Fraction expansions of analytic functions, SIAM J. Math. Anal. 14 (1983) 323-368.

8. Roland Louboutin, Sur un théorème de H. Poincaré relatif aux équations aux differences finies, C.R. Acad. Sc. Paris, t. 299, Serie I, n°12, (1984) 539-542.

9. Evar D. Nering, Linear Algebra and Matrix Theory, 2nd. ed., Wiley, New York, 1970.

10. Oskar Perron, Die Lehre von den Kettenbrüchen, Band II, Teubner, Stuttgart, 1957.

11. W.J. Thron and H. Waadeland, Accelerating convergence of limit periodic continued fractions $K(a_n/1)$, Num. Math. 34(1980) 155-170.

12. Haakon Waadeland, Tales about tails, Proc. A.M.S. 90 (1984) 57-64.

13. H.S. Wall, Analytic Theory of Continued Fractions, Van Nostrand, New York, 1948.

Continued Fraction Applications to Zero Location*

L.J. Lange

Department of Mathematics
University of Missouri
Columbia, Missouri 65211

1. Introduction.

In this paper we give a unified presentation of J- and δ-fraction criteria for ensuring zero counting, zero location, and stability properties of polynomials with real or complex coefficients. Given a polynomial $P(z)$ we investigate the question of whether there is an easily obtainable rational function $T(z)$ associated with $P(z)$, such that the elements of the J- or δ-fraction expansion of $T(z)$ give us the desired information about the zeros of $P(z)$ with little computation.

The subject of zero properties of polynomials is an old one and the literature on it is vast. Thus we have chosen not to attempt to present a complete historical documentation of the classical results that might have some bearing on our work here and who should be credited with them. To our knowledge, Wall [13] (1945) and Frank [1] (1946) were the first to give stability and half plane counting criteria in terms of continued fractions. In two more recent works, Jones and Steinhardt [5], [6] have shown how a class of continued fractions, called Schur fractions, can be applied to questions of D-stability of polynomials and to related problems in the area of digital filtering and signal processing. In 1983 Rogers [10]

* The research of L.J. Lange was supported in part by a grant from the Research Council of the Graduate School, University of Missouri - Columbia and a grant from the Nansen Foundation of Norway.

investigated applications of certain continued fractions to the location of zeros of polynomials. However, we doubt the validity of his so called "homotopy approach" in his sketches of proofs of the key theorems in the paper, especially when the given polynomial has multiple zeros.

Henrici [3], [4] treats the subject of zero location of polynomials in considerable detail in Chapter 6, Volume 1 of his 3-volume sequence of books on applied and computational complex analysis. He also takes up the topic in Chapter 12 on continued fractions in Volume 2 of this series. We recommend these two books as good modern references on the analytic theory of polynomials.

We now define some terms and concepts which will be used in various parts of this paper. We say that a polynomial $P(z)$ of degree $n \geq 1$ is stable, (D-stable) if $P(z_k) = 0 \Rightarrow \mathrm{Re}(z_k) < 0$, ($|z_k| < 1$). The paraconjugate $f^*(z)$ of a rational function $f(z)$ is defined by $f^*(z) = \overline{f(-\bar{z})}$. The function $f(z)$ is said to be para-odd if $f^*(z) = -f(z)$ and positive if $\mathrm{Re}(f(z)) > 0$ when $\mathrm{Re}(z) > 0$.

By a finite J-fraction we mean a continued fraction of the form

$$(1.1) \qquad \mathop{K}_{k=1}^{n} \frac{1}{a_k z + b_k}, \quad a_k, b_k \in \mathbb{C},\ a_k \neq 0,\ k = 1, 2, \ldots, n.$$

Wall [14] develops the theory of J-fraction expansions for rational functions in Chapter IX of his book. Unfortunately, not every rational function has a finite J-fraction representation.

Independent of a_k and b_k, set

$$\mathop{K}_{k=1}^{0} \frac{a_k}{b_k} = 0.$$

Then a finite or terminating regular δ-fraction is defined to be a continued fraction of the form

(1.2) $$b_0 - \delta_0 z + \sum_{k=1}^{n} \frac{d_k z}{1-\delta_k z}, \quad n \geq 0,$$

where

$$b_0 \in \mathbb{C};\ d_k \in \mathbb{C},\ d_k \neq 0,\ k = 1, 2, \ldots, n$$

$$\delta_n = \delta_{n-1} = 0,\ \delta_k = 0 \text{ or } 1 \text{ if } 0 \leq k \leq n-2,\ \delta_k = 1 \Rightarrow d_{k+1} = 1.$$

In 1983 Lange [8] introduced and developed the theory and application of δ-fractions. In this paper he proved that every rational function which is analytic at the origin has a unique terminating regular δ-fraction expansion of the form (1.2). For example

$$\frac{z}{1+z^4} = \frac{z}{(1-z)} + \frac{z}{(1-z)} + \frac{z}{(1-z)} + \frac{z}{1} - \frac{z}{1} + \frac{z}{1} + \frac{z}{(1-z)} + \frac{z}{1} - \frac{z}{1}.$$

This function has no J-fraction expansion of the form (1.1). We refer the reader to the recent book on continued fractions by Jones and Thron [7] for other basic definitions, formulas, and results in this field.

Finally, in our introduction, we resurrect two interesting theorems which can be used as side tools along with the results presented in this work. Descarte's ordinary Rule of Signs for real polynomials is of course very well known. However, we have found that the following extension of Descarte's Rule for real polynomials with real zeros is not so well known.

<u>Theorem</u> 1.1: <u>Let</u>

$$P(z) = \sum_{k=0}^{n} a_k z^{n-k}$$

<u>be a polynomial of degree</u> $n \geq 1$ <u>with real coefficients, and let</u> $V, (V')$ <u>be the number of variations in sign of the nonzero coefficients of</u>

$P(z)$, $((-1)^n P(-z))$. <u>If all zeros of</u> $P(z)$ <u>are real, then</u> (<u>counting multiplicities</u>) V <u>is exactly the number of positive zeros of</u> $P(z)$ <u>and</u> V' <u>is exactly the number of its negative zeros</u>.

A proof of Theorem 1.1 is given by Uspensky [12, p. 124]. This theorem becomes especially useful in conjunction with our continued fraction criteria in section 3 for deciding whether or not all zeros of a given polynomial are real.

Titchmarsh [11, section 8.61] in his book on the theory of functions states and proves the following theorem which he attributes to Laguerre:

<u>Theorem</u> (Laguerre) 1.2: <u>Let</u>

$$P(z) = \sum_{k=0}^{n} a_k z^k$$

<u>be a polynomial of degree</u> $n \geq 1$, <u>all of whose zeros are real; and let</u> $\Phi(w)$ <u>be an entire function of genus</u> 0 <u>or</u> 1, <u>which is real for real</u> w, <u>and all the zeros of which are real and negative</u>. <u>Then the polynomial</u>

$$Q(z) = \sum_{k=0}^{n} \Phi(k) a_k z^k$$

<u>has all its zeros real, and as many positive, zero, and negative zeros as</u> $P(z)$.

For example, one of the simplest choices for $\Phi(w)$ in Theorem 1.2 is $\Phi(w) = 1 + w$ which leads to an easy computation for the corresponding test polynomial $Q(z)$. We suggest that Laguerre's theorem can become useful when it is employed along with some of the other results in this paper such as Theorem 3.5 and the theorems in section 4. The idea is this: Suppose we know by one of our tests that the zeros of a given polynomial $P(z)$ are real. Next suppose we apply one of our tests for counting the number of positive zeros and the number of negative zeros, but this test breaks down for $P(z)$. We then transform P(z) à la Laguerre to Q(z) and apply our test to Q(z). If the test is successful when applied to Q(z) we will have our desired information. We shall illustrate this idea in section 4.

2. Two Fundamental Theorems.

The first result in this section, Theorem 2.1, is essentially equivalent to Theorem 43.1 in Wall's book [14], a theorem which Wall attributes to Grommer [2]. However, we give a new proof of this theorem based partly on some known results on positive para-odd rational functions. Theorem 2.1 will be employed directly in our proofs of Theorems 3.1 and 3.3 in the next section.

Each part of Theorem 2.2 is basically known and appears in some equivalent form in the literature. We have found it convenient for application to state the theorem in its present form, and we are especially proud of what we claim is a new proof of part (E) of this theorem. Part (E) is closely related to a result of Wall and Frank, [14, Theorem 48.1], which was derived by a different method. We shall derive from part (E) an equivalent form of the theorem of Wall and Frank as Theorem 4.1 in section 4. Theorem 2.2 will also play a direct role in our proof of Theorem 3.4 and an indirect role in the proofs of Theorems 4.1, 4.2, and 4.3.

Theorem 2.1: A necessary and sufficient condition for a rational function g to have the form

$$g(z) = \sum_{k=1}^{n} \frac{m_k}{z-x_k}, \tag{2.1}$$

where the x_k are real and distinct and the m_k are positive, is that g have a J-fraction representation of the form

(2.2) $$g(z) = \frac{1}{a_1z+b_1} + \frac{1}{a_2z+b_2} + \cdots + \frac{1}{a_nz+b_n},$$

where the a_k, b_k *are real and* $(-1)^{k-1}a_k > 0$, $k = 1, 2, \ldots, n$.

Proof: Let $h(z) = ig(iz)$. Then

$$h(z) = \sum_{k=1}^{n} \frac{m_k}{z+ix_k}.$$

Clearly, $\mathrm{Re}(h(z)) > 0$ if $\mathrm{Re}(z) > 0$. Also,

$$h^*(z) := \overline{h(-\bar{z})} = -h(z),$$

so $h(z)$ satisfies the requirements for being a positive para-odd rational function. It is easily seen that $1/h$ is also a positive para-odd function having the additional property

$$\lim_{z\to\infty} 1/h(z) = \infty.$$

Hence by [7, Theorem 7.30], there exists an integer m such that

(2.3) $$h(z) = \frac{1}{c_1z+d_1} + \frac{1}{c_2z+d_2} + \cdots + \frac{1}{c_mz+d_m}$$

$$= \mathop{K}_{k=1}^{m} 1/(c_kz+d_k)$$

where $c_k > 0$, $\mathrm{Re}(d_k) = 0$, $k = 1, 2, \ldots, m$. Since the denominator of $h(z)$ is a polynomial of degree n, an application of the difference equations for (2.3) shows that $m = n$. Therefore, in view of the relation $g(z) = -ih(-iz)$, we have

$$g(z) = -i\mathop{K}_{k=1}^{n} 1/(-c_kiz+d_k)$$

which by an equivalence transformation can be put into the form

$$g(z) = \mathop{K}_{k=1}^{n} 1/((-1)^{k-1}(c_kz+id_k)).$$

Thus $g(z)$ has a representation of the form (2.2) if it has a representation of the form (2.1).

Conversely, suppose g is given by (2.2). Then, after employing an equivalence transformation, we obtain

$$\text{(2.4)} \qquad ig(iz) = \sum_{k=1}^{n} 1/((-1)^{k-1}(a_k z - ib_k)).$$

Hence by [7, Theorems 7.29 and 7.30] $ig(iz)$ is positive para-odd. Since

$$\lim_{z\to\infty} ig(iz) = 0,$$

it follows from (2.4) and [7, Theorem 7.28] that

$$\text{(2.5)} \qquad ig(iz) = \frac{m_1}{z-iw_1} + \frac{m_2}{z-iw_2} + \dots + \frac{m_n}{z-iw_n},$$

where the w_k are distinct real numbers. The sum (2.5) has n terms because the denominator of $ig(iz)$ is of degree n. From (2.5) we obtain

$$g(z) = \frac{m_1}{z+w_1} + \frac{m_2}{z+w_2} + \dots + \frac{m_n}{z+w_n},$$

so $g(z)$ has a representation of the form (2.1). This completes our proof of Theorem 2.1.

Theorem 2.2: For each positive integer n let $P_n(z)$ and $Q_n(z)$ be the nth numerator and denominator, respectively, of the continued fraction

$$\text{(2.6)} \qquad f_n(z) = \frac{1}{a_n z + ib_n} + \frac{1}{a_{n-1} z + ib_{n-1}} + \dots + \frac{1}{a_1 z + ib_1},$$

where a_k, $b_k \in \mathbb{R}$, $a_k \neq 0$, $k = 1, 2, \dots, n$. Then

(A) Degree $(P_n) = n-1$ and degree $(Q_n) = n$.

(B) P_n and Q_n have no common zero.

(C) *If* $t \in \mathbb{R}$, *then* $\mathrm{Im}(P_n(it)) = \mathrm{Re}(Q_n(it)) = 0$ *if* n *is odd and* $\mathrm{Re}(P_n(it)) = \mathrm{Im}(Q_n(it)) = 0$ *if* n *is even.*

(D) *If* $t \in \mathbb{R}$, *then* $P_n(it) + Q_n(it) \neq 0$.

(E) *If* m *of the* a_k's *are positive and* $n-m$ *are negative, then* m *of the zeros of* $P_n(z) + Q_n(z)$ *have negative real part and* $n-m$ *have positive real part.*

Proof: Let sequences of polynomials $\{P_k\}$ and $\{Q_k\}$ be defined by

$$P_0 = 0, \qquad Q_0 = 1$$

$$P_k = Q_{k-1} \qquad Q_k = (a_k z + ib_k)Q_{k-1} + P_{k-1}, \quad k \geq 1, \tag{2.7}$$

where $\mathrm{Im}(a_k) = \mathrm{Im}(b_k) = 0$ and $\mathrm{Re}(a_k) \neq 0$ for all $k \geq 1$. Then Q_n is the nth denominator of the continued fraction

$$\overset{n}{\underset{k=1}{K}} \frac{1}{a_k z + ib_k},$$

and it is well known that $Q_{n-1} = P_n$ and Q_n are the nth numerator and denominator, respectively, of the continued fraction (2.6). It follows easily from (2.7) that for $k \geq 0$

$$P_{k+2}(z) = (a_{k+1}z + ib_{k+1})Q_k(z) + P_k(z)$$

$$Q_{k+2}(z) = (a_{k+2}z + ib_{k+2})((a_{k+1}z + ib_{k+1})Q_k(z) + P_k(z)) + Q_k(z). \tag{2.8}$$

Now (A) and (B) follow easily by induction using (14). Using (2.7) and (2.8), it is also not difficult to verify (C) by induction. (D) follows immediately from (B) and (C).

Now fix k and let

$$R(z) = (1 + a_{k+1}z + ib_{k+1})(P_k(z) + Q_k(z))$$

$$S(z) = (a_{k+1}z + ib_{k+1})P_k(z) \tag{2.9}$$

$$T(z,\lambda) = R(z) - \lambda S(z), \quad \lambda \in [0,1].$$

Then

$$T(z,0) = R(z) \tag{2.10}$$

and by (2.7)

$$T(z,1) = P_{k+1}(z) + Q_{k+1}(z). \tag{2.11}$$

We will now show that $T(z,0)$ and $T(z,1)$ have the same zero distributions with respect to the open left and right half planes.

Using (B), (C), and (D) it can be established after some simple algebraic calculations that

$$T(it,\lambda) \neq 0$$

if $t \in \mathbb{R}$ and $\lambda \in [0,1]$. Thus we can and do define the function $n(\lambda)$ on $[0,1]$ by

$$n(\lambda) = \frac{1}{2\pi i} \int_\gamma \frac{R'(z) - \lambda S'(z)}{R(z) - \lambda S(z)} dz,$$

where γ is the closed positively oriented path consisting of a semicircle in the left half plane with its center at the origin along with its diameter on the imaginary axis chosen so large that all zeros $T(z,\lambda)$ with negative real part are contained in the interior of γ for all $\lambda \in [0,1]$. This can be done because from (A) and the hypotheses on the a_k we have degree (R) > degree (S) and because $T(z,\lambda)$ is not zero on the imaginary axis. It is easily established that $n(\lambda)$ is continuous on $[0,1]$. Since it is also integer valued, it follows that $n(0) = n(1)$. This implies that $T(z,0)$ and $T(z,1)$ have the same number of zeros with negative real part as asserted. A similar argument with an obvious change of γ will show that these two polynomials have the same number of zeros with positive real part. In view of the relations (2.9), (2.10), and (2.11) it now follows that $P_{k+1} + Q_{k+1}$ has at least as many zeros with negative real part as $P_k + Q_k$ and at least as many with positive real part as $P_k + Q_k$. Since

$$\text{degree } (P_{k+1} + Q_{k+1}) = 1 + \text{degree } (P_k + Q_k),$$

$P_{k+1} + Q_{k+1}$ has one more zero than $P_k + Q_k$ and, by what we have shown, this zero must be in the same half plane in which the zero of the factor $(1 + a_{k+1}z + ib_{k+1})$ of $R(z)$ is located. Clearly, this zero is in the left or right half plane according as a_{k+1} is positive or negative. Hence, by the principle of mathematical induction, $P_n + Q_n$ has as many zeros with negative real part (positive real part) as there are positive (negative) a_k's in the expansion (2.6). With this we have proved part (E), so our proof of Theorem 2.2 is complete.

3. <u>Real Zero Criteria</u>.

In this section we state and prove five theorems giving continued fraction criteria for various real zero behavior patterns for polynomials. Theorems 3.1 and 3.2 give necessary and sufficient conditions in terms of finite J-fractions and terminating regular δ-fractions, respectively, for a polynomial with complex coefficients to have all its zeros real with a specified number of them distinct. In a way Theorem 3.3 goes a step further. It gives necessary and sufficient conditions, in terms of finite regular δ-fractions, for a complex polynomial to have all its zeros real and negative with a specified number of them distinct; thereby giving a sufficient condition for stability also.

Theorem 3.4 tells us that we can use finite J-fractions in some cases to count the number of distinct real zeros and the number of distinct conjugate pairs of complex zeros of a real polynomial.

Finally, Theorem 3.5 shows how, for real polynomials, δ-fractions can be used in some cases to count the number of distinct positive real zeros, the number of distinct negative real zeros, and the number of distinct conjugate pairs of complex zeros. We give examples illustrating these results. In most cases the zeros of the example polynomials are given or are easily found. These examples are not intended to show the full power of the theorems but only to give assurance to the reader that the theorem in question is being applied correctly.

<u>Theorem</u> 3.1: <u>A necessary and sufficient condition for a polynomial</u> P <u>of degree</u> $n \geq 1$ <u>with complex coefficients to have all its zeros real with</u> $m \leq n$ <u>of them distinct is that</u> P'/P <u>have a J-fraction expansion of the form</u>

$$\frac{P'(z)}{P(z)} = \frac{1}{a_1 z+b_1} + \frac{1}{a_2 z+b_2} + \cdots \frac{1}{a_m z+b_m}$$

<u>where the</u> a_k <u>and</u> b_k <u>are real and</u> $(-1)^{k-1} a_k > 0$, $k = 1, 2, \ldots, m$.

<u>Proof:</u> Let the distinct real zeros of P be denoted by x_k, $k = 1, 2, \ldots, m$, and let the corresponding multiplicity of x_k be denoted by m_k. Then

$$(3.1) \qquad \frac{P'(z)}{P(z)} = \sum_{k=1}^{m} \frac{m_k}{z - x_k} .$$

Hence, by Theorem 2.1,

$$(3.2) \qquad \frac{P'(z)}{P(z)} = \mathop{K}_{k=1}^{m} \frac{1}{a_k z+b_k} ,$$

where $a_k, b_k \in \mathbb{R}$, $(-1)^{k-1} a_k > 0$, $k = 1, 2, \ldots, m$.

Conversely, suppose (3.2) holds. Then by Theorem 2.1 there exists $\{m_k\}_{k=1}^m$, $m_k > 0$, and $\{x_k\}_{k=1}^m$, where the x_k are real and distinct, such that (3.1) holds. But then it is true that each x_k is a simple pole of P'/P with residue $m_k > 0$. Hence P must have a zero at each x_k. P can have no zeros other than the x_k, for if so, P'/P would have to have a pole of order 1 at such a place and clearly the representation (3.1) has no poles other than the x_k. With this our proof of Theorem 3.1 is complete.

<u>Example</u> 3.1: Let

$$P(z) = (z-1)^2(z+1)^2(z-2)$$

$$= z^5 - 2z^4 - 2z^3 + 4z^2 + x - 2.$$

Then

$$P'(z) = 5z^4 - 8z^3 - 6z^2 + 8z + 1.$$

By an Euclidean process similar to the one used to find the g.c.d. of two polynomials we obtain

$$\frac{P'(z)}{P(z)} = \cfrac{1}{\frac{1}{5}z - \frac{2}{25} + \cfrac{1}{\frac{-125}{36}z + \frac{25}{27} + \cfrac{1}{\frac{324}{625}z - \frac{432}{625}}}} .$$

So m = 3 and for k = 1, 2, 3

$$\text{sign }(a_k): \quad + \quad - \quad +$$

$$\text{sign}((-1)^{k-1}a_k): \quad + \quad + \quad + .$$

Thus by Theorem 3.1 all zeros of P(z) are real and 3 are distinct.

Theorem 3.2: A necessary and sufficient condition for a polynomial P of degree $n \geq 1$ with complex coefficients to have all its zeros real and $m \leq n$ of them distinct is that $P'(1/z)/P(1/z)$ have a regular δ-fraction expansion of the form

$$(3.3)\qquad \frac{P'(1/z)}{P(1/z)} = \frac{d_1 z}{1-\delta_1 z} + \frac{d_2 z}{1} + \frac{d_3 z}{1-\delta_3 z} + \frac{d_4 z}{1} + \dots + \frac{d_{2m-1} z}{1-\delta_{2m-1} z} + \frac{d_{2m} z}{1}$$

where $d_k \in \mathbb{R}$, $k = 1, 2, \dots, 2m$, and

(i) $\delta_{2m-1} = 0$, $\delta_{2k-1} = 0$ or 1, $k = 1, 2, \dots, 2m-1$.

(ii) $d_1 > 0$ and d_{2m} may or may not be $= 0$.

(iii) $\delta_{2k-1} = 1 \Rightarrow d_{2k} = 1$ and $d_{2k+1} > 0$

$k = 1, 2, \dots, m-1$

$\delta_{2k-1} = 0 \Rightarrow d_{2k} d_{2k+1} > 0$

Proof: We shall show that this theorem follows from Theorem 3.1 by proving that $P'(1/z)/P(1/z)$ has an expansion of the form (3.3) if and only if $P'(z)/P(z)$ has a representation of the form

$$(3.4)\qquad \frac{P'(z)}{P(z)} = \frac{1}{a_1 z + b_1} + \frac{1}{a_2 z + b_2} + \dots + \frac{1}{a_m z + b_m},$$

where the a_k and b_k are real and $(-1)^{k-1} a_k > 0$, $k = 1, 2, \dots, m$.

From (3.3), with the aid of an equivalence transformation, we obtain

$$(3.5)\qquad \frac{P'(z)}{P(z)} = \frac{d_1}{z-\delta_1} + \frac{d_2}{1} + \frac{d_3}{z-\delta_3} + \frac{d_4}{1} + \dots + \frac{d_{2m-1}}{z-\delta_{2m-1}} + \frac{d_{2m}}{1}$$

By contracting (3.5) we arrive at

$$(3.6)\qquad \frac{P'(z)}{P(z)} = \frac{a_1^*}{z+b_1^*} + \frac{a_2^*}{z+b_2^*} + \dots + \frac{a_m^*}{z+b_m^*}$$

where

$$a_1^* = d_1, \qquad b_1^* = d_2 - \delta_1$$

$$a_k^* = -d_{2k-2}d_{2k-1}, \; b_k^* = d_{2k-1} + d_{2k} - \delta_{2k-1}, \; k = 2, 3, \ldots, m.$$

Note that $a_1^* > 0$ and $a_k^* < 0$ if $k = 2, 3, \ldots, m$. By another equivalence transformation (3.6) can be put into the form (3.4).

Conversely, suppose $P'(z)/P(z)$ has a representation of the form (3.4). Then, by replacing z by $1/z$ in (3.4) and using equivalence transformations we obtain

$$\frac{P'(1/z)}{P(1/z)} = \frac{z}{a_1+b_1z} + \frac{z^2}{a_2+b_2z} + \cdots + \frac{z^2}{a_m+b_mz}$$

$$= \frac{z/a_1}{1+b_1z/a_1} + \frac{z^2/(a_1a_2)}{1+b_2z/a_2} + \cdots + \frac{z^2/(a_{m-1}a_m)}{1+b_mz/a_m}$$

$$(3.7) \qquad = \frac{a_1^*z}{1+b_1^*z} \; \frac{a_2^*z^2}{1+b_2^*z} + \cdots + \frac{a_m^*z^2}{1+b_m^*z},$$

where now

$$a_1^* = 1/a_1>0; \; a_k^* = 1/(a_{k-1}a_k), \; k=2,3,\ldots,m; \; b_k^* = b_k/a_k, \; k=1,2,\ldots,m.$$

By an extension the continued fraction (3.7) can be written as

$$(3.8) \qquad \frac{P'(1/z)}{P(1/z)} = \frac{a_1^*z}{1+(b_1^*-\rho_1)z} + \frac{\rho_1z}{1} + \frac{a_2^*z/\rho_1}{1+(b_2^*+a_2^*/\rho_1-\rho_2)z} + \frac{\rho_2z}{1}$$

$$- \frac{a_3^*z/\rho_2}{1+(b_3^*+a_3^*/\rho_2-\rho_3)z} + \cdots + \frac{\rho_{m-1}z}{1} - \frac{a_m^*z/\rho_{m-1}}{1+(b_m^*+a_m^*/\rho_{m-1})z},$$

where

$$\rho_1 = b_1^* \text{ if } b_1 \neq 0, \; \rho_1 = 1 \text{ if } b_1 = 0$$

and

$$\rho_k = b_k^* + a_k^*/\rho_{k-1} \text{ if } b_k^* + a_k^*/\rho_{k-1} \neq 0$$

$$k = 2, \ldots, m-1$$

$$\rho_k = 1 \text{ if } b_k^* + a_k^*/\rho_{k-1} = 0$$

Finally, (3.8) can be written in the form

$$(3.9) \quad \frac{P'(1/z)}{P(1/z)} = \frac{d_1 z}{1-\delta_1 z} + \frac{d_2 z}{1} + \frac{d_3 z}{1-\delta_3 z} + \ldots + \frac{d_{2m-2} z}{1} + \frac{d_{2m-1} z}{1-\delta_{2m-1} z} + \frac{d_{2m} z}{1},$$

where

$$d_1 = a_1^*, \; d_{2k} = \rho_k, \; d_{2k+1} = -a_{k+1}^*/\rho_k, \; k = 1, 2, \ldots, m-1$$

$$\delta_{2m-1} = 0, \; d_{2m} = b_m^* + a_m^*/\rho_{m-1}.$$

Thus it can be seen that (3.9) is in the form (3.3) of our theorem and the d_k's and δ_{2k-1}'s meet the requirements (i), (ii), and (iii). This completes the proof of Theorem 3.2.

<u>Examples</u> 3.2: We construct a table where for a given polynomial $P(z)$ in the left-hand column we give the unique regular δ-fraction expansion of $P'(1/z)/P(1/z)$ in the right-hand column. These examples illustrate the various forms that (3.3) can take in accordance with Theorem 3.2 in accurately describing the zero behavior of the given polynomial.

$P(z)$	$P'(1/z)/P(1/z)$
$(z+1)^2$	$\frac{2z}{1} + \frac{z}{1}$
$z(z-1)$	$\frac{2z}{1} - \frac{z/2}{1} - \frac{z/2}{1}$
$(z-1)(z+1)$	$\frac{2z}{1-z} + \frac{z}{1} + \frac{z}{1} - \frac{z}{1}$
$(z-1)(z+2)$	$\frac{2z}{1} + \frac{z/2}{1} + \frac{9z/2}{1} - \frac{4z}{1}$
$(z^2-1)^2(z+2)$	$\frac{5z}{1} + \frac{2z/5}{1} + \frac{18z/5}{1} - \frac{10z/3}{1} - \frac{z/6}{1} + \frac{3z/2}{1}$.

Theorem 3.3: *A polynomial P of degree $n \geq 1$ with complex coefficients has all its zeros real and negative with $m \leq n$ of them distinct iff $P'(1/z)/P(1/z)$ has a regular δ-fraction expansion of the form*

$$\frac{P'(1/z)}{P(1/z)} = \frac{d_1 z}{1} + \frac{d_2 z}{1} + \dots + \frac{d_{2m-1} z}{1} + \frac{d_{2m} z}{1}, \tag{3.10}$$

where the d_j are positive real numbers.

Proof: We first prove that condition (3.10) is necessary. Set

$$P(z) = b_0(z-x_1)^{m_1}(z-x_2)^{m_2} \dots (z-x_m)^{m_m}, \quad b_0(\neq 0) \in \mathbb{C},$$

where for $1 \leq i,j \leq m$

$$x_i = x_j \Rightarrow i = j, \quad x_j < 0, \quad \sum_{j=1}^{m} m_j = n$$

Then

$$P(-z^2) = b_0(-1)^n \prod_{j=1}^{m} (z - \sqrt{-x_j})^{m_j}$$

and

$$-\frac{2zP'(-z^2)}{P(-z^2)} = \sum_{j=1}^{m} \frac{m_j}{z-\sqrt{-x_j}} + \sum_{j=1}^{m} \frac{m_j}{z+\sqrt{-x_j}}.$$

Hence, by Theorem 2.1,

$$-\frac{2zP'(-z^2)}{P(-z^2)} = \mathop{K}_{j=1}^{2m} \frac{1}{a_j z + b_j}, \tag{3.11}$$

where the a_j, b_j are real and $(-1)^{j-1}a_j > 0$, $j = 1, 2, \dots, 2m$. Since

$$-2z\ P'(-z^2)/P(-z^2)$$

is an odd function of z, it follows from the uniqueness of the J-fraction expansion (3.11) that the b_j are equal to zero. Therefore,

$$-\frac{2zP'(-z^2)}{P(-z^2)} = \mathop{K}_{j=1}^{2m} \frac{1}{a_j z}$$

from which we derive

$$(3.12) \qquad \frac{P'(-z^2)}{P(-z^2)} = -(1/2) \mathop{K}_{j=1}^{2m} \frac{(a_j a_{j-1} z^2)^{-1}}{1}, \quad a_0 := 1$$

with the aid of an equivalence transformation. If we set

$$w = -1/z^2,\ c_1 = 1/(2a_1),\ c_j = -(a_j a_{j-1})^{-1},\ j = 2, 3, \ldots, 2m,$$

then (3.12) can be written as

$$(3.13) \qquad \frac{P'(1/w)}{P(1/w)} = \mathop{K}_{j=1}^{2m} \frac{c_j w}{1}$$

where $c_j > 0$, $j = 1, 2, \ldots, 2m$ since $(-1)^{j-1} a_j > 0$ for the same values of j. Since (3.13) is of the form (3.10), we are done with the first part of our proof.

We now prove the sufficiency of (3.10). Suppose

$$\frac{P'(1/w)}{P(1/w)} = \mathop{K}_{j=1}^{2m} \frac{d_j w}{1},$$

where the d_j are positive. Then

$$\frac{P'(-z^2)}{P(-z^2)} = \mathop{K}_{j=1}^{2m} \frac{-d_j z^{-2}}{1}$$

from which we obtain

(3.14) $$-\frac{2zP'(-z^2)}{P(-z)^2} = -2 \operatorname*{K}_{j=1}^{2m} \frac{-d_j}{z} .$$

Using an equivalence transformation we can write (3.14) in the form

$$\frac{\frac{d}{dz}[P(-z^2)]}{P(-z^2)} = \operatorname*{K}_{j=1}^{2m} \frac{1}{a_j z} ,$$

where $(-1)^{j-1}a_j > 0$, $j = 1, 2, \ldots, 2m$. Hence, by Theorem 2.1,

$$\frac{\frac{d}{dz}[P(-z^2)]}{P(-z^2)} = \sum_{j=1}^{2m} \frac{m_j}{a-x_j} ,$$

where the x_j are real and distinct and the m_j are positive. This implies that $P(-z^2)$ has 2m real and distinct zeros. But, if r is a zero of $P(-z^2)$, so is $-r$. Therefore, exactly m of the $-x_j^2$ are real, negative, and distinct zeros of $P(z)$. With this our proof is complete.

<u>Example</u> 3.3: If

$$P(z) = (z+1)^2(z+2) = z^3 + 4z^2 + 5z + 2,$$

then

$$P'(z) = 3z^2 + 8z + 5$$

and

$$\frac{P'(1/z)}{P(1/z)} = \frac{3z+8z^2+5z^3}{1+4z+5z^2+2z^3}$$

$$= \frac{3z}{1} + \frac{4z/3}{1} + \frac{z/6}{1} + \frac{3z/2}{1} .$$

The latter continued fraction is of the form (3.10) with $m = 2$ and $d_k > 0$, $k = 1, 2, 3, 4$. Hence, according to Theorem 3.3, $P(z)$ should have all its roots real and negative with two of them distinct, which of course is accurate.

Theorem 3.4: *Let* P *be a polynomial of degree* $n \geq 1$ *with real coefficients and suppose*

$$(3.15)\qquad \frac{P'(z)}{P(z)} = \frac{1}{a_1 z+b_1} + \frac{1}{a_2 z+b_2} + \dots + \frac{1}{a_m z+b_m},$$

where $1 \leq m \leq n$; $a_k, b_k \in \mathbb{R}$, $a_k \neq 0$, $k = 1, 2, \dots, m$. *Let*

$$q = \sum_{k=1}^{m} (1 - \operatorname{sgn}((-1)^{k-1} a_k))/2$$

Then P *has exactly* q *distinct pairs of conjugate complex zeros and* $m - 2q$ *distinct real zeros.*

Proof: Let ε be a positive real number which will be restricted in size later. With the aid of an equivalence transformation, it follows from (3.15) that

$$(3.16)\qquad \frac{i\varepsilon P'(iz)}{P(iz)} = \mathop{\mathrm{K}}_{k=1}^{m} \frac{1}{c_k z+id_k},$$

where

$$c_k = (-1)^{k-1} a_k/\varepsilon,\ d_k = (-1)^k \varepsilon b_k/\varepsilon,\ k \text{ odd}$$

$$c_k = (-1)^{k-1} \varepsilon a_k,\ d_k = (-1)^k \varepsilon b_k,\ k \text{ even}.$$

Now let $A_m(z)$ and $B_m(z)$ denote the mth numerator and denominator, respectively, of the continued fraction (3.16). It is well known that

P'/P has a representation of the form

$$P'(z)/P(z) = \sum_{k=1}^{q} \frac{m_k}{z-z_k}, \tag{3.17}$$

where the z_k are the distinct zeros of P, $m_k > 0$ is the multiplicity of z_k, and $\sum_{k=1}^{q} m_k = n$. From this we obtain

$$\frac{i\varepsilon P'(iz)}{P(iz)} = \sum_{k=1}^{q} \frac{\varepsilon m_k}{z+iz_k}. \tag{3.18}$$

By Theorem 2.2, A_m and B_m have no common zero, degree $(A_m) = m-1$ and degree $(B_m) = m$, so it must be true that $q = m$ in (3.18). Hence, from (3.17) we have

$$\frac{P'(z)}{P(z)} = \frac{A(z)}{B(z)} = \sum_{k=1}^{m} \frac{m_k}{z-z_k}, \tag{3.19}$$

where

$$B(z) := \prod_{k=1}^{m} (z-z_k)$$

and $A(z)$ is defined appropriately. Let us assume without loss of generality that the z_k are indexed so that $z_1, z_2, \ldots, z_r$ are real and $z_{r+1}, z_{r+2}, \ldots, z_m$ have nonzero imaginary part. From (3.16) and (3.19) we now have

$$\frac{i\varepsilon A(iz)}{B(iz)} = \sum_{k=1}^{m} \frac{1}{c_k z+id_k} \tag{3.20}$$

and

$$1 + \frac{i\varepsilon A(iz)}{B(iz)} = 1 + \varepsilon \sum_{k=1}^{r} \frac{m_k}{z+iz_k} + \varepsilon \sum_{k=r+1}^{m} \frac{m_k}{z+iz_k}. \tag{3.21}$$

For each k, $k=1, 2, \ldots, m$, let $D_k(-iz_k, \rho_k)$ denote the open disk with center $-iz_k$, radius ρ_k, and boundary γ_k, where the ρ_k are chosen small enough so none of the closed disks $\overline{D}_k(-iz_k, \rho_k)$ intersect and so that $\overline{D}_k(-z_k, \rho_k)$ does not intersect the imaginary axis for $k = r+1, r+2, \ldots, m$. Let

$$M := \max\left|\sum_{k=r+1}^{m} \frac{m_k}{z+iz_k}\right| \quad \text{for } z \in \bigcup_{k=1}^{r} \overline{D}(-iz_k, \rho_k)$$

and let

$$M^* := \max\left|\frac{A(iz)}{B(iz)}\right| \quad \text{for } z \in \bigcup_{k=1}^{m} \gamma_k \,.$$

Now we restrict $\varepsilon > 0$ so that

$$\varepsilon \max(M, M^*) \leq 1/2.$$

Then, since $\varepsilon M^* \leq 1/2$, it follows from Rouché's Theorem that there is exactly one zero of

$$H(z) := B(iz) + i\varepsilon A(iz)$$

in each $D_k(-iz_k, \rho_k)$, $k = 1, 2, \ldots, m$. If $\mathrm{Im}(z_k) > 0$ ($\mathrm{Im}(z_k) < 0$), then the corresponding zero of $H(z)$ in $D(-iz_k, \rho_k)$ has positive (negative) real part. Since $A(iz)$ and $B(iz)$ have no common zero and since by (3.20)

$$\frac{i\varepsilon A(iz)}{B(iz)} = \frac{A_m(z)}{B_m(z)} \,,$$

it follows that there exists a nonzero complex constant c such that

$$H(z) = c[A_m(z) + B_m(z)]. \tag{3.22}$$

Therefore, by Theorem 2.2, $H(z)$ has no zero on the imaginary axis. Hence, if z_k is real, the corresponding zero of $H(z)$ in $D(-iz_k, \rho_k)$

cannot have zero real part. Since $\varepsilon M \le 1/2$, it follows from (3.21) that $Re(H(z)) > 0$ if $Re(z) > 0$ and $z \in D(-iz_k, \rho_k)$ with z_k real. Hence the corresponding zero of $H(z)$ must have negative real part if $Im(z_k) = 0$ and $z \in D(-iz_k, \rho_k)$. Thus we have shown that the number of zeros of $H(z)$ with positive real part is equal to the number of zeros of $B(z)$ with positive imaginary part. By (3.22) and Theorem 2.2, the number of zeros of H with positive real part is equal to

$$\sum_{k=1}^{m} (1 - \mathrm{sgn}(c_k))/2 = \sum_{k=1}^{m} (1 - \mathrm{sgn}((-1)^{k-1}a_k))/2 = q$$

since $c_k = (-1)^{k-1}a_k\varepsilon^{(-1)^k}$, $k = 1, 2, \ldots, m$. Hence, noting that the coefficients of $B(z)$ are real, the number of nonreal zeros of $B(z)$ is $2q$; so that the number of real zeros of B is $m-2q$. Since B has the distinct zeros of P as its zeros, our proof of this theorem is now complete.

Examples 3.4: Suppose

$$P(z) = (z^2+1)(z-1)^2 = z^4 - 2z^3 + 2z^2 - 2z + 1.$$

Then after three divisions we obtain

$$\frac{P'(z)}{P(z)} = \frac{1}{z/4-1/8} + \frac{1}{16z+40} + \frac{1}{z/128-3/138}$$

which is an expansion of the form (3.15) with $m = 3$, $a_1 = 1/4$, $a_2 = 16$, and $a_3 = 1/128$. It follows that $q = 1$ and $m-2q = 1$ so that by Theorem 3.4 $P(z)$ has exactly 1 pair of conjugate complex zeros and 1 distinct real zero. A quick inspection of $P(z)$ tells us that this is indeed the case.

There are polynomials $Q(z)$ with real coefficients for which $Q'(z)/Q(z)$ does not have a representation of the form (3.15). For example

$$Q(z) = z^3 - 1$$

is such a polynomial. However, if we multiply $Q(z)$ by the simple factor $(z+1)$ to obtain

$$R(z) = (z+1)(z^3-1) = z^4 + z^3 - z - 1,$$

then

$$\frac{R'(z)}{R(z)} = \frac{1}{\frac{z}{4} + \frac{1}{16}} + \frac{1}{-\frac{64z}{3} + \frac{208}{3}} + \frac{1}{-\frac{3z}{512} - \frac{3}{256}} + \frac{1}{-\frac{512z}{3} - \frac{1024}{3}} .$$

Hence, according to Theorem 3.4, $R(z)$ has exactly 1 pair of conjugate complex zeros and 2 distinct real zeros. From this we easily deduce that $Q(z)$ has 1 pair of conjugate complex zeros and 1 real zero.

The last example suggests that, if the continued fraction test in Theorem 3.4 for determining the number of distinct real zeros of a real polynomial $P(z)$ breaks down, it might be practical to apply the test to $(z+\alpha)P(z)$ for some constant α, say α equal to an integer. Frank (see [1, Theorem 2.1]) has given necessary and sufficient conditions for the quotient of two polynomials to have a J-fraction expansion in terms of determinants whose entries are coefficients of the polynomials. This result of Frank may help lead us to a wise choice of a multiplicative factor in some cases.

<u>Theorem</u> 3.5: <u>Let</u> $P(z)$ <u>be a polynomial of degree</u> $n \geq 1$ <u>with real coefficients and suppose</u>

$$\frac{P'(1/z)}{P(1/z)} = \frac{a_1 z}{1} + \frac{a_2 z}{1} + \dots + \frac{a_{2m-1} z}{1} + \frac{a_{2m} z}{1}, \tag{3.30}$$

<u>where</u> $1 \leq m \leq n$ <u>and the</u> a_k <u>are nonzero real numbers</u>. <u>Let</u>

$$A := \sum_{k=1}^{m} \left(1 - \prod_{j=1}^{2k-1} \operatorname{sgn}(a_j)\right)/2$$

$$B := \sum_{k=1}^{m} \left(1 - \prod_{j=1}^{2k} \operatorname{sgn}(a_j)\right)/2. \tag{3.31}$$

<u>Then</u>

$B - A =$ <u>the number of distinct positive real zeros of</u> $P(z)$.

$m - A - B =$ <u>the number of distinct negative real zeros of</u> $P(z)$.

$A =$ <u>the number of pairs of conjugate complex zeros of</u> $P(z)$.

<u>Proof</u>: From (3.30), by equivalence transformations, we obtain

$$\frac{P'(z)}{P(z)} = \frac{a_1}{z} + \frac{a_2}{1} + \frac{a_3}{z} + \frac{a_4}{1} + \dots + \frac{a_{2m-1}}{z} \frac{a_{2m}}{1}$$

(3.32)

$$= \frac{1}{b_1 z} + \frac{1}{b_2} + \frac{1}{b_3 z} + \frac{1}{b_4} + \dots + \frac{1}{b_{2m-1} z} + \frac{1}{b_{2m}},$$

where, taking $a_0 := 1$,

$$b_{2k-1} = \prod_{j=1}^{k} \frac{a_{2j-2}}{a_{2j-1}}, \quad b_{2k} = \prod_{j=1}^{k} \frac{a_{2j-1}}{a_{2j}}, \quad k = 1, 2, \dots, m. \tag{3.33}$$

Let

$$\Phi(z) := P(z^2), \quad \Psi(z) := P(-z^2). \tag{3.34}$$

Using (3.32) and equivalence transformations we obtain

$$\frac{\Phi'(z)}{\Phi(z)} = \frac{1}{b_1 z/2} + \frac{1}{2b_2 z} + \frac{1}{b_3 z/2} + \frac{1}{2b_4 z} + \dots + \frac{1}{b_{2m-1} z/2} + \frac{1}{2b_{2m} z} \tag{3.35}$$

(3.36)

$$\frac{\Psi'(z)}{\Psi(z)} = \frac{1}{b_1 z/2} + \frac{1}{(-2b_2 z)} + \frac{1}{b_3 z/2} + \frac{1}{(-2b_4 z)} + \dots + \frac{1}{b_{2m-1} z/2} + \frac{1}{(-2b_{2m} z)}.$$

The number α of distinct real zeros of $Q(z)$ is equal to twice the number of distinct positive real zeros of $P(z)$. According to Theorem 3.4

$$\alpha = 2m - 2\left(\sum_{k=1}^{m}(1 - \operatorname{sgn}(b_{2k-1}))/2 + \sum_{k=1}^{m}(1 + \operatorname{sgn}(b_{2k}))/2\right)$$

$$= 2\left(\sum_{k=1}^{m}(1 - \operatorname{sgn}(b_{2k}))/2 - \sum_{k=1}^{m}(1 - \operatorname{sgn}(b_{2k-1}))/2\right.$$

$$= 2\,(B - A),$$

where B and A are given by (3.31). The number ß of distinct real zeros of $\psi(z)$ is equal to twice the number of distinct negative real zeros of P(z). Again according to Theorem 3.4

$$ß = 2m - 2\left(\sum_{k=1}^{m}(1 - \mathrm{sgn}(b_{2k-1}))/2 + \sum_{k=1}^{m}(1 - \mathrm{sgn}(b_{2k}))/2\right)$$

$$= 2m - 2(A + B),$$

where A and B are defined by (3.31). Hence the number of distinct complex zeros of P(z) is

$$m - (B - A + m - A - B) = 2A,$$

so there are A pairs of conjugate complex zeros of P. This completes our proof.

<u>Examples</u> 3.5: For our first example we choose

$$P(z) = 1.44z^6 - 5.76z^5 - 4.8z^4 + 19.2z^3 + 23.2z^2 + 7.04z + 0.64$$

It is not difficult to calculate that

$$\frac{P'(1/z)}{P(1/z)} = \frac{6z}{1} - \frac{2z/3}{1} - \frac{5z}{1} + \frac{29z/4}{1} + \frac{35z/261}{1} + \frac{9z/29}{1}$$

which is an expansion of the form (3.30) with m = 3. It is easy to calculate that A = 0 and B = 1. Hence, according to Theorem 3.5, P(z) has B - A = 1 distinct positive real zeros, m - A - B = 2 distinct negative real zeros and A = 0 distinct pairs of conjugate complex zeros. We can look at a factored form of P(z)

$$P(z) = 0.16(z+1)^2 (3z^2 - 9z - 2)^2$$

to see that this zero information is accurate.

For our second example we choose

$$P(z) = z^3 + 1$$

Then the regular δ-fraction expansion of $P'(1/z)/P(1/z)$ is given by

$$\frac{P'(1/z)}{P(1/z)} = \frac{3z}{1-z} + \frac{z}{1-z} + \frac{z}{1} + \frac{z}{1} - \frac{z}{1} + \frac{z}{1} - \frac{z}{1}\ .$$

Since $\delta_1 = \delta_2 = 1$ in this representation, it is not of the form (3.30). Though the zero behavior of $P(z)$ may be buried some way in the elements of this continued fraction, we as yet do not know how to unscramble this information. However, we try a method of attack similar to the one used in the preceding example section which is to multiply $P(z)$ by the factor $(z+1)$ to obtain

$$Q(z) = (z+1)P(z) = (z+1)(z^3 + 1) = z^4 + z^3 + z + 1.$$

It then turns out that

$$\frac{Q'(1/z)}{Q(1/z)} = \frac{4z}{1} + \frac{z/4}{1} + \frac{3z/4}{1} + \frac{4z}{1} - \frac{6z}{1} + \frac{z}{1}$$

which is an expansion of the form (3.30) with $m = 3$. For this continued fraction we obtain $A = B = 1$, so by Theorem 3.5 $Q(z)$ has no positive real zeros, 1 distinct negative real zero, and 1 pair of conjugate complex zeros. Clearly from this information about $Q(z)$ we can easily deduce the desired real zero information about $P(z)$.

4. Stability and Half Plane Zero Counting Criteria.

In this section we state and prove four two part theorems. The first part of each theorem gives sufficient conditions in terms of a continued fraction for a polynomial of degree $n \geq 1$ to have m zeros with negative real part and $n-m$ zeros with positive real part. The second part of each theorem gives necessary and sufficient conditions in terms of a continued fraction for a polynomial to be a stable polynomial. The test continued fractions used in Theorem 4.1 are finite J-fractions, whereas those used in the other three theorems are terminating regular δ-fractions of various types.

Theorem 4.1 which deals with polynomials with complex coefficients is essentially a combination of two results of Wall and Frank (see [14], Theorems 47.1 and 48.1). However, we give a new proof of part (A) of this theorem and our proof of part (B) is a modification of the proof of Theorem 7.32 presented by Jones and Thron [7] in their book on continued fractions. Theorem 4.3 is our δ-fraction counterpart to Theorem 4.1. We would like to suggest that a δ-fraction test for stability is at least aesthetically better than a J-fraction test in the following sense. A polynomial $P(z)$ may be determined unstable because the J-fraction expansion of an associated rational function $T(z)$ does not exist, whereas $P(z)$ is unstable by a δ-fraction test because the δ-fraction expansion of $T(z)$ exists but is not of the correct form.

Theorems 4.2 and 4.4 deal with polynomials with real coefficients. They show what effects different rational functions associated with a given polynomial have on the corresponding δ-fraction whose elements bear the desired zero information about $P(z)$. As we might suspect, the test continued fractions are simpler in form in the real case.

Let $P(z)$ be the given polynomial in any one of the theorems in this section. Though we shall not do so in this paper, it can be proved (using a result of Frank [1, Theorem 5.1]) that there exists a $\delta > 0$ such that for all $\eta \in (-\delta, 0)$ the test fraction in part (A) exists and is of the right form for giving the number of zeros of $P(z+\eta)$ in each of the half planes $\mathrm{Re}(z) > 0$ and $\mathrm{Re}(z) < 0$. For the sake of length we will not expand in this paper on all the ramifications of this result. Nor, for the same reason, will we expand on how one can obtain other desirable zero information from the results in this section by applying them to appropriately transformed versions of the given polynomial. We will give examples after the proof of each theorem illustrating the results just proved.

Theorem 4.1: *Let*

$$P(z) = \sum_{k=0}^{n} c_{n-k} z^{n-k}$$

be a polynomial of degree $n \geq 1$ *with complex coefficients*

$$c_{n-k} = \alpha_{n-k} + i\beta_{n-k}, \ \alpha_n \neq 0, \ \beta_n = 0, \ k = 0, 1, \ldots, n.$$

Let

$$Q(z) = \sum_{k=1}^{n} d_{n-k} z^{n-k},$$

where $d_{n-k} = \alpha_{n-k}$ or $i\beta_{n-k}$ *according as to whether* k *is odd or even, respectively*. (A) *If*

$$\frac{Q(z)}{P(z) - Q(z)} = \mathop{K}_{k=1}^{n} \frac{1}{a_k z + i b_k}, \tag{4.1}$$

where the a_k *and* b_k *are real*, $b_k \equiv 0$ if $\beta_{n-k} \equiv 0$, m *of the* a_k *are positive and* $n-m$ *are negative, then* m *of the zeros of* $P(z)$ *have negative real part and* $n-m$ *have positive real part*. (B) $P(z)$ *is a stable polynomial iff* $Q(z)/(P(z) - Q(z))$ *has an expansion of the form* (4.1), *where*

$$a_k, \ b_k \in \mathbb{R}, \ a_k > 0, \ k = 1, 2, \ldots, n.$$

Proof: Let $A_n(z)$ and $B_n(z)$ be the nth numerator and denominator, respectively, of the continued fraction (4.1). By Theorem 2.2, degree $(A_n(z)) = n-1$, degree $(B_n(z)) = n$, and $A_n(z)$ and $B_n(z)$ have no common zeros. It follows that $Q(z)$ and $P(z) - Q(z)$ have no common zero, so that the same is true for $P(z)$ and $Q(z)$. Hence there exists a complex constant $c \neq 0$ such that

$$Q(z) = c\, A_n(z),\ P(z) - Q(z) = c\, B_n(z)$$

which lead to

$$P(z) = c(A_n(z) + B_n(z)).$$

Part (A) now follows from Theorem 2.2.

We shall now prove part (B). Let

$$T(z) := \frac{Q(z)}{P(z) - Q(z)}$$

and suppose $T(z)$ has an expansion of the form (4.1), where $a_k, b_k \in \mathbb{R}$, $a_k > 0$, $k = 1, 2, \ldots, n$. Then by part A, $P(z)$ is stable.

Conversely, suppose P is stable. Then P and its paraconjugate P^* have no common zeros. It is easily verified that

$$Q(z) = \frac{P(z) + (-1)^{n-1} P^*(z)}{2},$$

so

$$T(z) = \frac{P(z) + (-1)^{n-1} P^*(z)}{P(z) - (-1)^{n-1} P^*(z)}. \tag{4.2}$$

Since

$$T^*(z) = -T(z),$$

we have that $T(z)$ is para-odd.

If we write $P(z)$ in the form

$$P(z) = c_n \prod_{k=1}^{n} (z - z_k),\ c_n \in \mathbb{R},$$

where the z_k are the zeros of $P(z)$, then

$$P^*(z) = (-1)^n c_n \prod_{k=1}^{n} (z + \bar{z}_k).$$

Since $z_k < 0$ for $k = 1, 2, \ldots, n$, it is true that

$$|z+\overline{z}\ | < |z-z_k|$$

for the same values of k. Hence

$$|P^*(z)| < |P(z)|.$$

Using this inequality and (4.2), it is easy to see that $T(z)$ is a positive rational function. Since the degree of $P(z)$ exceeds the degree of Q, it is also true that

$$\lim_{z\to\infty} T(z) = 0. \tag{4.3}$$

The facts that $T(z)$ is a positive para-odd rational function satisfying (4.3) now guarantee, by [7, Theorem 7.30], that $T(z)$ has a representation of the form

$$T(z) = \mathop{K}_{k=1}^{m} \frac{1}{a_k z + ib_k}, \tag{4.4}$$

where a_k, $b_k \in \mathbb{R}$, $a_k > 0$, $k = 1, 2, \ldots, m$, for some integer m. Noting that the denominator of $T(z)$ is of degree n, it follows from the recurrence formulas for the denominators of the continued fraction (4.4) that $m = n$.

If the coefficients of $P(z)$ are real, then $T(z)$ is an odd function. Hence, if $T(z)$ is represented by a continued fraction of the form (4.1), then

$$-T(z) = \mathop{K}_{k=1}^{n} \frac{1}{-a_k z - ib_k}$$

and

$$T(-z) = \mathop{K}_{k=1}^{n} \frac{1}{-a_k z + ib_k}.$$

The uniqueness of the expansion (4.1) now guarantees that $b_k = 0$, $k = 1, 2, \ldots, n$. This concludes our proof of Theorem 4.1.

Examples 4.1:

We give two examples illustrating Theorem 4.1. In our first example we let P(z) be the complex polynomial

$$P(z) = z^3 + (4+2i)z^2 + (4+5i)z + (1+3i).$$

Then

$$Q(z) = 4z^2 + 5iz + 1$$

and

$$\frac{Q(z)}{P(z) - Q(z)} = \frac{1}{\frac{z}{4} + \frac{3}{16}i} + \frac{1}{\frac{64}{75}z + \frac{208}{375}i} + \frac{1}{\frac{1875}{1024}z + \frac{1125}{1024}i}.$$

Since the J-fraction is of length 3 and the coefficients of z in its partial denominators are all positive, P(z) is a stable polynomial by Theorem 4.1.

In our second example we let

$$P(z) = z^5 + 3z^4 - 9z^3 + 27z^2 - 32z + 30.$$

Then

$$Q(z) = 3z^4 + 27z^2 + 30$$

and

$$\frac{Q(z)}{P(z) - Q(z)} = \frac{1}{z/3} + \frac{1}{-z/6} + \frac{1}{-9z/10} + \frac{1}{-4z/3} + \frac{1}{-z/2}.$$

Since the J-fraction on the right of the last equation is of length 5 and there are 4 negative coefficients and 1 positive coefficient of z in its partial denominators, it follows from Theorem 4.1 that P(z) has 1 negative real zero and 4 zeros with positive real part.

Theorem 4.2: *Let*

$$P(z) = \sum_{k=0}^{n} a_{n-k} z^{n-k}$$

be a polynomial of degree $n \geq 1$ *with real coefficients, and let*

$$r(z) := z\left(\sum_{k=0}^{[(n-1)/2]} a_{2k+1} z^k\right) / \left(\sum_{k=0}^{[n/2]} a_{2k} z^k\right) . \tag{4.5}$$

(A) *If* $r(z)$ *has a terminating regular* δ*-fraction expansion of the form*

$$r(z) = \frac{d_1 z}{1} + \frac{d_2 z}{1} + \dots + \frac{d_n z}{1} , \tag{4.6}$$

where $d_k \in \mathbb{R}$, $d_k \neq 0$, $k = 1, 2, \dots, n$, *and*

$$\varepsilon_k := \prod_{j=1}^{k} \operatorname{sgn} d_j , \quad k = 1, 2, \dots, n, \tag{4.7}$$

then the number m *of positive* ε_k *is the number of zeros of* $P(z)$ *with negative real part and* $n-m$ *is the number with positive real part.*
(B) $P(z)$ *is a stable polynomial iff* $r(z)$ *has a representation of the form* (4.6), *where* $d_k > 0$, $k = 1, 2, \dots, n$.

Proof: Assume $r(z)$ has an expansion of the form (4.6), where $d_k \neq 0$, $k = 1, 2, \dots, n$. Then, using (4.6) and an equivalence transformation, we obtain

$$zr(1/z^2) = \frac{d_1}{z} + \frac{d_2}{z} + \dots + \frac{d_n}{z} . \tag{4.8}$$

By another equivalence transformation, we can write (4.8) in the form

$$zr(1/z^2) = \frac{1}{b_1 z} + \frac{1}{b_2 z} + \dots + \frac{1}{b_n z} \tag{4.9}$$

where

$$b_1 = 1/d, \quad b_{2k} = \prod_{j=1}^{k} d_{2j-1}/d_{2j}, \quad k = 1, 2, \dots, [n/2]$$

$$b_{2k+1} = 1/(d_{2k+1} b_{2k}), \quad k = 1, 2, \dots, [(n-1)/2].$$

If ε_k is given by (4.7), it is not difficult to verify that

$$\varepsilon_k = \operatorname{sgn}(b_k), \quad k = 1, 2, \ldots, n.$$

Now let $R(z)$ be the polynomial defined by

$$R(z) = z^n P(1/z).$$

Then it is easy to check that

$$(4.10) \qquad zr(1/z^2) = \frac{Q(z)}{R(z) - Q(z)} := T(z),$$

where $T(z)$ is the test rational function in Theorem 4.1 for $R(z)$. Since $T(z)$ has the expansion (4.9), it now follows from Theorem 4.1 that if m of the ε_k are positive and $n-m$ are negative then m of the zeros of $R(z)$ have negative real part and $n-m$ have positive real part. Thus, since $P(z)$ and $R(z)$ have the same left and right half plane zero distributions, we have that part (A) is true and also that $P(z)$ is stable if the d_k are positive in (4.6).

To finish our proof of part (B), let us now assume that $P(z)$ is stable. Then $R(z)$ is stable. So by (4.10) and Theorem 4.1 we have

$$(4.11) \qquad zr(1/z^2) = \frac{1}{c_1 z} + \frac{1}{c_2 z} + \cdots + \frac{1}{c_n z},$$

where the c_k are positive.

From (4.11) we use a substitution and an equivalence transformation to derive

$$(4.12) \qquad r(z^2) = \frac{z^2}{c_1} + \frac{z^2}{c_2} + \cdots + \frac{z^2}{c_n}.$$

Finally, from (4.12), after replacing z^2 by z, we obtain

$$(4.13) \qquad r(z) = \frac{d_1^* z}{1} + \frac{d_2^* z}{1} + \cdots + \frac{d_n^* z}{1},$$

where

$$d_1^* = 1/c_1,\ d_k^* = 1/(c_{k-1}c_k),\ k = 2, \ldots, n .$$

Since (4.13) is of the form (4.6) and the d_k^* are positive, our proof of part B is now complete.

Examples 4.2: Let

$$P(z) = z^4 - 9z^3/4 + z^2 - 9z/4 + 5/2.$$

Then it can be shown that the J-fraction expansion of the associated rational function $Q(z)/(P(z) - Q(z))$ does not exist, so $P(z)$ is unstable by that Theorem. However, Theorem 4.1 does not tell us anything else directly about the zeros of $P(z)$. So we turn to Theorem 4.2 for some possible help. The associated rational function $r(z)$ in Theorem 4.2 is given by

$$r(z) = \frac{z(-9/4 - 9z/4)}{5/2 + z + z^2} .$$

The regular δ-fraction expansion (which exists because $r(z)$ is analytic at $z = 0$) is given by

$$r(z) = \frac{-9z/10}{1} - \frac{3z/5}{1} + \frac{5z/3}{1} - \frac{2z/3}{1} .$$

This is an expansion of the form (4.6) with $n = 4$ and $d_1 < 0$, $d_2 < 0$, $d_3 > 0$, $d_4 < 0$. Hence, $\varepsilon_1 = -1$, $\varepsilon_2 = 1$, $\varepsilon_3 = 1$, and $\varepsilon_4 = -1$. Thus Theorem 4.2 tells us that $P(z)$ has 2 zeros with negative real part and 2 zeros with positive real part.

We now apply Theorem 4.2 to the simple polynomial

$$P(z) = (z-1)(z+1)^2 = z^3 + z^2 - z - 1.$$

Hence, $r(z) = z$, so that the δ-fraction expansion of $r(z)$ is

$$r(z) = \frac{z}{1}\ .$$

Since this δ-fraction is not of length 3, $P(z)$ is unstable by Theorem 4.2, but this theorem does not tell us directly how the zeros are distributed with respect to the left and right half planes. We will now show how Theorem 4.2 can be used indirectly to give more information about the zeros of $P(z)$. The regular δ-fraction expansion of $P'(1/z)/P(1/z)$ is given by

$$\frac{P'(1/z)}{P(1/z)} = \frac{3z}{1} + \frac{z/3}{1} + \frac{8z/3}{1} - \frac{3z}{1}\ .$$

Hence, by Theorem 3.2, $P(z)$ has 2 distinct zeros and both are real. So we can apply Laguerre's Theorem from section 1 with $\phi(w) = 1+w$ to $P(z)$ to obtain

$$Q(z) = -1 - 2z + 3z^2 + 4z^3\ .$$

For this polynomial, $r(z)$ of Theorem 4.2 is given by

$$r(z) = z(-2 + 4z)/(-1 + 3z).$$

Simple calculations will show that

$$r(z) = \frac{2z}{1} - \frac{z}{1} - \frac{2z}{1}$$

which is of the form (4.6). Part (A) of Theorem 4.2 now tells us that $Q(z)$ has 2 negative real zeros and 1 positive real zero. Since Theorem 3.2 told us there are only two distinct real zeros of Q, we now know with the aid of Laguerre's Theorem that $P(z)$ has exactly two distinct zeros, both real, one positive and one negative.

Theorem 4.3: *Let*

$$P(z) = \sum_{k=0}^{n} a_{n-k} z^{n-k}$$

be a polynomial of degree $n \geq 1$ *with complex coefficients*

$$a_{n-k} = \alpha_{n-k} + i\beta_{n-k},\ \alpha_n \neq 0,\ \beta_n = 0,\ k = 0, 1, \ldots, n.$$

Let

(4.14) $$R(z) := \sum_{k=0}^{n} a_{n-k} z^k,\ Q(z) := \sum_{k=1}^{n} b_{n-k} z^k,$$

where $b_{n-k} = \alpha_{n-k}$ *or* $i\beta_{n-k}$ *according as to whether* k *is odd or even, respectively.* (A) *If* $Q(z)/R(z)$ *can be expressed as a terminating regular* δ*-fraction of the form*

(4.15)

$$\frac{Q(z)}{R(z)} = \frac{(d_1/\sigma_0)z}{1} + \frac{d_1\sigma_1 z}{1} - \frac{(d_2/\sigma_1)z}{1} + \frac{d_2\sigma_2 z}{1} - \cdots - \frac{(d_n/\sigma_{n-1})z}{1} \quad \frac{d_n\sigma_n z}{1},$$

where

$$\sigma_0 = 1,\ \sigma_k \in \mathbb{C},\ \mathrm{Re}(\sigma_k - 1/\sigma_{k-1}) = 0,\ \beta_{n-k} \equiv 0 \Rightarrow \mathrm{Im}(\sigma_k) \equiv 0$$

(4.16)

$$d_k \in \mathbb{R},\ d_k \neq 0,\ k = 1, 2, \ldots, n,$$

then the number m *of positive* d_k *is the number of zeros of* $P(z)$ *with negative real part and* $n-m$ *is the number with positive real part.*
(B) $P(z)$ *is a stable polynomial iff* $Q(z)/R(z)$ *has an expansion of the form* (4.15) *meeting conditions* (4.16) *and the condition* $d_k > 0$, $k = 1, 2, \ldots, n$.

Proof: Let the polynomial $\hat{Q}(z)$ be defined by

$$\hat{Q}(z) := \sum_{k=1}^{n} b_{n-k} z^{n-k}$$

In view of Theorem 4.1, to prove Theorem 4.3 it is sufficient to prove that $Q(z)/R(z)$ has a representation of the form (4.15) with (4.16) holding iff $\hat{Q}/(P - \hat{Q})$ has a representation of the form

$$(4.17) \qquad \frac{\hat{Q}(z)}{P(z) - \hat{Q}(z)} = \frac{1}{a_1 z + ib_1} + \frac{1}{a_2 z + ib_2} + \dots + \frac{1}{a_n z + ib_n},$$

where

$$a_k \in \mathbb{R},\ b_k \in \mathbb{R},\ a_k \neq 0,\ a_k = 1/d_k,\ k = 1, 2, \dots, n.$$

So suppose first that $Q(z)/R(z)$ has an expansion (4.15) meeting conditions (4.16). Then by contracting (4.15) to its even part we obtain

(4.18)

$$\frac{Q(z)}{R(z)} = \frac{(d_1/\sigma_0)z}{1+d_1\sigma_1 z} + \frac{d_1 d_2 z^2}{1+d_2(\sigma_2 - 1/\sigma_1)z} + \frac{d_2 d_3 z^2}{1+d_3(\sigma_3 - 1/\sigma_2)z} + \dots + \frac{d_{n-1} d_n z^2}{1+d_n(\sigma_n - 1/\sigma_{n-1})z}.$$

After replacing z by $1/z$, noting that $\sigma_0 = 1$, and using an equivalence transformation we derive from (4.18) that

$$(4.19) \qquad \frac{Q(1/z)}{R(1/z)} = \frac{1}{z/d_1 + \sigma_1} + \frac{1}{z/d_2 + (\sigma_2 - 1/\sigma_1)} + \dots + \frac{1}{z/d_n + (\sigma_n - 1/\sigma_{n-1})}.$$

Since

$$(4.20) \qquad z^n Q(1/z) = \hat{Q}(z),\ z^n R(1/z) = P(z)$$

it follows from (4.19) that

(4.21)

$$\frac{\hat{Q}(z)}{P(z) - \hat{Q}(z)} = \frac{1}{z/d_1 + \sigma_1 - 1} + \frac{1}{z/d_2 + (\sigma_2 - 1/\sigma_1)} + \dots + \frac{1}{z/d_n + (\sigma_n - 1/\sigma_{n-1})}$$

But $\mathrm{Re}(\sigma_1) = 1$, so (4.21) is of the form (4.17) where $a_k = 1/d_k$, $k = 1, 2, \ldots, n$.

Now suppose $\hat{Q}/(P(z)-\hat{Q}(z))$ has a representation of the form (4.17), where $a_k, b_k \in \mathbb{R}$, $a_k \neq 0$, $k = 1, 2, \ldots, n$. Then

$$(4.22) \qquad \frac{\hat{Q}(z)}{P(z)} = \frac{1}{a_1z+1+ib_1} + \frac{1}{a_2z+ib_2} + \cdots + \frac{1}{a_nz+ib_n}.$$

By (4.20) and an equivalence transformation, it follows from (4.22) that

$$(4.23) \qquad \frac{Q(z)}{R(z)} = \frac{z}{a_1+z(1+ib_1)} + \frac{z^2}{a_2+ib_2z} + \frac{z^2}{a_3+ib_3z} + \cdots + \frac{z^2}{a_n+ib_nz}.$$

If we set $\rho_0 = 1$, $\rho_k = ib_k + 1/\rho_{k-1}$, $k = 1, 2, \ldots, n$, then (4.23) can be extended to

$$(4.24) \qquad \frac{Q(z)}{R(z)} = \frac{z}{a_1} + \frac{\rho_1 z}{1} - \frac{z/\rho_1}{a_2} + \cdots + \frac{\rho_{n-1}z}{1} - \frac{z/\rho_{n-1}}{a_n} + \frac{\rho_n z}{1}.$$

By an equivalence transformation, (4.24) can be written as

(4.25)
$$\frac{Q(z)}{R(z)} = \frac{(d_1/\rho_0)}{1} + \frac{d_1\rho_1 z}{1} - \frac{(d_2/\rho_1)z}{1} + \frac{d_2\rho_2 z}{1} - \cdots - \frac{(d_n/\rho_{n-1})z}{1} + \frac{d_n\rho_n z}{1},$$

where $d_k = 1/a_k$, $k = 1, 2, \ldots, n$. If we now set $\sigma_k \equiv \rho_k$, then (4.25) is of the form (4.15) and meets conditions (4.16). With this our proof of Theorem 4.3 is complete.

<u>Example</u> 4.3: Let

$$P(z) = z^3 + (-1+i)z^2 - 2iz + (1-i)$$

and let $Q(z)$, $R(z)$ be its associated polynomials in Theorem 4.3. Then

$$Q(z) = -z - 2iz^2 + z^3$$

$$R(z) = 1 + (-1+i)z - 2iz^2 + (1-i)z^3$$

and

$$\frac{Q(z)}{R(z)} = -\frac{z}{1} - \frac{z(1+i)}{1} - \frac{(1-i)z/2}{1} + \frac{(1-i)z/2}{1} + \frac{(1+i)z}{1} - \frac{(1-i)z}{1}$$

$$= -\frac{z}{1} - \frac{z(1+i)}{1} - \frac{z/(1+i)}{1} + \frac{(1-i)z/2}{1} - \frac{-(1+i)z}{1} + \frac{-(1-i)z}{1}.$$

Hence, $Q(z)/R(z)$ has an expansion of the form (4.15) with

$$d_1 = -1,\ d_2 = 1,\ d_3 = -1;\ \sigma_0 = 1,\ \sigma_1 = 1+i,\ \sigma_2 = (1-i)/2,\ \sigma_3 = 1-i.$$

Thus, according to Theorem 4.3, $P(z)$ has 1 zero with negative real part and 2 zeros with positive real part.

Theorem 4.4: *Let*

$$P(z) = \sum_{k=0}^{n} a_{n-k} z^{n-k}$$

be a polynomial of degree $n \geq 1$ *with real coefficients. Let*

$$Q(z) = \sum_{k=0}^{[(n-1)/2]} a_{n-2k-1} z^{2k+1},\quad R(z) = \sum_{k=0}^{[n/2]} a_{n-2k} z^{2k},$$

and let q^+, q^- *denote the number of zeros of* $P(z)$ *with positive and negative real parts, respectively.* (A) *If* $Q(z)/P(z)$ *has a terminating regular* δ-*fraction of the form*

$$(4.26)\qquad \frac{Q(z)}{R(z)} = \frac{d_1 z}{1-z} + \frac{z}{1} - \frac{d_2 z}{1} + \frac{d_2 z}{1} - \frac{d_3 z}{1} + \frac{d_3 z}{1} - \cdots - \frac{d_n z}{1} + \frac{d_n z}{1},$$

where $d_k \in \mathbb{R}$, $d_k \neq 0$, $k = 1, 2, \ldots, n$, *and* m *denotes the number of positive* d_k, *then*

$$q^- = m, \qquad q^+ = n-m \text{ if } d_1 > 0$$

$$q^- = n-m-1, \; q^+ = m+1 \text{ if } d_1 < 0$$

(B) $P(z)$ <u>is a stable polynomial iff</u> $Q(z)/R(z)$ <u>has a representation of the form</u> (4.26), <u>where</u> $d_k > 0$, $k = 1, 2, \ldots, n$.

<u>Proof:</u> Let

$$\hat{Q}(z) = \sum_{k=0}^{[n-1/2]} a_{n-2k-1} z^{n-2k-1}, \quad \hat{R}(z) = \sum_{k=0}^{[n/2]} a_{n-2k} z^{n-2k},$$

and suppose

$$(4.27) \qquad \frac{\hat{Q}(z)}{\hat{R}(z)} = \frac{1}{a_1 z} + \frac{1}{a_2 z} + \cdots + \frac{1}{a_n z},$$

where $a_k \in \mathbb{R}$, $a_k \neq 0$, $k = 1, 2, \ldots, n$. Then by replacing z by $1/z$ in the continued fraction (4.27) and using an equivalence transformation it can be seen that

$$\frac{Q(z)}{R(z)} = \frac{z}{a_1} + \frac{z^2}{a_2} + \cdots + \frac{z^2}{a_n}$$

$$(4.28) \qquad = \frac{z/a_1}{1} + \frac{z^2}{b_2} + \frac{z^2}{b_3} + \cdots + \frac{z^2}{b_n},$$

where $b_k = a_k(a_1)^{(-1)^k}$, $k = 2, 3, \ldots, n$. By extending the continued fraction (4.28), we obtain

$$\frac{Q(z)}{R(z)} = \frac{z/a_1}{1-z} + \frac{z}{1} - \frac{z}{b_2} + \frac{z}{1} - \frac{z}{b_3} + \cdots + \frac{z}{1} - \frac{z}{b_n} + \frac{z}{1}$$

which, after an equivalence transformation, leads us to

$$(4.29) \qquad \frac{Q(z)}{R(z)} = \frac{d_1 z}{1-z} + \frac{z}{1} - \frac{d_2 z}{1} + \frac{d_2 z}{1} - \frac{d_3 z}{1} + \frac{d_3 z}{1} - \cdots - \frac{d_n z}{1} + \frac{d_n z}{1},$$

where $d_1 = 1/a_1$, $d_k = 1/b_k = (a_1)^{(-1)^{k+1}}/a_k$.

Conversely, suppose $Q(z)/R(z)$ has an expansion of the form (4.29), where now our conditions on the d_k are $d_k \in \mathbb{R}$, $d_k \neq 0$, $k = 1, 2, \ldots, n$. Then, after replacing z by 1/z in (4.29) and using an equivalence transformation, we arrive at

$$(4.30) \qquad \frac{\hat{Q}(z)}{\hat{R}(z)} = \frac{d_1}{z-1} + \frac{1}{1} - \frac{d_2}{z} + \frac{d_2}{1} - \cdots - \frac{d_n}{z} + \frac{d_n}{z} .$$

By contracting the continued fraction (4.30) we get

$$\frac{\hat{Q}(z)}{\hat{R}(z)} = \frac{d_1}{z} + \frac{d_2}{z} + \frac{d_2 d_3}{z} + \frac{d_3 d_4}{z} + \cdots + \frac{d_n d_{n-1}}{z}$$

which, through an equivalence transformation, can be put into the form

$$(4.31) \qquad \frac{\hat{Q}(z)}{\hat{R}(z)} = \frac{1}{a_1 z} + \frac{1}{a_1 z} + \cdots + \frac{1}{a_n z} ,$$

where $a_1 = 1/d_1$, $a_k = (d_1)^{(-1)^k}/d_k$, $k = 2, \ldots, n$. Now, since the expansion (4.27) implies (4.29) and the expansion (4.26) with $d_k \in \mathbb{R}$, $d_k \neq 0$, $k = 1, 2, \ldots, n$ implies (4.31), our theorem follows from Theorem 4.1.

Example 4.4: Let

$$P(z) = z^4 - z^3 - z^2 + 1$$

Then the associated polynomials Q(z) and R(z) in Theorem 4.4 are given by

$$Q(z) = -z; \quad R(z) = 1 - z^2 + z^4.$$

Using the algorithm for obtaining regular δ-fraction expansions of rational functions we obtain

$$\frac{Q(z)}{R(z)} = \frac{-z}{1-z} + \frac{z}{1} + \frac{z}{1} - \frac{z}{1} + \frac{z}{1} - \frac{z}{1} - \frac{z}{1} + \frac{z}{1}$$

The latter expansion is of the form (4.26) with

$$d_1 = -1,\ d_2 = -1,\ d_3 = -1,\ d_4 = 1.$$

Clearly $m = 1$ since only $d_4 > 0$. Since $d_1 < 0$, we have

$$q^- = n - m - 1 = 4 - 1 - 1 = 2;\ q^+ = m + 1 = 1 + 1 = 2.$$

Thus, by Theorem 4.4, $P(z)$ has 2 zeros with negative real part and 2 zeros with positive real part. If we desire more knowledge about the zeros of $P(z)$ we can appeal to Theorem 3.5. It can be verified that

$$\frac{P'(1/z)}{P(1/z)} = \frac{4z}{1} - \frac{z/4}{1} - \frac{11z/4}{1} + \frac{20z/11}{1} - \frac{34z/55}{1}$$

$$+ \frac{341z/170}{1} - \frac{1265z/11594}{1} - \frac{374z/341}{1}.$$

By Theorem 3.5 we now know that $P(z)$ has 2 distinct positive real zeros and 1 pair of conjugate complex zeros. From our information derived from Theorem 4.4 we know that these complex zeros have negative real part. Thus we have illustrated how a combination of our results in this paper can be used to pin down the zero behavior of a polynomial.

REFERENCES

1. Evelyn Frank, On the zeros of polynomials with complex coefficients, Bull. Amer. Math. Soc., 52 (1946) 144-157.

2. J. Grommer, Ganze transcendente Functionen mit lauter reellen Nullstellen, Jour. für Math., 44 (1914) 212-238.

3. Peter Henrici, Applied and Computational Complex Analysis, Vol. 1, Power Series, Integration, Conformal Mapping, and Location of Zeros, Wiley, New York, 1974.

4. Peter Henrici, Applied and Computational Complex Analysis, Vol. 2, Special Functions, Integral Transforms, Asymptotics, and Continued Fractions, Wiley, New York, 1977.

5. William B. Jones and Allan Steinhardt, Digital filters and continued fractions, Analytic Theory of Continued Fractions, Lecture Notes in Mathematics 932, Springer-Verlag, New York (1982), 129-151.

6. William B. Jones and Allan Steinhardt, Applications of Schur fractions to digital filtering and signal processing, Rational Approximation and Interpolation, (eds. P. Graves-Morris, E.B. Saff, R.S. Varga) Lecture Notes in Mathematics 1105 (1984), 210-226.

7. William B. Jones and W.J. Thron, Continued Fractions: Analytic Theory and Applications, Encyclopedia of Mathematics and its Applications, Vol. II, Addison-Wesley, Reading, MA, 1980, distributed now by Cambridge U. Press, New York.

8. L.J. Lange, δ-Fraction expansions of analytic functions, SIAM J. Math. Anal., 14 (1983) 323-368.

9. Oscar Perron, Die Lehre von den Kettenbrüchen, Band II, Teubner, Stuttgart, 1957.

10. Joseph W. Rogers, Location of roots of polynomials, SIAM Review 25 (1983) 327-342.

11. E.C. Titchmarsh, The Theory of Functions, 2nd ed., Oxford U. Press, London 1939.

12. J.V. Uspensky, Theory of Equations, McGraw-Hill, New York, 1948.

13. H.S. Wall, Polynomials whose zeroes have negative real parts, Amer. Math. Monthly, 52 (1945) 308-322.

14. H.S. Wall, Analytic Theory of Continued Fractions, Van Nostrand, 1948.

A MULTI-POINT PADÉ APPROXIMATION PROBLEM

Olav Njåstad
Department of Mathematics
University of Trondheim-NTH
N-7034 Trondheim
Norway

1. Introduction

In [7] and [8] we studied systems of sequences $\{c_j^{(i)}\}$ which may in a natural way be said to be positive definite. We introduced orthogonal <u>R-functions</u> $Q_n(z)$ and their associated R-functions $P_n(z)$, and used them to investigate an extended Hamburger moment problem. (R-functions are rational functions having no poles in the extended complex plane outside a given finite set of real numbers $\{a_1,\ldots,a_p\}$.) For details see Section 2. We shall now show that the Padé approximant of type (n-1,n) in the situation described below is exactly $-\frac{P_n(z)}{Q_n(z)}$.

Analogous results are known for one-point Padé approximants of type (n-1,n) (see e.g. [1, Section 5.3]), and also for two-point Padé approximants of type (n-1,n), see e.g., [9].

A general multi-point Padé approximation problem is the following (see [5], [6], [10], cf. also [2]): For a pair (m,n) of natural numbers, let there be given $m+n+1$ points $\beta_k^{(m,n)}$, $k=1,\ldots,m+n+1$, in the complex plane, and $m+n+1$ formal power series

$$L_{m,n,k}(z) = \sum_{j=0}^{\infty} \alpha_j^{(m,n,k)} \cdot (z-\beta_k^{(m,n)})^j, \quad k=1,\ldots,m+n+1.$$

The rational function $\frac{A_m(z)}{B_n(z)}$ is called the <u>Padé approximant of type (m,n)</u> if A_m and B_n are polynomials such that $\deg A_m \leq m$, $\deg B_n \leq n$, and

$$A_m(z)-B_n(z)\cdot L_{m,n,k}(z) = \prod_{s=1}^{m+n+1} (z-\beta_s^{(m,n)}) \cdot \sum_{j=0}^{\infty} d_j^{(m,n,k)}(z-\beta_k^{(m,n)})^j,$$

$k=1,\ldots,m+n+1$.

(In [5], [6] is considered the case that the series $L_{m,n,k}$ are Taylor series expansions of an analytic function f. See also [3], where the case that the points $\beta_k^{(m,n)}$ are given as initial sections of fixed sequence $\{\beta_k : k=1,2,\ldots\}$, is treated. Cf. [1].)

Let $a_1,\dots,a_p$ be given real numbers. Every natural number n can be written in a unique way as $n = p\cdot q_n + r_n$, $1 \leq r_n \leq p$. Let the $2n$ points $\beta_i^{(n-1,n)}$, $i = 1,\dots,2n$, be determined as follows: $2q_n+2$ of the points are equal to a_i, for $i = 1,\dots,r_n-1$, $2q_n$ of the points are equal to a_i, for $i = r_n,\dots,p$, and 2 of the points are equal to ∞. (Note that $(r_n-1)(2q_n+2)+(p-r_n+1)\cdot 2q_n+2 = 2n$.)

Let $\{c_j^{(i)}: j = 1,2,\dots\}$, $i = 1,\dots,p$, be given sequences of real numbers. Let the power series $L_{n-1,n,i}$ associated with the points $\beta_i^{(n-1,n)}$ that are equal to a_i be

$$\sum_{j=0}^{\infty} c_{j+1}^{(i)}(z-a_i)^j \;, \quad i = 1,\dots,p \;,$$

and let the series associated with those two points that are equal to ∞ be $\{-\frac{1}{z}\}$. The situation can be expressed in terms of a given fixed sequence $\{\beta_i\}$ as follows: Let $\beta_1 = \beta_2 = \infty$, $\beta_3 = \beta_4 = a_1,\dots,\beta_{2p+1} = \beta_{2p+2} = a_p$, $\beta_{2p+3} = \beta_{2p+4} = a_1,\dots,\beta_{4p+1} = \beta_{4p+2} = a_p$, $\beta_{4p+3} = \beta_{4p+4} = a_1$, and so on until $2n$ is reached. Then the series $-\frac{1}{z}$ is associated with β_1 and β_2, the series $\sum_{j=0}^{\infty} c_{j+1}^{(1)}(z-a_1)^j$ is associated with $\beta_3,\beta_4,\beta_{2p+3},\beta_{2p+4},\beta_{2p+3},\beta_{2p+4},\dots$, and so on. The condition for $\frac{A_{n-1}(z)}{B_n(z)}$ (where $\deg A_{n-1} \leq n-1$, $\deg B_n \leq n$) to be the Padé approximant of type $(n-1,n)$ can be written as follows (we write q for q_n and r for r_n):

$$A_{n-1}(z) - B_n(z)\cdot\sum_{j=0}^{2q+1} c_{j+1}^{(i)}(z-a_i)^j = \sum_{j=2q+2}^{\infty} \gamma_j^{(i)}(z-a_i)^j \quad \text{for } i = 1,\dots,r-1,$$

$$A_{n-1}(z) - B_n(z)\cdot\sum_{j=0}^{2q-1} c_{j+1}^{(i)}(z-a_i)^j = \sum_{j=2q}^{\infty} \gamma_j^{(i)}(z-a_i)^j \quad \text{for } i = r,\dots,p \;,$$

$$A_{n-1}(z) - B_n(z)\cdot(-\frac{1}{z}) = \gamma_{n-2}z^{n-2} + \gamma_{n-3}z^{n-3} + \dots + \gamma_0 + \gamma_{-1}z^{-1} + \dots \;.$$

2. Preliminaries on R-functions

As in Section 1, let $a_1,\dots,a_p$ be given real numbers. Let $\mathcal{R}$ denote the linear space consisting of all functions of the form

$$R(z) = \alpha_0 + \sum_{i=1}^{p}\sum_{j=1}^{N_i} \frac{\alpha_{ij}}{(z-a_i)^j} \;, \quad \alpha_0,\alpha_{ij} \in \mathbb{C}.$$

Elements of $\mathcal{R}$ are called R-functions. We note that a function R belongs to $\mathcal{R}$ iff it can be written in the form $R(z) = \frac{A(z)}{B(z)}$, where B is a polynomial with all its zeros among the points $a_1,\ldots,a_p$, and A is a polynomial of $\deg A \leq \deg B$.

Every natural number n has a unique decomposition $n = q_n \cdot p + r_n$, $1 \leq r_n \leq p$. We write $q = q_n$, $r = r_n$. We denote by $\mathcal{R}_n$ the space of R-functions of the form

$$R(z) = \alpha_0 + \sum_{i=1}^{r} \sum_{j=1}^{q+1} \frac{\alpha_{ij}}{(z-a_i)^j} + \sum_{i=r+1}^{p} \sum_{j=1}^{q} \frac{\alpha_{ij}}{(z-a_i)^j} .$$

Let $\{c_j^{(i)}: j = 1,2,\ldots\}$, $i = 1,\ldots,p$, be given sequences of real numbers. Let Φ be the linear functional defined on $\mathcal{R}$ by

$$\Phi\left(\alpha_0 + \sum_{i=1}^{p} \sum_{j=1}^{N_i} \frac{\alpha_{ij}}{(z-a_i)^j}\right) = 1 + \sum_{i=1}^{p} \sum_{j=1}^{N_i} \alpha_{ij}\, c_j^{(i)}.$$

This functional gives rise to a bilinear form $\langle .,. \rangle$ on $\mathcal{R}\times\mathcal{R}$, defined by $\langle A,B \rangle = \Phi(A\cdot B)$. (Note that $\mathcal{R}$ is closed under multiplication.) We shall assume that the sequences $\{c_j^{(i)}\}$ have the property that the bilinear form is positive definite, i.e. $\Phi(R^2) > 0$ when $R(z) \not\equiv 0$, $\alpha_0, \alpha_{ij} \in \mathbf{R}$.

By applying the Gram-Schmidt orthonormalization procedure to the sequence

$$\{1, \frac{1}{(z-a_1)}, \ldots, \frac{1}{(z-a_p)}, \frac{1}{(z-a_1)^2}, \ldots \frac{1}{(z-a_p)^2}, \frac{1}{(z-a_1)^3}, \ldots\}$$

(in the indicated order) we obtain an orthonormal sequence $\{Q_n\}$ of R-functions. We note that $Q_n \in \mathcal{R}_n - \mathcal{R}_{n-1}$, and that $\{Q_0,\ldots,Q_{n-1}\}$ is a base for $\mathcal{R}_{n-1}$ so that $\langle Q_n, A\rangle = 0$ for every $A \in \mathcal{R}_{n-1}$. Furthermore Q_n may be written as $Q_n(z) = \frac{V_n(z)}{N_n(z)}$, where

$$N_n(z) = (z-a_1)^{q+1} \ldots (z-a_r)^{q+1} (z-a_{r+1})^q \ldots (z-a_p)^q,$$

and $V_n(z)$ is a polynomial of degree at most n. (Actually $\deg V_n = n$ or $\deg V_n = n-1$, see [7].)

The function $t \to \dfrac{Q_n(t)-Q_n(z)}{t-z}$ is an R-function. We define

$$P_n(z) = \Phi_t\left(\frac{Q_n(t)-Q_n(z)}{t-z}\right).$$

(We write Φ_t to emphasize that the functional Φ is applied to $\frac{Q_n(t)-Q_n(z)}{t-z}$ as a function of t.) We can write $P_n(z) = \frac{U_n(z)}{N_n(z)}$, where $U_n(z)$ is a polynomial, $\deg U_n \leq \deg V_n - 1$. The function P_n is called the <u>R-function associated with</u> Q_n.

For more information on R-functions, see [7], [8].

Our aim is to show that $-\frac{U_n(z)}{V_n(z)}$ (or equivalently: $-\frac{P_n(z)}{Q_n(z)}$) is the Padé approximant of type $(n-1,n)$ for the situation desribed in Section 1. The argument which will be given is related to the proof of a similar result in [4] for two-point approximants. It is possible to give an alternative proof by utilizing Gaussian quadrature formulas associated with the orthogonal R-functions. (For such quadrature formulas, see [7], [8].)

3. The interpolation result

<u>Theorem</u>. <u>The following formulas are valid</u>:

$$U_n(z) + V_n(z) \cdot \sum_{j=0}^{2q+1} c_{j+1}^{(i)}(z-a_i)^j = \sum_{j=2q+2}^{\infty} \gamma_j^{(i)}(z-a_i)^j \quad, \text{ for } i = 1,\ldots,r-1,$$

$$U_n(z) + V_n(z) \cdot \sum_{j=0}^{2q-1} c_{j+1}^{(i)}(z-a_i)^j = \sum_{j=2q}^{\infty} \gamma_j^{(i)}(z-a_i)^j \quad, \text{ for } i = r,\ldots,p,$$

$$U_n(z) - \frac{1}{z} V_n(z) = \sum_{j=2}^{\infty} \gamma_{n-j}\, z^{n-j}.$$

<u>Proof</u>: Set $\delta_i = q+1$ when $i < r$, $\delta_i = q$ when $i \geq r$. The function $\psi_i(t) = \frac{1}{(t-z)}\left[\frac{(z-a_i)^{\delta_i} - (t-a_i)^{\delta_i}}{(t-a_i)^{\delta_i}}\right]$ belongs to R_{n-1}. Therefore we may write

$$\Phi_t\left(\frac{1}{(t-z)}\left[\frac{(z-a_i)^{\delta_i}}{(t-a_i)^{\delta_i}} Q_n(t) - Q_n(z)\right]\right) = \Phi_t\left(\frac{1}{(t-z)}\left[Q_n(t)-Q_n(z)\right]\right)$$

$$+ \Phi_t\left(\psi_i(t) \cdot Q_n(t)\right) = P_n(z) + \langle \psi_i, Q_n \rangle = P_n(z).$$

We set $\Delta_i(z) = P_n(z)+Q_n(z)\cdot\sum_{j=0}^{2\delta_i-1} c_{j+1}^{(i)}(z-a_i)^j$, $i=1,\ldots,p$.

By using the expression for $P_n(z)$ above and the definition of Φ we get

$$\Delta_i(z) = \Phi_t\left(\frac{1}{(t-z)}\left[\frac{(z-a_i)^{\delta_i}}{(t-a_i)^{\delta_i}} Q_n(t)-Q_n(z)\right]\right) + Q_n(z)\cdot\sum_{j=0}^{2\delta_i-1}(z-a_i)^j\Phi_t\left(\frac{1}{(t-a_i)^{j+1}}\right)$$

$$= \Phi_t\left(\frac{1}{(t-z)}\left[\frac{(z-a_i)^{\delta_i}}{(t-a_i)^{\delta_i}} Q_n(t)-Q_n(z)\right]\right) + \Phi_t\left(Q_n(z)\left[\frac{(t-a_i)^{2\delta_i}-(z-a_i)^{2\delta_i}}{(t-a_i)^{2\delta_i}(t-z)}\right]\right)$$

$$= \Phi_t\left(\frac{1}{(t-z)}\left[\frac{(z-a_i)^{\delta_i}(t-a_i)^{\delta_i}Q_n(t)-(z-a_i)^{2\delta_i}Q_n(z)}{(t-a_i)^{2\delta_i}}\right]\right).$$

We set $E_i(z,t) = \frac{1}{(t-z)}\left[\frac{(z-a_i)^{\delta_i}(t-a_i)^{\delta_i}Q_n(t)-(z-a_i)^{2\delta_i}Q_n(z)}{(t-a_i)^{2\delta_i}}\right]$.

The function $t \to E_i(z,t)$ is an R-function. It follows from the definition of δ_i and the form of Q_n that $\Delta_i(z) = \Phi_t(E_i(z,t))$ has a Taylor series expansion about a_i, $i=1,\ldots,p$, of the following form:

$$\Delta_i(z) = g_{q+1}^{(i)}(z-a_i)^{q+1} + \ldots , \text{ for } i=1,\ldots,r-1,$$

$$\Delta_r(z) = g_{q-1}^{(r)}(z-a_r)^{q-1} + \ldots ,$$

$$\Delta_i(z) = g_q^{(i)}(z-a_i)^q + \ldots , \text{ for } i=r+1,\ldots,p.$$

Furthermore the function $t \to \frac{z\,Q_n(t) - t\,Q_n(z)}{z(t-z)}$ is an R-function, and we find by using the properties of Q_n described in Section 2 and the definition of Φ (we set $\Delta(z) = P_n(z) - \frac{1}{z}Q_n(z)$):

$$\Delta(z) = \Phi_t\left(\frac{Q_n(t)-Q_n(z)}{t-z} - \frac{1}{z}Q_n(z)\right)$$

$$= \Phi_t\left(\frac{z\,Q_n(t) - t\,Q_n(z)}{(t-z)\cdot z}\right) = g_{-2}\,z^{-2} + \ldots .$$

Multiplying the expressions for $\Delta(z)$, $\Delta_i(z)$ on both sides by $N(z)$, taking into account that the factor $(z-a_i)$ occurs $q+1$ times for $i \leq r$ and q times for $i > r$, we get the desired formulas.

□

References

1. Baker, George A. Jr. and Graves-Morris, Peter, Padé Approximants I, II, Encyclopedia of Mathematics and its Applications 13, 14. Addison Wesley (1980).

2. Chui, C.K., Recent results on Padé approximants and related problems, Approximation Theory II, Editors: C.C. Lorentz, C.K. Chui and L.L. Schumaker, Academic Press (1976), 79-115.

3. Gallucci, M.A., and William B. Jones, Rational approximations corresponding to Newton series (Newton-Padé approximants), Journal of Approximation Theory 17 (1976), 366-392.

4. Jones, William B., Olav Njåstad and W.J. Thron, Szegö polynomials and Perron-Caratheodory fractions, in preparation.

5. Karlsson, J., Rational interpolation and best rational approximation, Preprint University of Umeå, Department of Mathematics, No. 1, 1974.

6. Karlsson, J., Rational Hermite interpolation procedures, Preprint University of Umeå, Department of Mathematics, No. 2, 1974.

7. Njåstad, Olav, An extended Hamburger moment problem, Proc. Edinb. Math. Soc., (Series II) 28 (1985), 167-183.

8. Njåstad, Olav, Unique solvability of an extended Hamburger moment problem, submitted.

9. Njåstad, Olav and W.J. Thron, The theory of sequences of orthogonal L-polynomials, Padé approximants and continued fractions, Editors: Haakon Waadeland and Hans Wallin, Det Kongelige Norske Videnskabers Selskab, Skrifter, No. 1 (1983), 54-91.

10. Warner, D.D., Hermite interpolation with rational functions, Ph.D. Thesis, University of California at San Diego, 1974.

$\hat{J}$-FRACTIONS AND STRONG MOMENT PROBLEMS

A. Sri Ranga
Universidade de São Paulo
Instituto de Ciências Matematicas de São Carlos
Av. Dr. Carlos Botelho, 1465
13560 - São Carlos, SP, Brazil

1. Introduction

A $\hat{J}$-fraction is a continued fraction of the form

$$\frac{a_1}{Z+b_1} - \frac{\{a_2(Z)\}^2}{Z+b_2} - \frac{\{a_3(Z)\}^2}{Z+b_3} - \frac{\{a_4(Z)\}^2}{Z+b_4} - \dots, \qquad [1.1]$$

in which the functions $a_n(Z)$ are given by $a_n Z^{\delta_n}$. Here, each δ_n is either 0 or 1, and all $a_n, b_n \in \mathbb{C}$.

If all the $a_n(Z)$ in this fraction are constants "a_n", then this fraction is a J-fraction. We can, therefore, consider the set of all J-fractions as a sub-class of the set of all $\hat{J}$-fractions. This means that some of the properties of J-fractions may extend to certain other $\hat{J}$-fractions, and also that $\hat{J}$-fractions which are not J-fractions may possess various other properties.

For example, there exists a class of $\hat{J}$-fractions (which can be called positive definite $\hat{J}$-fractions) with many properties analogous to those of the positive definite J-fractions of Wall [1948]. A detailed study of these $\hat{J}$-fractions was made by the author [1983].

Secondly, it is well known that under appropriate conditions a J-fraction corresponds to a formal power series expansion

$$f_0(Z) = \frac{C_0}{Z} + \frac{C_1}{Z^2} + \frac{C_2}{Z^3} + \dots,$$

where $C_n \in \mathbb{C}$ for $n \geq 0$. McCabe [1983] has shown that a $\hat{J}$-fraction may be made to correspond to the formal power series $f_0(Z)$ and to the second formal power series

$$g_0(Z) = -C_{-1} - C_{-2}Z - C_{-3}Z^2 - \dots,$$

where $C_{-n} \in \mathbb{C}$ for $n \geq 1$, in such a way that the convergents of this fraction form a certain sawtooth sequence in the two-point Padé table of these two power series. In this article we undertake a detailed study of certain $\hat{J}$-fractions, so that their correspondence properties and conditions of existence are fully understood.

Finally, the connections between the real J-fractions and the ordinary moment problems are well known. See for example Wall [1948]. Here we look at the connections between real $\hat{J}$-fractions and strong moment problems. We do this by using the properties of the real $\hat{J}$-fractions to find conditions for the existence of solutions to these moment problems.

2. $\hat{J}$-fractions and Their Correspondence Properties

Let us consider first the regular $\hat{J}$-fraction, that is a $\hat{J}$-fraction where $a_{2n}(Z) = a_{2n}$ and $a_{2n+1}(Z) = a_{2n+1}Z$, for all $n \geq 1$. This fraction can be written as

$$\frac{a_1}{Z+b_1} - \frac{a_2^2}{Z+b_2} - \frac{a_3^2Z^2}{Z+b_3} - \frac{a_4^2}{Z+b_4} - \frac{a_5^2Z^2}{Z+b_5} - \dots .$$

This fraction can also be expressed in the equivalent form

$$\frac{\alpha_1}{Z+\beta_1} - \frac{\alpha_2}{Z+\beta_2} - \frac{\alpha_3Z^2}{(1+\alpha_3)Z+\beta_3} - \frac{\alpha_4}{Z+\beta_4} - \frac{\alpha_5Z^2}{(1+\alpha_5)Z+\beta_5} - \dots, \qquad [2.1]$$

provided $a_{2n+1} \neq 1$ for $n \geq 1$.

This latter form is much more convenient for the analysis, because in this case the denominator polynomials of its convergents are monic polynomials. We also refer to this latter continued fraction as a $\hat{J}$-fraction.

We denote the nth convergent of the $\hat{J}$-fraction [2.1] by $L_n(Z) = P_n(Z)/Q_n(Z)$. Then the numerator $P_n(Z)$ and the denominator $Q_n(Z)$ both satisfy the following three term relations.

$$R_{2n}(Z) = (Z+\beta_{2n})R_{2n-1}(Z) - \alpha_{2n}R_{2n-2}(Z),$$

$$R_{2n+1}(Z) = \{(1+\alpha_{2n+1})Z + \beta_{2n+1}\}R_{2n}(Z) - \alpha_{2n+1}Z^2R_{2n-1}(Z), \qquad [2.2a]$$

for $n \geq 1$, with initial conditions

$$Q_0(Z) = 1, \quad P_0(Z) = 0, \quad Q_1(Z) = Z + \beta_1 \quad \text{and} \quad P_1(Z) = \alpha_1. \qquad [2.2b]$$

From [2.2] it can be easily seen that all the $P_n(Z)$ and $Q_n(Z)$ are polynomials of the form

$$P_n(Z) = \alpha_1Z^{n-1} + \text{lower order terms}, \ n \geq 1, \qquad [2.3a]$$

and

$$Q_{2n}(Z) = Z^{2n} + \dots + q_0^{(2n)},$$

$$Q_{2n+1}(Z) = Z^{2n+1} + \dots + (1 + \alpha_{2n+1})q_0^{(2n)}Z + \beta_{2n+1}q_0^{(2n)}, \qquad [2.3b]$$

for all $n \geq 1$ where

$$q_0^{(2n)} = Q_{2n}(0) = \prod_{r=1}^{n} (\beta_{2r-1}\beta_{2r} - \alpha_{2r}),\ n \geq 1. \qquad [2.3c]$$

From [2.2] it also follows that

$$L_{2n}(Z)-L_{2n-2}(Z) = \frac{(Z+\beta_{2n})\alpha_{2n-1}\alpha_{2n-2}\ \cdots\ \alpha_2\alpha_1 Z^{2n-2}}{Q_{2n}(Z)Q_{2n-2}(Z)},$$

$$L_{2n+1}(Z)-L_{2n-1}(Z) = \frac{\{(1+\alpha_{2n+1})Z+\beta_{2n+1}\}\alpha_{2n}\alpha_{2n-1}\cdots\alpha_2\alpha_1 Z^{2n-2}}{Q_{2n+1}(Z)Q_{2n-1}(Z)}, \qquad [2.4]$$

$$L_{2n}(Z)-L_{2n-1}(Z) = \frac{\alpha_{2n}\alpha_{2n-1}\ \cdots\ \alpha_2\alpha_1 Z^{2n-2}}{Q_{2n}(Z)Q_{2n-1}(Z)},$$

for $n \geq 1$. Now, since $Q_n(Z)$ is a monic polynomial of degree n, expanding the right-hand sides of these equations in power of 1/Z yields

$$L_{2n}(Z) - L_{2n-2}(Z) = \gamma_{2n-2}^{(0)}(1/Z)^{2n-1} + O(1/Z)^{2n},$$

$$L_{2n+1}(Z) - L_{2n-1}(Z) = \gamma_{2n-1}^{(1)}(1/Z)^{2n+1} + O(1/Z)^{2n+2},$$

$$L_{2n}(Z) - L_{2n-1}(Z) = \gamma_{2n-1}^{(0)}(1/Z)^{2n+1} + O(1/Z)^{2n+2},$$

for all $n \geq 1$, where

$$\gamma_n^{(0)} = \alpha_{n+1}\alpha_n\ \cdots\ \alpha_2\alpha_1,$$

$$\gamma_{2n+1}^{(1)} = (1 + \alpha_{2n+3})\gamma_{2n+1}^{(0)}, \qquad n \geq 0. \qquad [2.5]$$

From this we realize that there exists a series in negative powers corresponding to the $\hat{J}$-fraction (2.1). If this power series is $f_0(Z)$ then

$$f_0(Z)-L_{2n}(Z)=\gamma_{2n}^{(0)}(1/Z)^{2n+1}+O(1/Z)^{2n+2},$$

$$f_0(Z)-L_{2n+1}(Z)=\gamma_{2n+1}^{(1)}(1/Z)^{2n+3}+O(1/Z)^{2n+4}, \qquad n \geq 0. \qquad [2.6]$$

By multiplying the above relations by $Q_{2n}(Z)$ and $Q_{2n+1}(Z)$, respectively, they can also be written as follows:

$$Q_{2n}(Z)f_0(Z)-P_{2n}(Z) = \gamma_{2n}^{(0)}/Z + O(1/Z)^2,$$

$$Q_{2n+1}(Z)f_0(Z)-P_{2n+1}(Z) = \gamma_{2n+1}^{(1)}/Z^2 + O(1/Z)^3, \qquad n \geq 0. \qquad [2.7]$$

We now consider the expansions of the expressions [2.4] in terms of powers of Z. For this, we first make the assumption

$$Q_{2n}(0) \neq 0, \quad \text{for } n \geq 1. \qquad [2.8]$$

This assumption is less restrictive than $Q_n(0) \neq 0$ since, as can be seen from [2.3b], $Q_{2n+1}(0)$ can be zero even when $Q_{2n}(0) \neq 0$. From [2.8] it immediately follows that

$$L_{2n}(Z)-L_{2n-2}(Z) = \frac{\beta_{2n}\alpha_{2n-1}\alpha_{2n-2}\cdots\alpha_2\alpha_1}{Q_{2n}(0)Q_{2n-2}(0)} Z^{2n-2} + O(Z)^{2n-1} \qquad [2.9]$$

and

$$\begin{aligned} Q_{2n-2}(Z)L_{2n}(Z)-P_{2n-2}(Z) &= \rho^{(0)}_{2n-2}Z^{2n-2} + O(Z)^{2n-1} \\ Q_{2n-1}(Z)L_{2n}(Z)-P_{2n-1}(Z) &= \rho^{(1)}_{2n-1}Z^{2n-2} + O(Z)^{2n-1}, \end{aligned} \qquad [2.10]$$

for all $n \geq 1$, where

$$\begin{aligned} \rho^{(0)}_{2n} &= \beta_{2n+2}\alpha_{2n+1}\alpha_{2n}\cdots\alpha_2\alpha_1/Q_{2n+2}(0), \\ \rho^{(1)}_{2n+1} &= \alpha_{2n+2}\alpha_{2n+1}\cdots\alpha_2\alpha_1/Q_{2n+2}(0), \end{aligned} \qquad n \geq 0. \qquad [2.11]$$

Relation [2.9] indicates that the $2n^{th}$ convergent of the fraction [2.1] corresponds to a power series about the origin. Hence, for this power series to be $g_0(Z)$, it follows from (2.10) that

$$\begin{aligned} Q_{2n}(Z)g_0(Z) - P_{2n}(Z) &= \rho^{(0)}_{2n}Z^{2n} + O(Z)^{2n+1}, \\ Q_{2n+1}(Z)g_0(Z) - P_{2n+1}(Z) &= \rho^{(1)}_{2n+1}Z^{2n} + O(Z)^{2n+1}, \end{aligned} \qquad n \geq 0. \qquad [2.12]$$

The equations [2.7] and [2.12] provide a system of equations in the coefficients of $Q_n(Z)$. Here, using the fact that $Q_n(Z)$ is a monic polynomial the following results are derived.

$$\begin{aligned} \gamma^{(0)}_{2n} &= H^{(-2n)}_{2n+1}/H^{(-2n)}_{2n}, \quad \gamma^{(1)}_{2n+1} = H^{(-2n)}_{2n+2}/H^{(-2n)}_{2n+1}, \\ \rho^{(0)}_{2n} &= H^{(-(2n+1))}_{2n+1}/H^{(-2n)}_{2n}, \quad \rho^{(1)}_{2n+1} = H^{(-(2n+1))}_{2n+2}/H^{(-2n)}_{2n+1}, \end{aligned} \qquad [2.13]$$

for $n \geq 0$, and

$$Q_{2n}(Z) = \frac{1}{H^{(-2n)}_{2n}} \begin{vmatrix} C_{-2n} & C_{-2n+1} & \cdots & C_0 \\ \vdots & \vdots & & \vdots \\ C_{-1} & C_0 & \cdots & C_{2n-1} \\ 1 & Z & \cdots & Z^{2n} \end{vmatrix}, \quad n \geq 1, \qquad [2.14a]$$

$$Q_{2n+1}(Z) = \frac{1}{H^{(-2n)}_{2n+1}} \begin{vmatrix} C_{-2n} & C_{-2n+1} & \cdots & C_1 \\ \vdots & \vdots & & \vdots \\ C_0 & C_1 & \cdots & C_{2n+1} \\ 1 & Z & \cdots & Z^{2n+1} \end{vmatrix}, \quad n \geq 0. \qquad [2.14b]$$

Here, the Hankel determinants $H^{(k)}_m$ are given by

$$H^{(k)}_{-1} = 0, \quad H^{(k)}_0 = 1,$$

$$H_m^{(k)} = \begin{vmatrix} C_k & C_{k+1} & \cdots & C_{k+m-1} \\ C_{k+1} & C_{k+2} & \cdots & C_{k+m} \\ \vdots & \vdots & & \vdots \\ C_{k+m-1} & C_{k+m} & \cdots & C_{k+2m-2} \end{vmatrix},$$

for all k and all positive m.

The condition required for obtaining the above results from [2.7] and [2.12] is

$$H_{2n+1}^{(-2n)} \neq 0 \quad \text{and} \quad H_{2n}^{(-2n)} \neq 0, \quad n \geq 0.$$

Thus, by remembering that [2.12] was arrived at under the assumption that $Q_{2n}(0) \neq 0$ for all $n \geq 1$, we find

$$H_{2n+1}^{(-2n)} \neq 0, \quad H_{2n}^{(-2n)} \neq 0, \quad \text{and} \quad H_{2n}^{(-(2n-1))} \neq 0, \; n \geq 0 \qquad [2.15]$$

as the required condition for the existence of the $\hat{J}$-fraction [2.1], with correspondence to the power series $f_0(Z)$ given by [2.6] and to the power series $g_0(Z)$ given by

$$g_0(Z)-L_{2n}(Z) = \frac{\rho_{2n}^{(0)}}{Q_{2n}(0)}Z^{2n} + O(Z)^{2n+1}, \quad n \geq 0.$$

Using the relations [2.3c], [2.5] and [2.11], we find that under [2.15] the coefficients α_r and β_r are uniquely determined by

$$\alpha_1 = C_0 \qquad \beta_1 = -C_{-1}/C_0,$$

$$\alpha_{2n}=\left[\frac{H_{2n}^{(-(2n-1))}}{H_{2n-1}^{(-(2n-2))}}\right]^2 \frac{H_{2n-2}^{(-(2n-2))}}{H_{2n}^{(-2n)}}, \quad \beta_{2n}=\frac{-H_{2n}^{(-(2n-1))}\; H_{2n-1}^{(-(2n-1))}}{H_{2n-1}^{(-(2n-2))}\; H_{2n}^{(-2n)}}$$

$$\alpha_{2n+1}=\frac{H_{2n+1}^{(-2n)}\; H_{2n-1}^{(-(2n-2))}}{\{H_{2n}^{(-(2n-1))}\}^2}, \qquad \beta_{2n+1}=\frac{-H_{2n+1}^{(-(2n-1))}\; H_{2n}^{(-2n)}}{H_{2n+1}^{(-2n)}\; H_{2n}^{(-(2n-1))}}, \qquad [2.16]$$

for all $n \geq 1$. To arrive at these results, we have made extensive use of the well-known Jacobi identity (Henrici [1974]).

So far as the correspondence of the odd convergents of the fraction [2.1] to the power series $g_0(Z)$ has not been looked at. From [2.12] it follows that, if $Q_{2n+1}(0) \neq 0$ for any particular value of n, then for that n, the convergent $L_{2n+1}(Z)$ must also correspond to 2n terms of the series $g_0(Z)$. Using [2.14b] the additional condition required for this is found to be $H_{2n+1}^{(-(2n-1))} \neq 0$. Thus the following theorem is established.

THEOREM 1. *For the given two formal power series* $f_0(Z)$ *and* $g_0(Z)$ *let the associated Hankel determinants satisfy condition* [2.15]. *Then there exists a unique* $\hat{J}$*-fraction of the form* [2.1], *with coefficients given by* [2.16], *such that its* $2n^{th}$ *convergent corresponds to exactly* 2n *terms of* $f_0(Z)$ *and at least* 2n *terms of* $g_0(Z)$, *while its* $(2n+1)^{th}$ *convergent corresponds to at least* (2n+2) *terms of* $f_0(Z)$. *In addition, if the Hankel determinants of* $f_0(Z)$ *and* $g_0(Z)$ *satisfy the condition*

$$H_{2n+1}^{(-(2n-1))} \neq 0, \quad n \geq 0,$$

then the $(2n+1)^{st}$ *convergent of this fraction also corresponds to exactly* 2n *terms of* $g_0(Z)$.

We now consider the case in which the coefficients of the expansions $f_0(Z)$ and $g_0(Z)$ are all real. Then, by including in [2.15] the additional restriction that the first two determinants are positive, it follows that all the α_r are positive and all the β_r are real. Furthermore, it follows from the Jacobi identity that this restriction also implies that the determinants $H_{2n}^{(-(2n-2))}$ are all positive. Hence, from [2.3] and [2.16] we note that this implies that $Q'_{2n+1}(0) \neq 0$ for all $n \geq 0$. Here, $Q'_n(Z)$ is the derivative of $Q_n(Z)$. Thus, we obtain

THEOREM 2. *If the coefficients of the formal power series expansions* $f_0(z)$ *and* $g_0(Z)$ *are real and if the associated Hankel determinants satisfy the condition*

$$H_{2n+1}^{(-2n)} > 0, \quad H_{2n}^{(-2n)} > 0, \text{ and } H_{2n}^{(-(2n-1))} \neq 0, \quad n \geq 0, \qquad [2.17]$$

then there exists a unique real $\hat{J}$*-fraction of the form* [2.1] *with positive partial numerators and such that its* $2n^{th}$ *convergent corresponds to exactly* 2n *terms of* $f_0(Z)$ *and at least* 2n - 1 *terms of* $g_0(Z)$.

Similarly, by considering the regular $\hat{J}$-fraction

$$\frac{\alpha_1^*}{Z+\beta_1^*} - \frac{\alpha_2^* Z^2}{(1+\alpha_2^*)Z+\beta_2^*} - \frac{\alpha_3^*}{Z+\beta_3^*} - \frac{\alpha_4^* Z^2}{(1+\alpha_4^*)Z+\beta_4^*} - \cdots, \qquad [2.18]$$

with

$$\alpha_1^* = \{C_{-1}\}^2/C_{-2}, \qquad \beta_1^* = -C_{-1}/C_{-2},$$

$$\alpha^*_{2n} = \frac{H_{2n}^{(-(2n))} H_{2n-2}^{(-(2n-2))}}{\{H_{2n-1}^{(-(2n-1))}\}^2}, \qquad \beta^*_{2n} = \frac{-H_{2n-1}^{(-2n)} H_{2n}^{(-(2n-1))}}{H_{2n-1}^{(-(2n-1))} H_{2n}^{(-2n)}},$$

[2.19]

$$\alpha^*_{2n+1} = \left\{\frac{H_{2n+1}^{(-(2n+1))}}{H_{2n}^{(-2n)}}\right\}^2 \frac{H_{2n-1}^{(-2n)}}{H_{2n+1}^{(-(2n+2))}}, \qquad \beta^*_{2n+1} = \frac{-H_{2n+1}^{(-(2n+1))} H_{2n}^{(-(2n+1))}}{H_{2n}^{(-2n)} H_{2n+1}^{(-(2n+2))}},$$

for all $n \geq 1$, we have

THEOREM 3. <u>If the Hankel determinants of the formal power series expansions</u> $f_0(Z)$ <u>and</u> $g_0(Z)$ <u>satisfy the condition</u>

$$H_{2n-1}^{(-2n)} \neq 0,\ H_{2n}^{(-2n)} \neq 0,\ \text{and}\ H_{2n-1}^{(-(2n-1))} \neq 0,\ n \geq 1,$$

<u>then there exists a unique $\hat{J}$-fraction of the form</u> [2.18] <u>with coefficients given by</u> [2.19]. <u>The denominator of the</u> $(2n+1)^{st}$ <u>convergent of this $\hat{J}$-fraction does not contain zero as one of its roots.</u>

<u>With the more restricted condition</u>

$$C_n,\ C_{-n-1} \in R,\ H_{2n+1}^{(-2n)} > 0,\ H_{2n}^{(-2n)} > 0,\ \text{and}\ H_{2n+1}^{(-(2n+1))} \neq 0, \qquad [2.20]$$

<u>for all</u> $n \geq 0$, <u>this fraction is real, has all its partial numerators positive and its</u> $2n^{th}$ <u>convergent corresponds to exactly</u> $2n$ <u>terms of</u> $f_0(Z)$ <u>and at least</u> $2n - 1$ <u>terms of</u> $g_0(Z)$, <u>while its</u> $(2n + 1)^{st}$ <u>convergent corresponds to exactly</u> $2n$ <u>terms of</u> $f_0(Z)$ <u>and at least</u> $2n + 2$ <u>terms of</u> $g_0(Z)$.

We note that for both regular real $\hat{J}$-fractions so far considered the condition

$$C_n,\ C_{-n-1} \in \mathbb{R},\ H_{2n+1}^{(-2n)} > 0 \text{ and } H_{2n}^{(-2n)} > 0,\ n \geq 0, \qquad [2.21]$$

appears as part of the required condition for their existence and correspondence. Under this condition it follows: If for any value of $r \geq 1$, $H_{2r}^{(-(2r-1))} = 0$ then

$$H_{2r-1}^{(-(2r-1))} \neq 0 \quad \text{and} \quad H_{2r+1}^{(-(2r+1))} \neq 0,$$

while if $H_{2r+1}^{(-(2r+1))} = 0$ then

$$H_{2r}^{(-(2r-1))} \neq 0 \quad \text{and} \quad H_{2r+2}^{(-(2r+1))} \neq 0.$$

Using these results we find that under the condition [2.21] alone, it is always possible to construct a real $\hat{J}$-fraction corresponding to both the power series $f_0(Z)$ and $g_0(Z)$. There may be more than one such

$\hat{J}$-fraction which can be constructed. To understand this we look at one possible construction.

Suppose that for some $r \geq 1$, $H_{2r}^{(-(2r-1))} = 0$ and $H_{2s}^{(-(2s-1))} \neq 0$ for $s = 1, 2, \ldots, r-1$. Then we can start the continued fraction with partial quotients as in the $\hat{J}$-fraction [2.1] and at the $(2r-1)^{th}$ stage switch to partial quotients as in the $\hat{J}$-fraction [2.18]. At the point of switching the continued fraction takes the form

$$\ldots - \frac{\alpha_{2r-3}Z^2}{(1+\alpha_{2r-3})Z+\beta_{2r-3}} - \frac{\alpha_{2r-2}}{Z+\beta_{2r-2}} - \frac{\alpha_{2r-1}^{(-)}Z^2}{(1+\alpha_{2r-1}^{(-)})Z+\beta_{2r-1}^{(-)}} \qquad [2.22a]$$

$$- \frac{\alpha_{2r}^{*}Z^2}{(1+\alpha_{2r}^{*})Z+\beta_{2r}^{*}} - \frac{\alpha_{2r+1}^{*}}{Z+\beta_{2r+1}^{*}} - \ldots .$$

Again if $H_{2m+1}^{(-(2m+1))} = 0$ for some $m > r$, then we switch back to partial quotients as in the $\hat{J}$-fraction [2.1] and at this stage the fraction takes the form

$$\frac{\alpha_{2m-1}^{*}}{Z+\beta_{2m-1}^{*}} - \frac{\alpha_{2m}^{*}Z^2}{(1+\alpha_{2m}^{*})Z+\beta_{2m}^{*}} - \frac{\alpha_{2m+1}^{(+)}}{Z+\beta_{2m+1}^{(+)}} - \frac{\alpha_{2m+2}}{Z+\beta_{2m+2}} - \ldots . \qquad [2.22b]$$

Such changeovers are made whenever they are necessary. It can be shown that the coefficients α_r, β_r, α_r^*, β_r^*, whenever they appear, take the values given by [2.16] and [2.19], and the coefficients $\alpha_{2s+1}^{(-)}$, $\beta_{2s+1}^{(-)}$, $\alpha_{2s+1}^{(+)}$, $\beta_{2s+1}^{(+)}$, which are used for the switching, take the values given by

$$\alpha_{2s+1}^{(-)} = \left\{\frac{H_{2s+1}^{(-(2s+1))}}{H_{2s}^{(-(2s-1))}}\right\}^2 \frac{H_{2s-1}^{(-(2s-2))}}{H_{2s+1}^{(-(2s+2))}},$$

$$\beta_{2s+1}^{(-)} = \frac{-H_{2s}^{(-2s)}H_{2s+1}^{(-(2s+1))}}{H_{2s}^{(-(2s-1))}H_{2s+1}^{(-(2s+2))}},$$

$$\alpha_{2s+1}^{(+)} = \frac{H_{2s-1}^{(-2s)}H_{2s+1}^{(-2s)}}{\left\{H_{2s}^{(-2s)}\right\}^2},$$

$$\beta_{2s+1}^{(+)} = \frac{-H_{2s+1}^{(-2s)}H_{2s}^{(-(2s+1))}}{H_{2s+1}^{(-(2s+1))}H_{2s}^{(-2s)}} + \frac{-H_{2s+2}^{(-(2s+1))}H_{2s}^{(-2s)}}{H_{2s+1}^{(-2s)}\,H_{2s+1}^{(-(2s+1))}}.$$

These results, and some further results which can be easily derived are summarized as follows:

THEOREM 4. *If the coefficients of the formal power series* $f_0(Z)$ *and* $g_0(Z)$ *are all real and if their associated Hankel determinants satisfy the condition*

$$H_{2n+1}^{(-2n)} > 0 \quad \text{and} \quad H_{2n}^{(-2n)} > 0 \quad \text{for } n \geq 1,$$

then there always exists a real $\hat{J}$*-fraction satisfying the following properties:*

1. *The partial numerators are positive,*
2. *The convergents correspond to both power series* $f_0(Z)$ *and* $g_0(Z)$.
3. *Zero is not a common root of the denominator polynomials of any two successive convergents,*
4. *If zero is a root of the denominator polynomial of one of the convergents, then it is a simple root.*

We consider as an example the two power series expansions $f_0(Z)$ and $g_0(Z)$ for which the coefficients C_n are given by

$$C_n = \int_{-\infty}^{\infty} t^n w(t)dt, \qquad n = \ldots, -2, -1, 0, 1, 2, \ldots,$$

where

$$w(t) = \{e/\sqrt{2\pi}\}e^{-(1/2)(t^2+1/t^2)}.$$

Using integration by parts, these coefficients can be shown to satisfy

$$C_0 = 1, \quad C_{2s+1} = 0, \quad C_{-s-2} = C_s, \; C_{2s+2} = (2s+1)C_{2s} + C_{2s-2},$$

for all $s \geq 0$.

Here, since $w(t)$ is a positive function in the interval $(-\infty, \infty)$, the associated Hankel determinants do satisfy the condition of Theorem 4. Hence, there must exist a real $\hat{J}$-fraction corresponding to the power series $f_0(Z)$ and $g_0(Z)$. We find such a fraction, which is

$$\frac{1}{Z} - \frac{1}{Z} - \frac{Z^2}{2Z} - \frac{1}{Z} - \frac{2Z^2}{3Z} - \frac{1}{Z} - \frac{3Z^2}{4Z} - \frac{1}{Z} - \cdot \qquad [2.23]$$

This is a regular $\hat{J}$-fraction of the form [2.1], with coefficients α_r, β_r taking the values

$$\alpha_1 = 1, \quad \alpha_{2r} = 1, \quad \alpha_{2r+1} = r, \text{ and } \beta_r = 0, \; r \geq 1.$$

Since $\beta_r = 0$ for all $r \geq 1$, then from [2.16] it follows that

$$H_{2n+1}^{(-(2n+1))} = 0 \quad \text{and} \quad H_{2n+1}^{(-(2n-1))} = 0 \text{ for } n \geq 0.$$

This indicates that [2.23] is the only $\hat{J}$-fraction of any of the forms that we have considered, which exists and corresponds to these series expansions.

3. $\hat{J}$-fractions and the Strong Moment Problem

The strong moment problem which was first considered by Jones, Thron and Waadeland [1980] can be stated as follows:

Given a doubly infinite sequence of real numbers $\{C_n\}_{n=-\infty}^{\infty}$, *find conditions to ensure the existence of a bounded nondecreasing function* $\phi(t)$ *in an interval* [a, b] *such that*

$$\int_a^b t^n d\phi(t) = C_n,\ n = \ldots, -2, -1, 0, 1, 2, \ldots . \qquad [3.1]$$

Such a function $\phi(t)$ is referred to as a solution of the strong moment problem.

This moment problem has now been solved by jones, Thron and Waadeland [1980] for the case where the interval [a, b] is the positive (or negative) half of the real axis, and by Jones, Njåstad and Thron [1984] for the case where it is the whole real axis. In these papers these two cases are called the strong Stieltjes moment problem and the strong Hamburger moment problem, respectively.

The strong Stieltjes moment problem was solved by making the problem equivalent to the existence of a certain corresponding T-fraction. On the other hand, the strong Hamburger moment problem was solved with the use of a certain sequence of functions which the authors called orthogonal Laurent polynomials. Jones, Njåstad and Thron [1983] gave a partial solution of the strong Hamburger moment problem using T-fractions.

Here we sketch a means of obtaining a complete solution to the above moment problem using $\hat{J}$-fractions. We also give conditions for the function ϕ to vanish outside a finite interval [a, b].

We first consider the regular $\hat{J}$-fraction [2.1]. For this fraction we obtain from the three term relations [2.2],

$$\begin{aligned} S_{2n}(Z) &= \alpha_{2n}\alpha_{2n-1} \cdots \alpha_2\alpha_1 Z^{2n-2}, \quad n \geq 1, \\ S_{2n+1}(Z) &= \alpha_{2n+1}\alpha_{2n} \cdots \alpha_2\alpha_1 Z^{2n}, \quad n \geq 1, \end{aligned} \qquad [3.2]$$

where $S_n = \{P_n(Z)Q_{n-1}(Z)-P_{n-1}(Z)Q_n(Z)\}$, and

$$\begin{aligned} T_{2n}(Z) &= \{Q_{2b-1}(Z)\}^2 + \alpha_{2n}T_{2n-1}(Z), \quad n \geq 1, \\ T_{2n+1}(Z) &= \{Q_{2n}(Z)\}^2 + \alpha_{2n+1}\{Q_{2n}(Z) - ZQ_{2n-1}(Z)\}^2 \\ &\quad + \alpha_{2n+1}\alpha_{2n}Z^2T_{2n-1}(Z), \quad n \geq 1, \end{aligned} \qquad [3.3]$$

where $T_n(Z) = \{Q_n'(Z)Q_{n-1}(Z) - Q_{n-1}'(Z)Q_n(Z)\}$. Hence, we have the theorem,

<u>THEOREM 5.</u> <u>If the $\hat{J}$-fraction [2.1] satisfies the condition</u>

$$\alpha_n,\ \beta_n \in \mathbb{R},\ \alpha_n > 0 \ \underline{\text{and}}\ Q_{2n}(0) \neq 0,\ \underline{\text{for}}\ n \geq 1, \qquad [3.4]$$

<u>then for any</u> $r > 1$ <u>the roots of the denominator polynomial</u> $Q_r(Z)$ <u>are all real, distinct and different from those of</u> $Q_{r-1}(Z)$. <u>In addition, if for any</u> a <u>and</u> b $(a < b)$ <u>in</u> $\mathbb{R}$ <u>the conditions</u>

$$(-1)^n Q_n(a) > 0 \quad \underline{\text{and}} \quad Q_n(b) > 0 \quad \underline{\text{for}}\ n \geq 1, \qquad [3.5]$$

<u>are satisfied then all the roots of</u> $Q_r(Z)$ <u>are inside</u> (a, b).

<u>Proof</u>: We note from [3.3] that under the condition [3.4] all $T_n(Z)$ are positive. Hence, as in the case of ordinary orthogonal polynomials (Szeg [1959]), this immediately leads to the first of the results of the theorem. Now, in $T_n(Z)$, using also the conditions of [3.5], we see that the largest and smallest roots of $Q_n(Z)$ are contained inside (a, b). Hence, the theorem is proved.

Now, from [3.2] it follows that under the condition [3.4], if $Z_r^{(n)}$ is a non-zero root of $Q_n(Z)$ then it is not a root of $P_n(Z)$, while if it is a zero it is also a root of $P_n(Z)$ $(n > 1)$. Thus, the quotient $P_n(Z)/Q_n(Z)$ must have a partial decomposition of the form

$$\frac{P_n(Z)}{Q_n(Z)} = \sum_{r=1}^{n} \frac{\ell_r^{(n)}}{Z - Z_r^{(n)}}, \qquad n \geq 1 \qquad [3.6]$$

where

$$\ell_r^{(n)} = P_n(Z_r^{(n)})/Q_n'(Z_r^{(n)}), \qquad r = 1, 2, \ldots, n.$$

Since $Z_r^{(n)}$ is a root of $Q_n(Z)$, it can be seen that $\ell_r^{(n)}$ can also be give an

$$\ell_r^{(n)} = S_n(Z_r^{(n)})/T_n(Z_r^{(n)}), \qquad r = 1, 2, \ldots, n.$$

This implies that for all $n \geq 1$,

$$\ell_r^{(n)} \geq 0 \quad \text{for } r = 1, 2, \ldots, n.$$

Since we have assumed that $Q_{2n}(0) \neq 0$ for all $n \geq 1$, equality here can hold only in the case of $\ell_r^{(2n+1)}$ when $Z_r^{(2n+1)} = 0$.

Furthermore, from [2.3] and [3.6] we find that

$$\sum_{r=1}^{n} \ell_r^{(n)} = \lim_{Z\to\infty} \frac{ZP_n(Z)}{Q_n(Z)} = C_0, \qquad n \geq 1.$$

Hence, if we define a sequence of step functions $\psi_n(t)$, by

$$\psi_n(t) = \begin{cases} 0, & \text{for } -\infty < t \leq Z_1^{(n)}, \\ \sum_{s=1}^{r} \ell_s^{(n)}, & \text{for } Z_r^{(n)} < t \leq Z_{r+1}^{(n)},\ r = 1, 2, \ldots, n-1, \\ C_0, & \text{for } Z_n^{(n)} < t < \infty, \end{cases}$$

for $n \geq 1$, then we can write

$$\frac{P_n(Z)}{Q_n(Z)} = \int_{-\infty}^{\infty} \frac{1}{Z - t}\, d\psi_n(t) = \int_a^b \frac{1}{Z - t}\, d\psi_n(t), \qquad [3.7]$$

where (a, b) is the interval inside which all the roots of the polynomials $Q_n(Z)$, $(n \geq 1)$ lie. This result follows from the definition of the Stieltjes integrals (see for example Widder [1952]).

Here, using the same reasoning as that of, for example, Jones, Thron and Waadeland [1980], the following result is arrived at.

THEOREM 6. *If the $\hat{J}$-fraction [2.1] satisfies the condition [3.4], then there exists a bounded non-decreasing function $\psi(t)$ (with infinitely many points of increase) in the interval $(-\infty, \infty)$ such that*

$$\lim_{n\to\infty} \frac{P_n(Z)}{Q_n(Z)} = \int_{-\infty}^{\infty} \frac{1}{Z - t}\, d\psi(t).$$

In particular, for any real numbers a, b satisfying the condition [3.5] the function $\psi(t)$ is constant outside the interval (a, b) and satisfies

$$\lim_{n\to\infty} \frac{P_n(Z)}{Q_n(Z)} = \int_a^b \frac{1}{Z - t}\, d\psi(t).$$

Let us now consider the given doubly infinite sequence $\{C_n\}_{n=-\infty}^{\infty}$ of real numbers. Using this we can construct two formal power series of the form $f_0(Z)$ and $g_0(Z)$. Hence, from Theorem 2 it follows that under condition [2.17] there exists a real $\hat{J}$-fraction of the form [2.1] which not only corresponds to these power series expansions but also satisfies the condition [3.4]. Hence, from the above theorem this real $\hat{J}$-fraction has a bounded non-decreasing function $\psi(t)$ (with infinitely many points of increase) in the interval $(-\infty, \infty)$ such that

$$\int_{-\infty}^{\infty} \frac{1}{Z - t}\, d\psi(t) = \frac{C_0}{Z} + \frac{C_1}{Z^2} + \frac{C_2}{Z^3} + \ldots, \qquad Z \sim \infty,$$

$$= -C_{-1} - C_{-2}Z - C_{-3}Z^2 - \ldots, \qquad Z \sim 0.$$

For any real numbers a, b if the denominator polynomials of this fraction also satisfies the condition [3.5] this function $\psi(t)$ is constant outside the interval (a, b) and satisfies

$$\int_a^b \frac{1}{Z-t}\, d\varphi(t) = \frac{C_0}{Z} + \frac{C_1}{Z^2} + \frac{C_2}{Z^3} + \ldots, \qquad Z \sim \infty,$$

$$= -C_{-1} - C_{-2}Z - C_{-3}Z^2 - \ldots, \qquad Z \sim 0.$$

From this the following is obtained.

THEOREM 7. *Given a doubly infinite sequence* $\{C_n\}_{n=-\infty}^{\infty}$ *of real numbers such that*

$$H_{2n+1}^{(-2n)} > 0, \quad H_{2n}^{(-2n)} > 0, \quad \text{and } H_{2n}^{(-(2n-1))} \neq 0, \quad n \geq 0.$$

Then there exists a bounded non-decreasing function $\varphi(t)$ *(with infinitely many points of increase) in the interval* $(-\infty, \infty)$ *such that*

$$C_m = \int_{-\infty}^{\infty} t^m d\varphi(t), \quad m = \ldots, -2, -1, 0, 1, 2, \ldots .$$

If in addition the conditions

$$\begin{vmatrix} C_{-2n} & C_{-2n+1} & \cdots & C_0 \\ \vdots & \vdots & & \vdots \\ C_{-1} & C_0 & \cdots & C_{2n-1} \\ 1 & a & \cdots & a^{2n} \end{vmatrix} > 0, \quad \begin{vmatrix} C_{-2n} & C_{-2n+1} & \cdots & C_0 \\ \vdots & \vdots & & \vdots \\ C_{-1} & C_0 & & C_{2n-1} \\ 1 & b & \cdots & b^{2n} \end{vmatrix} > 0,$$

$$\begin{vmatrix} C_{-2n} & C_{-2n+1} & \cdots & C_1 \\ \vdots & \vdots & & \vdots \\ C_0 & C_1 & \cdots & C_{2n+1} \\ 1 & a & \cdots & a^{2n+1} \end{vmatrix} < 0, \quad \begin{vmatrix} C_{-2n} & C_{-2n+1} & \cdots & C_1 \\ \vdots & \vdots & & \vdots \\ C_0 & C_1 & \cdots & C_{2n+1} \\ 1 & b & \cdots & b^{2n+1} \end{vmatrix}$$

also hold for $n \geq 0$, *then this function* $\varphi(t)$ *is constant outside* (a, b) *and satisfies*

$$C_m = \int_a^b t^m d\varphi(t), \quad m = \ldots, -2, -1, 0, 1, 2, \ldots .$$

The first of the conditions is a sufficient condition for the existence of a solution $\varphi(t)$ to the strong Hamburger moment problem. In this theorem, by taking $a = 0$ and $b = \infty$, we find

$$H_{2n+1}^{(-2n)} > 0, \quad H_{2n}^{(-2n)} > 0, \quad H_{2n}^{(-(2n-1))} > 0, \quad H_{2n+1}^{(-(2n-1))} > 0,$$

for $n \geq 0$, as a sufficient condition for the strong Stieltjes moment problem, and by taking $a = 0$ and $b = 1$, we obtain

$$H_{2n+1}^{(-2n)} > 0, \quad H_{2n}^{(-2n)} > 0, \quad H_{2n}^{(-(2n-1))} > 0, \quad H_{2n+1}^{(-(2n-1))} > 0,$$

$$\begin{vmatrix} c_{-2n} & c_{-2n+1} & \cdots & c_0 \\ \vdots & \vdots & & \vdots \\ c_{-1} & c_0 & \cdots & c_{2n-1} \\ 1 & 1 & \cdots & 1 \end{vmatrix} > 0, \text{ and } \begin{vmatrix} c_{-2n} & c_{-2n+1} & \cdots & c_1 \\ \vdots & \vdots & & \vdots \\ c_0 & c_1 & \cdots & c_{2n+1} \\ 1 & 1 & \cdots & 1 \end{vmatrix} > 0,$$

for $n \geq 0$, as a sufficient condition for the moment problem which can be called the strong Hausdorff moment problem.

It is also easily verified that these sufficient conditions for the strong Stieltjes and the strong Hausdorff moment problems are also their respective necessary conditions.

By considering the regular real $\hat{J}$-fraction [2.16], the condition

$$H_{2n+1}^{(-2n)} > 0, \quad H_{2n}^{(-2n)} > 0, \quad H_{2n+1}^{(-(2n+1))} \neq 0, \quad n \geq 0,$$

is found to be another sufficient condition for the existence of a solution to the strong Hamburger moment problem.

However, Jones, Njåstad and Thron [1984] have shown that a necessary and sufficient condition for the existence of a solution to this problem is

$$H_{2n+1}^{(-2n)} > 0 \quad \text{and} \quad H_{2n}^{(-2n)} > 0, \quad n \geq 0.$$

To arrive at this solution we need the use of a corresponding $\hat{J}$-fraction which exists under this condition alone. From Theorem 4, it follows that there exists such a $\hat{J}$-fraction and therefore, it is required here only to show that there exist also integral representations for the convergents of this $\hat{J}$-fraction.

Since it may be possible to have more than one such $\hat{J}$-fraction under the above condition, we assume here the continued fraction constructed according to the rules of [2.22].

Let $\alpha_n^{\cdot}$ and $\beta_n^{\cdot}$ be the partial quotients of this $\hat{J}$-fraction. Furthermore, let $A_n(Z)$ and $B_n(Z)$ be the numerator and denominator of the nth convergent of this fraction. Then, unlike [3.3] for the function

$$T_n^{\cdot}(Z) = \{B_n'(Z)B_{n-1}(Z) - B_{n-1}'(Z)B_n(Z)\},$$

there may be up to three different types of recursive relations. This can be seen as follows:

Case I:

$$\cdots - \frac{\alpha_n^{\cdot}}{Z + \beta_n^{\cdot}} - \cdots,$$

then $T_n^{\cdot}(Z) = \{B_{n-1}(Z)\}^2 + \alpha_n^{\cdot} T_{n-1}^{\cdot}(Z)$.

Case II:

$$\cdots - \frac{\alpha_{n-1}}{Z + \beta_{n-1}^{\cdot}} - \frac{\alpha_n^{\cdot} Z^2}{(1 + \alpha_n^{\cdot})Z + \beta_n^{\cdot}} - \cdots$$

then $T_n^{\cdot}(Z) = \{B_{n-1}(Z)\}^2 + \alpha_n^{\cdot}\{B_{n-1}(Z) - ZB_{n-2}(Z)\}^2 + \alpha_n^{\cdot}\alpha_{n-1}^{\cdot} Z^2 T_{n-2}(Z)$.

Case III:

$$\cdots - \frac{\alpha_{n-2}^{\cdot}}{Z+\beta_{n-2}} - \frac{\alpha_{n-1}^{\cdot} Z^2}{(1+\alpha_{n-1}^{\cdot})Z+\beta_{n-1}^{\cdot}} - \frac{\alpha_n^{\cdot} Z^2}{(1+\alpha_n^{\cdot})Z+\beta_n} - \cdots,$$

then

$$\begin{aligned} T_n^{\cdot}(Z) = {} & \{B_{n-1}(Z)\}^2 + \alpha_n^{\cdot}\{B_{n-1}(Z) - ZB_{n-2}(Z)\}^2 \\ & + \alpha_n^{\cdot}\alpha_{n-1}^{\cdot} Z^2\{B_{n-2}(Z) - ZB_{n-3}(Z)\}^2 \\ & + \alpha_n^{\cdot}\alpha_{n-1}^{\cdot}\alpha_{n-2}^{\cdot} Z^4 T_{n-3}(Z). \end{aligned}$$

Since $\alpha_n^{\cdot}$ are positive for all $n \geq 0$ (Theorem 4), the functions $T_n^{\cdot}(Z)$ are easily found to be positive for all real values of Z other than zero. To show that $T_n^{\cdot}(0)$ are also positive, it is required in cases II and III that $B_{n-1}(0) \neq 0$. The way this continued fraction was constructed is also found to hold. Consequently, we have the results that all the roots of the denominator polynomials $B_r(Z)$ are real distinct and different from those of $B_{r-1}(Z)$.

Furthermore, the relation

$$\{A_n(Z)B_{n-1}(Z) - A_{n-1}(Z)B_n(Z)\} = \alpha_n^{\cdot}\alpha_{n-1}^{\cdot} \cdots \alpha_2^{\cdot}\alpha_1 Z^{2r},$$

$r > n$, and $n \geq 1$ also gives that if $Z_r^{(n)}$ is a root of $B_n(Z)$ then it may be a root of $A_n(Z)$ only when it is zero.

These results are sufficient for the quotients $A_n(Z)/B_n(Z)$ $(n \geq 1)$ to have integral representations as in the case of the $\hat{J}$-fraction [2.1]. Hence, using this $\hat{J}$-fraction the results of Jones, Njåstad and Thron [1984] can be easily established.

REFERENCES

1. Henrici, P., Applied and Computational Complex Analysis, Vol. 1, Power Series, Integration, Conformal Mapping and Location of Zeros, Wiley, New York, 1974.

2. Jones, W. B., Njåstad, O., and Thron, W. J., "Continued fractions and the strong Hamburger moment problems", Proc. London Math. Soc. 47 (1983), 363-384.

3. Jones, W. B., Njåstad, O., and Thron, W. J., "Orthogonal Laurent Polynomials and the strong Hamburger moment problem", J. Math. Anal. & Appl. 98 (1984), 528-554.

4. Jones, W. B., Thron, W. J., and Waadeland, H., "A Strong Stieltjes Moment Problem", Trans. Amer. Math. Soc. 261 (1980), 503-528.

5. McCabe, J. H., "A Formal Extension of the Padé Table to Include Two Point Padé Quotients", J. Inst. Math. Appl. 15 (1975), 363-372.

6. McCabe, J. H., "The Quotient-Difference Algorithm and the Padé Table: An Alternative Form and a General Continued Fraction", Math. Comp. Vol. 41 (1983).

7. Sri Ranga, A., "Continued fractions which correspond to two series expansions and the strong Hamburger moment problem", Ph.D. Thesis, University of St. Andrews, Scotland, 1983.

8. Szeg, G., Orthogonal Polynomials, Colloquium Publications, Vol. 23, Amer. Math. Soc., New York, 1959.

9. Wall, H. S., Analytic Theory of Continued Fractions, Van Nostrand, New York, 1948.

10. Widder, D. V., Advanced Calculus, Prentice Hall Inc., New York, 1952.

ON THE CONVERGENCE OF A CERTAIN CLASS OF CONTINUED FRACTIONS $K(a_n/1)$ WITH $a_n \to \infty$

Ellen Sørsdal and Haakon Waadeland
Department of Mathematics and Statistics
University of Trondheim
N-7055 Dragvoll, Norway

1. Introduction.

The periodic continued fraction $K(a/1)$, $a \in \mathbb{C}$, $a \neq 0$, diverges (by oscillation) for all $a \in (-\infty, -\frac{1}{4})$, and converges to a finite value for any other a [5, Thm. 3.1 and 3.2]. If the continued fraction

$$(1.1) \qquad \underset{n=1}{\overset{\infty}{K}} \left(\frac{a_n}{1}\right) = \frac{a_1}{1} + \frac{a_2}{1} + \cdots + \frac{a_n}{1} + \cdots$$

is limit periodic with $a_n \to a$, it converges if a is off the closed ray $[-\infty, -\frac{1}{4}]$, possibly to ∞. If $a \in (-\infty, -\frac{1}{4})$, some special results are known [1].

If $a = -\frac{1}{4}$, i.e. $a_n \to -\frac{1}{4}$ in (1.1), it is known, as a simple consequence of the parabola theorem [7,5, Thm. 4.42], that the continued fraction converges if all a_n from a certain n on are located in some $(\pi-2\varepsilon)$-sector, $0 < \varepsilon < \pi/2$, with vertex at $-\frac{1}{4}$ and not containing -1. We furthermore know, that

$$(1.2) \qquad |a_n + \tfrac{1}{4}| \leq \frac{1}{4(4n^2-1)} \leq \frac{1}{16(n-\frac{1}{2})(n+\frac{1}{2})}$$

suffices for convergence [8]. An open question for a long time was whether 1/16 is the best constant in (1.2). This was affirmatively answered in [3] by proving that the continued fraction

$$(1.3) \qquad \underset{n=1}{\overset{\infty}{K}} \left(\left(-\frac{1}{4} - \frac{C}{16(n+\theta)(n+\theta+1)} \right) / 1 \right),$$

$\theta \neq -1,-2,-3,\ldots$, $C \in \mathbf{R}$, diverges for all $C > 1$. For $C \leq 1$ the values of (1.3) and all the tails are known [8]. The interest in convergence questions for continued fractions with $a_n \to -\frac{1}{4}$ is justified in the introductions of [2] and [4].

If $a=\infty$, i.e. $a_n \to \infty$ in (1.1), it is known from the parabola theorem that if all a_n are in a certain parabolic region and the speed at which $a_n \to \infty$ satisfies certain conditions, then (1.1) converges to a finite number [5, Thm. 4.42].

In the paper [4] a "bridge" is built between certain continued fractions (1.1) with $a_n \to \infty$ and certain continued fractions (1.1) with $a_n \to -\frac{1}{4}$. There the following proposition is proved:

Proposition 1.1. *If a continued fraction* $K(a_n/1)$ *is such that*

$$(1.4) \qquad a_n \to \infty\,, \quad \frac{a_{n+1}}{a_n} \to 1\,, \quad 1+a_n \neq 0\,, \quad 1+a_n+a_{n+1} \neq 0\,, \quad n \geq 2\,,$$

then its even and odd parts, $K(c_n/1)$ *and* $d_0 + K(d_n/1)$, *are limit periodic with* $c_n \to -\frac{1}{4}$ *and* $d_n \to -\frac{1}{4}$.

This proposition can be used in investigations of convergence of continued fractions $K(a_n/1)$ with $a_n \to \infty$ (if the rest of the conditions (1.4) hold). If even *or* odd part diverges, then $K(a_n/1)$ diverges. If even *and* odd parts both converge, we need to find out more before we can conclude convergence or divergence. If for instance (in addition) $f_n - f_{n-1} \to 0$ as $n \to \infty$, $\{f_n\}$ being the sequence of approximants, it follows that $K(a_n/1)$ converges. The purpose of the present note is to give an example, in the form of a 1-parameter family of continued fractions, for which the proposition can be used. This example will also produce one of the border points between divergence and convergence of continued fractions (1.1) with $a_n \to \infty$.

2. The continued fraction, even and odd parts.

In the paper [4] the example

$$(2.1) \qquad \frac{-1^2}{1} + \frac{-2\cdot 4}{1} + \frac{-3^2}{1} + \frac{-4\cdot 6}{1} + \frac{-5^2}{1} + \frac{-6\cdot 8}{1} + \cdots$$

is used to illustrate Proposition 1.1. Even and odd parts are there explicitely computed. They both have (after a couple of partial fractions) the form (1.3) with $C=1$, $\theta=2$ and $\theta=1$, and can therefore be exactly determined. Both of them have the value 1. Hence (2.1) converges and has the value 1. This, however, is known, and easily seen by using directly results from [10].

Since for any fixed θ the continued fraction (1.3) with $C = 1$ is a "border point" between convergent and divergent continued fractions in the class of continued fractions (1.3), it is a natural question to ask for a similar 1-parameter family, where (2.1) is a border point. The class of continued fractions

$$\text{(2.2)} \qquad \mathop{K}_{n=1}^{\infty} \frac{-n(n+x(1+(-1)^n))}{1}, \quad x \geq 0$$

is a good candidate: For $x = 0$ we get $K(-n^2/1)$, where numerical evidence strongly suggests divergence, for $x = 1$ we get the continued fraction (2.1), and for $x > 1$ numerical experiments strongly suggest convergence.

In the rest of the paper we shall study (2.2). It clearly satisfies the conditions (1.4) in Proposition 1.1.

From [4, (2.3) and (2.6)] we know that the continued fraction (1.1) has an even part $K(c_n/1)$ and an odd part $d_0 + K(d_n/1)$, where

$$\text{(2.3)} \qquad \begin{cases} c_1 = \dfrac{a_1}{1+a_2}, \quad c_2 = \dfrac{-a_2 a_3}{(1+a_2)(1+a_3+a_4)}, \\[2ex] c_n = \dfrac{-a_{2n-2}\, a_{2n-1}}{(1+a_{2n-3}+a_{2n-2})(1+a_{2n-1}+a_{2n})}, \quad n \geq 3, \end{cases}$$

and

$$\text{(2.4)} \qquad \begin{cases} d_0 = a_1, \quad d_1 = \dfrac{-a_1 a_2}{1+a_2+a_3} \\[2ex] d_n = \dfrac{-a_{2n-1}\, a_{2n}}{(1+a_{2n-2}+a_{2n-1})(1+a_{2n}+a_{2n+1})}, \quad n \geq 2, \end{cases}$$

provided that $1+a_2 \neq 0$ and $1+a_n+a_{n+1} \neq 0$ for $n \geq 2$. The formulas are proved by using formulas in [5, Sec. 2.4], followed by an equivalence transformation.

For the continued fraction (2.2) we find

$$(2.5)\quad \begin{cases} c_1 = \dfrac{1}{3+4x}\,, \quad c_2 = -\dfrac{9}{2}\,\dfrac{(1+x)}{(3+4x)(3+x)}\,, \\[2ex] c_n = -\dfrac{1}{4} - \dfrac{(2-x^2) + \dfrac{x-1}{n}}{16(n+\frac{x-1}{2})(n+\frac{x-3}{2})}\,, \quad n \geq 3, \end{cases}$$

$$(2.6)\quad \begin{cases} d_0 = -1\,, \quad d_1 = \dfrac{1+x}{3+x}\,, \\[2ex] d_n = -\dfrac{1}{4} - \dfrac{(2-x^2) + \dfrac{x+1}{n-1}}{16(n+\frac{x-1}{2})(n+\frac{x+1}{2})}\,, \quad n \geq 2. \end{cases}$$

An immediate observation is that for $x > \sqrt{2}$ we have $c_n + \frac{1}{4} > 0$ and $d_n + \frac{1}{4} > 0$ from a certain n on, depending upon x. This means that from a certain n on c_n and d_n approach $-\frac{1}{4}$ from the right, and from the parabola theorem it follows that even and odd parts both converge (not necessarily to a finite number). For $1 < x \leq \sqrt{2}$ it follows from (2.5) and (2.6) that $|c_n + \frac{1}{4}| \leq \dfrac{1}{16(n-\frac{1}{2})(n+\frac{1}{2})}$, and the same for d_n from some n on. Hence <u>the even part and the odd part of</u> (2.2) <u>both converge for</u> $x \geq 1$. For $x = 1$ we even know, from [4], that they converge to the same value 1.

For $0 \leq x < 1$ we would have had a continued fraction of the type (1.3) with $C > 1$, and hence divergence, if we could have neglected the terms $\frac{x-1}{n}$ in c_n and $\frac{x+1}{n-1}$ in d_n. In the paper [2] in the present proceedings L. Jacobsen has extended the result of [3] to the case when C is replaced by $C + \varepsilon_n$, $\varepsilon_n \to 0$ sufficiently fast, in (1.3) which holds in the present case. This implies that <u>the continued fraction</u> (2.2) <u>diverges for</u> $0 \leq x < 1$.

3. Convergence investigation in the case $x > 1$.

Since even and odd parts of (2.2) both converge, the only thing that can ruin convergence of (2.2) itself, would be that they converge to different values. Numerical computation indicates strongly that they converge to the same value, and hence that (2.2) converges. See the tables below. Observe in particular the different behavior of even and odd order approximants f_n.

$x = 1.1$

n	f_n
1000	0.457308
1001	0.418966
1002	0.457332
1003	0.419026
1004	0.457357
1005	0.419085
1006	0.457381
1007	0.419145
1008	0.457405
10000	0.474660
10001	0.461416

$x = 6.0$

n	f_n
20	4.457450
21	4.382003
22	4.457741
23	4.406928
24	4.457920
25	4.422834
26	4.458035
27	4.433270
28	4.458109
200	4.458288
201	4.458287

$x = 10.0$

n	f_n
2	$2.325581 \cdot 10^{-2}$
3	$-15.384615 \cdot 10^{-2}$
4	$2.551521 \cdot 10^{-2}$
5	$-1.785714 \cdot 10^{-2}$
6	$2.588503 \cdot 10^{-2}$
7	$1.299468 \cdot 10^{-2}$
8	$2.596715 \cdot 10^{-2}$
9	$2.155195 \cdot 10^{-2}$
10	$2.598957 \cdot 10^{-2}$
50	$2.600107 \cdot 10^{-2}$
51	$2.600101 \cdot 10^{-2}$

In the present section we shall see how observations on the computer lead to the basic idea of the proof of a result. In order to prove convergence, it suffices to prove that $f_{n+1} - f_n \to 0$, where f_n is the nth approximant. We shall use the formula

$$(3.1) \qquad f_{n+1} - f_n = (-1)^{n+1} \cdot \prod_{\nu=2}^{n+1} (1 - \frac{1}{h_\nu}) a_1 ,$$

[6,9], where $h_1 = 1$, $h_n = 1 + \frac{a_n}{1} + \frac{a_{n-1}}{1} + \dots + \frac{a_2}{1}$, $n \geq 2$.

Observe the recurrence relation

$$(3.2) \qquad h_{n+1} = 1 + \frac{a_{n+1}}{h_n} .$$

For our continued fraction (2.2) we find

$$(3.3) \qquad h_1 = 1, \quad h_2 = -(3+4x), \quad h_3 = \frac{4(3+x)}{3+4x} .$$

Inspired by observations on the computer we study the h_n's of even and odd order separately. (See the tables below.)

$x = 1.1$

h_3	2.22	
h_4		-10.19
h_5	3.45	
h_6		-13.25
h_7	4.70	
h_8		-16.37
h_9	5.95	
h_{10}		-19.51
h_{11}	7.20	
h_{12}		-22.66

$x = 10.0$

h_3	1.21	
h_4		-78.38
h_5	1.32	
h_6		-117.28
h_7	1.42	
h_8		-157.00
h_9	1.52	
h_{10}		-196.90
h_{11}	1.61	
h_{12}		-236.84

Simple computation gives the reccurrence relations:

(3.4) $$h_{2n+1} = \frac{\frac{m+1}{m+x} h_{2m-1} + 1}{1 - \frac{1}{4m(m+x)} h_{2m-1}}$$

(3.5) $$(-h_{2m+2}) = \frac{\left(1 + \frac{4(x+1)m+2+4x}{(2m+1)^2}\right)(-h_{2m}) - 1}{1 + \frac{1}{(2m+1)^2}(-h_{2m})}$$

We shall use some inequalities for the h_n's. For those with even indices we shall keep the somewhat awkward notation $(-h_{2m+2})$, in order to emphasize that we handle $(-h_n)$ for even n and h_n for odd n.

Lemma 3.1.

For all $m \geq 1$ and $x \geq 1$ we have

(3.6) $$1 \leq h_{2m-1} \leq 2m.$$

Proof:

Use of (3.3) and (3.4). Straightforward induction.

<u>Lemma 3.2</u>.

<u>For all sufficiently small</u> $\varepsilon > 0$ <u>the following holds</u>: <u>For all</u> $m \geq 1$ <u>and all</u> $x \geq 1 + \frac{\varepsilon(2+5\varepsilon)}{8(1+\varepsilon)}$ <u>we have</u>

(3.7) $\quad (-h_{2m}) \geq (1+\varepsilon)(2m+1).$

<u>Proof</u>:

We will show this by induction. To start the induction, we must show that (3.7) holds for $m=1$:

(3.8) $\quad (-h_2) = 3+4x \geq (1+\varepsilon)3 \quad \text{for} \quad x \geq \frac{3}{4}\varepsilon .$

Suppose that m is such that (3.7) holds, then we get from (3.5) that

$$(-h_{2m+2}) \geq \frac{\left(1 + \frac{4(x+1)m+2+4x}{(2m+1)^2}\right)(1+\varepsilon)(2m+1) - 1}{1 + \frac{1}{(2m+1)^2}(1+\varepsilon)(2m+1)}$$

$$= (1+\varepsilon)(2m+3) + \frac{(4x(1+\varepsilon) - 2(1+\varepsilon)^2 - 2)m + 4x(1+\varepsilon) - 3(1+\varepsilon)^2 - 1}{2m+2+\varepsilon}$$

since $\left(\frac{Au-1}{1+Bu}\right)' > 0$ for $A,B > 0,\ u > 0$.

It remains to prove that

(3.9) $\quad (4x(1+\varepsilon) - 2(1+\varepsilon)^2 - 2)m + 4x(1+\varepsilon) - 3(1+\varepsilon)^2 - 1 \geq 0$

for all $m \geq 1$. This we will do in two steps:

a) Show that it holds for $m=1$.

b) Show that the coefficient of m on the lefthand side of (3.9) is positive.

a) For $m=1$ (3.9) takes the form

(3.10) $\quad 8x(1+\varepsilon) - 5(1+\varepsilon)^2 - 3 \geq 0 \quad \text{for} \quad x \geq 1 + \frac{\varepsilon(5\varepsilon+2)}{8(1+\varepsilon)} .$

b) The coefficient of m is positive iff

$$4x(1+\varepsilon) - 2(1+\varepsilon)^2 - 2 \geq 0,$$

i.e.

$$x \geq 1 + \frac{\varepsilon^2}{2(1+\varepsilon)} . \tag{3.11}$$

Since the bound in (3.10) is larger (and larger than (3.8) for all sufficiently small $\varepsilon > 0$), the lemma is proved.

We are now ready to complete the proof of the convergence of (2.2) for $x > 1$. From Lemma 3.1 we have

$$\prod_{k=2}^{m}\left(1 - \frac{1}{h_{2k-1}}\right) \leq \prod_{k=2}^{m}\left(1 - \frac{1}{2k}\right) \leq \exp\left[-\frac{1}{2}\left(\frac{1}{2} + \ldots + \frac{1}{m}\right)\right]$$

$$\leq \exp\left[-\frac{1}{2}\log\left(\frac{m+1}{2}\right)\right] = \frac{\sqrt{2}}{\sqrt{m+1}} < \sqrt{\frac{2}{m}} .$$

From Lemma 3.2 we have, with an $\varepsilon > 0$ such that $1 + \frac{\varepsilon(2+5\varepsilon)}{8(1+\varepsilon)} \leq x$:

$$\prod_{k=1}^{m}\left(1 - \frac{1}{h_{2k}}\right) \leq \prod_{k=1}^{m}\left(1 + \frac{1}{(1+\varepsilon)(2k+1)}\right)$$

$$\leq \exp\left[\frac{1}{2(1+\varepsilon)}\left(1 + \frac{1}{2} + \cdots + \frac{1}{m}\right)\right]$$

$$\leq \exp\left[\frac{1}{2(1+\varepsilon)}(1 + \log m)\right] = e^{\frac{1}{2(1+\varepsilon)}} \cdot m^{\frac{1}{2+2\varepsilon}} .$$

Since

$$m^{-\frac{1}{2}} \cdot m^{\frac{1}{2+2\varepsilon}} = m^{-\frac{\varepsilon}{2(1+\varepsilon)}} \to 0$$

when $m \to \infty$, we have from (3.1) and the preceding discussion that (2.2) converges. In conclusion we thus have:

<u>The continued fraction</u>

$$\underset{n=1}{\overset{\infty}{K}} \frac{-n(n+x(1+(-1)^n))}{1} , \quad x \geq 0,$$

converges for $x \geq 1$ and diverges for $0 \leq x < 1$.

Remark: It is not hard to prove estimates, better than the ones in the two lemmas, if we want them to depend upon x. We have for instance that $h_{2m-1} \leq m+x$ for $x \geq 2$.

References

1. Gill, J., Infinite compositions of Möbius transformations, Trans. Amer. Math. Soc. 176 (1973), 479-487.

2. Jacobsen, L., On the convergence of limit periodic continued fractions $K(a_n/1)$, where $a_n \to -\frac{1}{4}$, Part II, Proceedings of Seminar-Workshop in Pitlochry, Scotland, 1985, these Lecture Notes.

3. Jacobsen, L. and Magnus, A., On the convergence of limit periodic continued fractions $K(a_n/1)$, where $a_n \to -\frac{1}{4}$, Proceedings of a conference in Tampa 1983, Lecture Notes in Mathematics 1105, Springer-Verlag (1984), 243-248.

4. Jacobsen, L. and Waadeland, H., Even and odd parts of limit periodic continued fractions, Journal of Computational and Applied Mathematics. To appear.

5. Jones, W.B. and Thron, W.J., Continued Fractions: Analytic Theory and Applications. Encyclopedia of Mathematics and its Applications, Addison-Wesley, 1980.

6. Overholt, M., A class of element and value regions for continued fractions, Lecture Notes in Mathematics 932, Proceedings of a Seminar-Workshop, Loen, Norway, 1981, Springer-Verlag (1982), 194-205.

7. Thron, W.J., On parabolic convergence regions for continued fractions, Math. Zeitschr. 69 (1958), 173-182.

8. Thron, W.J. and Waadeland, H., On a certain transformation of continued fractions, Lecture Notes in Mathematics 932, Proceedings of a Seminar-Workshop, Loen, Norway, 1981, Springer-Verlag (1982), 225-240.

9. Thron, W.J. and Waadeland, H., Truncation error bounds for limit periodic continued fractions, Mathematics of Computation, 40, #162, April 1963, 589-597.

10. Waadeland, H., Tales about tails, Proc. Amer. Math. Soc. 90, #1, January 1984, 57-64.

A NOTE ON PARTIAL DERIVATIVES OF CONTINUED FRACTIONS

Haakon Waadeland
Department of Mathematics and Statistics
University of Trondheim
N-7055 Dragvoll, Norway

For subsets A of $\mathbb{C}^\infty$ for which the continued fraction $K(z_n/1)$ is defined and converges it defines a function from A to $\hat{\mathbb{C}}$. This function is rational in each z_n. The purpose of the present note is to draw attention to a quite trivial, but useful observation on the partial derivatives of such continued fractions.

Observation: Let $A \subseteq \mathbb{C}^\infty$ be such that the continued fraction

$$\text{(1)} \qquad \operatorname*{K}_{n=1}^{\infty}\left(\frac{z_n}{1}\right) = \frac{z_1}{1} + \frac{z_2}{1} + \cdots + \frac{z_n}{1} + \cdots$$

is defined and converges to a finite value for $\bar{z} = (z_1, z_2, \ldots, z_n, \ldots) \in A$, and let $f^{(N)}$ be the functions, defined by the tails as follows

$$\text{(2)} \qquad f^{(N)}(z_1, z_2, \ldots, z_{N+1}, z_{N+2}, \ldots) = \operatorname*{K}_{n=N+1}^{\infty}\left(\frac{z_n}{1}\right),$$

where the tail values are assumed to be finite. Then the following formula holds:

$$\text{(3)} \qquad \frac{\partial f^{(0)}(\bar{z})}{\partial z_{n+1}} = \frac{f^{(0)}(\bar{z})}{z_{n+1}} \cdot \prod_{k=1}^{n}\left(\frac{-f^{(k)}(\bar{z})}{1+f^{(k)}(\bar{z})}\right), \quad n = 1,2,3,\ldots$$

Observe that $z_{n+1} \neq 0$ by the definition of a continued fraction, and that $f^{(k)} \neq -1$, since $f^{(k-1)} = z_k/(1+f^{(k)})$ is finite. Observe furthermore that we, without loss of generality, may assume that A has the property: For any fixed n there is an $h > 0$, such that $(z_1, z_2, \ldots, z_n, \ldots) \in A \Rightarrow (z_1, z_2, \ldots, z_n', \ldots) \in A$ for $|z_n' - z_n| < h$.

<u>Proof</u>: Since $f^{(k)} = z_{k+1}/(1+f^{(k+1)})$, $k = 0,1,2,\ldots$, the chain rule gives

$$(4) \qquad \frac{\partial f^{(0)}}{\partial z_{n+1}} = \frac{\partial f^{(0)}}{\partial f^{(1)}} \cdot \frac{\partial f^{(1)}}{\partial f^{(2)}} \cdots \frac{\partial f^{(n)}}{\partial z_{n+1}} ,$$

where

$$\frac{\partial f^{(k)}}{\partial f^{(k+1)}} = - \frac{z_{k+1}}{(1+f^{(k+1)})^2} = - \frac{f^{(k)}}{1+f^{(k+1)}}$$

and

$$\frac{\partial f^{(n)}}{\partial z_{n+1}} = \frac{1}{1+f^{(n+1)}} .$$

From this follows immediately the formula (3). □

<u>Remark</u>: Let A_k and B_k denote the normalized numerator and denominator of the kth approximant

$$f_k = \frac{z_1}{1} + \frac{z_2}{1} + \cdots + \frac{z_k}{1} = \frac{A_k}{B_k} ,$$

see e.g. [3, Ch. 2]. Then the following well known equality holds:

$$(5) \qquad f^{(0)} = \frac{A_n + A_{n-1}\, f^{(n)}}{B_n + B_{n-1}\, f^{(n)}} = \frac{A_n + A_{n-1} \cdot \dfrac{z_{n+1}}{1+f^{(n+1)}}}{B_n + B_{n-1} \cdot \dfrac{z_{n+1}}{1+f^{(n+1)}}}$$

Here A_n, A_{n-1}, B_n, B_{n-1} and $f^{(n+1)}$ are independent of z_{n+1}. We can use (5) for a more direct computation of the partial derivative, but in order to arrive at (3) we need in addition formulas from the analytic theory of continued fractions: We find from (5)

$$(6) \qquad \frac{\partial f^{(0)}}{\partial z_{n+1}} = \frac{A_{n-1}B_n - A_n B_{n-1}}{(B_n + B_{n-1} f^{(n)})^2} \cdot \frac{1}{1+f^{(n+1)}} .$$

Using in turn the determinant formula [3, (2.1.9)],

$$A_{n-1}B_n - A_nB_{n-1} = (-1)^n \cdot z_1 \cdot z_2 \cdots z_n ,$$

the formula (3) in the paper [2],

$$B_n + B_{n-1}\, f^{(n)} = \prod_{k=1}^{n} (1 + f^{(k)}),$$

and $z_k = f^{(k-1)}(1+f^{(k)})$, we easily arrive at the formula (3).

Example 1. Let $a \notin (-\infty, -\frac{1}{4})$. Then it is well known that the value of $K(a/1)$ is

$$\Gamma = \tfrac{1}{2}[\sqrt{1+4a} - 1] \ , \quad \operatorname{Re}\sqrt{\ } \geq 0 \ (= \text{only for } a = -\tfrac{1}{4}).$$

Formula (3), evaluated at $\bar{z} = (a,a,a,\ldots)$, gives

$$\text{(7)} \qquad \left(\frac{\partial f^{(0)}}{\partial z_{n+1}}\right)_0 = \frac{1}{1+\Gamma} \cdot \left(\frac{-\Gamma}{1+\Gamma}\right)^n .$$

Here and in the following $(\ )_0$ shall mean evaluation at the particular $\bar{z}$ in question. Another way of writing (7) is

$$\left(\frac{a}{1} + \frac{a}{1} + \cdots + \underset{\substack{\uparrow \\ (n+1)\text{th}}}{\frac{a+h}{1}} + \cdots\right) - \Gamma = \frac{1}{1+\Gamma}\left(\frac{-\Gamma}{1+\Gamma}\right)^n \cdot h + o(h),$$

in particular

$$\left(\frac{-\frac{1}{4}}{1} + \frac{-\frac{1}{4}}{1} + \cdots + \frac{-\frac{1}{4}+h}{1} + \cdots\right) + \frac{1}{2} = 2h + o(h).$$

For $a \notin (-\infty,-\frac{1}{4}]$ we have $|\Gamma| < |1+\Gamma|$. The weakness in these results is that we merely have terms $o(h)$ instead of remainder terms where the size is under control. This deficiency is taken care of in [5] if $a \notin (-\infty, -\frac{1}{4}]$.

Example 2. The continued fraction

$$\frac{1\cdot 3}{1} + \frac{2\cdot 4}{1} + \frac{3\cdot 5}{1} + \cdots + \frac{n(n+2)}{1} + \cdots$$

has the sequence of tail values

$$1,2,3,4,5,\ldots$$

starting with the value of the continued fraction. The formula (3), evaluated at $(z_1,z_2,\ldots,z_n,\ldots) = (1\cdot3,2\cdot4,\ldots,n(n+2),\ldots)$ is

$$\left(\frac{\partial f^{(0)}}{\partial z_{n+1}}\right)_0 = \frac{2\cdot(-1)^n}{(n+1)(n+2)(n+3)} .$$

Example 3. The continued fraction

$$\frac{-1^2}{1} \; + \; \frac{-2\cdot4}{1} \; + \; \frac{-3^2}{1} \; + \; \frac{-4\cdot6}{1} \; + \; \frac{-5^2}{1} \; + \cdots$$

converges [4, Thm. 1], and the sequence of tail values is

$$1,-2,3,-4,5,-6,\ldots$$

The formula (3) takes in this case the value (for odd order z_{n+1})

$$\left(\frac{\partial f^{(0)}}{\partial z_{2m+1}}\right)_0 = \frac{-1}{(m+1)(2m+1)} .$$

Comments.

1) In the unpublished paper [1] the formula (4.20) gives the logarithmic derivative of a function, represented by a C-fraction, in terms of what is essentially the tails. Partial derivatives are not explicitely mentioned, but they are contained in the formula, and the formula can be proved by using them. Apart from this the question about partial derivatives of continued fractions with respect to the elements apparently has not been raised. This is rather surprising for such a natural question. One reason is perhaps that the interest in tails until recently has been rather moderate. Another possible reason is that the problem may have been regarded as uninteresting. The results of [5] indicate that this point of view is not justified. We shall touch upon this later.

2) The formula (3) is competely useless, unless we know the values of the tails, or at least something about their location in the complex plane. The simplest case is the 1-periodic case, illustrated in Example 1. Other examples are the k-periodic continued fractions, regular

C-fraction expansions of hypergeometric or confluent hypergeometric functions, the continued fractions in Examples 2 and 3, to name but a few.

3) In the paper [5] are shown two examples of applications of the formula (3):

(i) A method of finding a good approximation to the set of all possible values of $K(a_n/1)$, $a_n \in E$ contained in $|a_n - a| \leq \rho$, $a \notin (-\infty, -\frac{1}{4}]$, ρ small.

(ii) Convergence accelertation of limit periodic continued fractions.

Both are based upon the inequality

$$(8) \qquad \left| \underset{n=1}{\overset{\infty}{K}} \frac{a+\delta_n}{1} - \Gamma - \frac{1}{1+\Gamma} \sum_{n=1}^{\infty} \left(\frac{-\Gamma}{1+\Gamma}\right)^n \delta_n \right| \leq K \cdot \rho^2 ,$$

which is proved by a combination of convergence theory for continued fractions and basic results in the theory of functions of one complex variable. This result justifies to call the expression

$$\frac{1}{1+\Gamma} \sum_{n=1}^{\infty} \left(\frac{-\Gamma}{1+\Gamma}\right)^n \delta_n$$

the total differential of (1) at $(z_1, z_2, \ldots) = (a, a, a, \ldots)$. It is likely that (8) and the applications can be extended to much more general cases. For convergence acceleration this is strongly supported by numerical experiments. See [5], [6].

4) Formulas for partial derivatives of $K(z_n/u_n)$ are easily established.

Acknowledgement. The author is indebted to professor Stephan Ruscheweyh for calling his attention to the paper [1].

References.

1. Atkinson, F.V., A value-region problem occuring in the theory of continued fractions, MRC Technical Summary Report # 4/9, December 1963 Madison, Wisconsin.

2. Jacobsen, L. and Waadeland, H., Some useful formulas involving tails of continued fractions, Proceedings of a Seminar-Workshop, Loen, Norway, 1981, Lecture Notes in Mathematics, Vol. 932, Springer-Verlag, Berlin, Heidelberg, New York 1982, pp. 99-105.

3. Jones, W.B. and Thron, W.J., Continued Fractions: Analytic Theory and Applications, Encyclopedia of Mathematics and its Applications, Vol. 11, Addison-Wesley, Reading, Mass., 1980.

4. Waadeland, H., Tales about Tails, Proceedings of the American Mathematical Society, Vol. 90, Number 1, January 1984.

5. Waadeland, H., Local properties of continued fractions, Proceedings of the International Conference on Rational Approximation and its Applications in Mathematics and Physics, Łańcût, Polen 1985, Lecture Notes in Mathematics, Springer-Verlag. To appear.

6. Waadeland, H., Derivatives of continued fractions with applications to hypergeometric functions, Proceedings of the Congress on Extrapolation and Padé Approximation, Luminy, France 1985. Journal of Comp. and Appl. Math. To appear.

Vol. 1090: Differential Geometry of Submanifolds. Proceedings, 1984. Edited by K. Kenmotsu. VI, 132 pages. 1984.

Vol. 1091: Multifunctions and Integrands. Proceedings, 1983. Edited by G. Salinetti. V, 234 pages. 1984.

Vol. 1092: Complete Intersections. Seminar, 1983. Edited by S. Greco and R. Strano. VII, 299 pages. 1984.

Vol. 1093: A. Prestel, Lectures on Formally Real Fields. XI, 125 pages. 1984.

Vol. 1094: Analyse Complexe. Proceedings, 1983. Edité par E. Amar, R. Gay et Nguyen Thanh Van. IX, 184 pages. 1984.

Vol. 1095: Stochastic Analysis and Applications. Proceedings, 1983. Edited by A. Truman and D. Williams. V, 199 pages. 1984.

Vol. 1096: Théorie du Potentiel. Proceedings, 1983. Edité par G. Mokobodzki et D. Pinchon. IX, 601 pages. 1984.

Vol. 1097: R.M. Dudley, H. Kunita, F. Ledrappier, École d'Éte de Probabilités de Saint-Flour XII – 1982. Edité par P.L. Hennequin. X, 396 pages. 1984.

Vol. 1098: Groups – Korea 1983. Proceedings. Edited by A.C. Kim and B.H. Neumann. VII, 183 pages. 1984.

Vol. 1099: C.M. Ringel, Tame Algebras and Integral Quadratic Forms. XIII, 376 pages. 1984.

Vol. 1100: V. Ivrii, Precise Spectral Asymptotics for Elliptic Operators Acting in Fiberings over Manifolds with Boundary. V, 237 pages. 1984.

Vol. 1101: V. Cossart, J. Giraud, U. Orbanz, Resolution of Surface Singularities. Seminar. VII, 132 pages. 1984.

Vol. 1102: A. Verona, Stratified Mappings – Structure and Triangulability. IX, 160 pages. 1984.

Vol. 1103: Models and Sets. Proceedings, Logic Colloquium, 1983, Part I. Edited by G.H. Müller and M.M. Richter. VIII, 484 pages. 1984.

Vol. 1104: Computation and Proof Theory. Proceedings, Logic Colloquium, 1983, Part II. Edited by M.M. Richter, E. Börger, W. Oberschelp, B. Schinzel and W. Thomas. VIII, 475 pages. 1984.

Vol. 1105: Rational Approximation and Interpolation. Proceedings, 1983. Edited by P.R. Graves-Morris, E.B. Saff and R.S. Varga. XII, 528 pages. 1984.

Vol. 1106: C.T. Chong, Techniques of Admissible Recursion Theory. IX, 214 pages. 1984.

Vol. 1107: Nonlinear Analysis and Optimization. Proceedings, 1982. Edited by C. Vinti. V, 224 pages. 1984.

Vol. 1108: Global Analysis – Studies and Applications I. Edited by Yu.G. Borisovich and Yu.E. Gliklikh. V, 301 pages. 1984.

Vol. 1109: Stochastic Aspects of Classical and Quantum Systems. Proceedings, 1983. Edited by S. Albeverio, P. Combe and M. Sirugue-Collin. IX, 227 pages. 1985.

Vol. 1110: R. Jajte, Strong Limit Theorems in Non-Commutative Probability. VI, 152 pages. 1985.

Vol. 1111: Arbeitstagung Bonn 1984. Proceedings. Edited by F. Hirzebruch, J. Schwermer and S. Suter. V, 481 pages. 1985.

Vol. 1112: Products of Conjugacy Classes in Groups. Edited by Z. Arad and M. Herzog. V, 244 pages. 1985.

Vol. 1113: P. Antosik, C. Swartz, Matrix Methods in Analysis. IV, 114 pages. 1985.

Vol. 1114: Zahlentheoretische Analysis. Seminar. Herausgegeben von E. Hlawka. V, 157 Seiten. 1985.

Vol. 1115: J. Moulin Ollagnier, Ergodic Theory and Statistical Mechanics. VI, 147 pages. 1985.

Vol. 1116: S. Stolz, Hochzusammenhängende Mannigfaltigkeiten und ihre Ränder. XXIII, 134 Seiten. 1985.

Vol. 1117: D.J. Aldous, J.A. Ibragimov, J. Jacod, Ecole d'Été d Probabilités de Saint-Flour XIII – 1983. Édité par P.L. Hennequir IX, 409 pages. 1985.

Vol. 1118: Grossissements de filtrations: exemples et applications Seminaire, 1982/83. Edité par Th. Jeulin et M. Yor. V, 315 pages. 1985

Vol. 1119: Recent Mathematical Methods in Dynamic Programming Proceedings, 1984. Edited by I. Capuzzo Dolcetta, W.H. Fleminc and T. Zolezzi. VI, 202 pages. 1985.

Vol. 1120: K. Jarosz, Perturbations of Banach Algebras. V, 118 pages 1985.

Vol. 1121: Singularities and Constructive Methods for Their Treatmen Proceedings, 1983. Edited by P. Grisvard, W. Wendland and J.F Whiteman. IX, 346 pages. 1985.

Vol. 1122: Number Theory. Proceedings, 1984. Edited by K. Alladi VII, 217 pages. 1985.

Vol. 1123: Séminaire de Probabilités XIX 1983/84. Proceedings. Edit par J. Azéma et M. Yor. IV, 504 pages. 1985.

Vol. 1124: Algebraic Geometry, Sitges (Barcelona) 1983. Proceec ings. Edited by E. Casas-Alvero, G.E. Welters and S. Xambo Descamps. XI, 416 pages. 1985.

Vol. 1125: Dynamical Systems and Bifurcations. Proceedings, 1984 Edited by B.L.J. Braaksma, H.W. Broer and F. Takens. V, 129 pages 1985.

Vol. 1126: Algebraic and Geometric Topology. Proceedings, 1983 Edited by A. Ranicki, N. Levitt and F. Quinn. V, 523 pages. 1985.

Vol. 1127: Numerical Methods in Fluid Dynamics. Seminar. Edited by F. Brezzi, VII, 333 pages. 1985.

Vol. 1128: J. Elschner, Singular Ordinary Differential Operators anc Pseudodifferential Equations. 200 pages. 1985.

Vol. 1129: Numerical Analysis, Lancaster 1984. Proceedings. Editec by P.R. Turner. XIV, 179 pages. 1985.

Vol. 1130: Methods in Mathematical Logic. Proceedings, 1983. Editec by C.A. Di Prisco. VII, 407 pages. 1985.

Vol. 1131: K. Sundaresan, S. Swaminathan, Geometry and Nonlinear Analysis in Banach Spaces. III, 116 pages. 1985.

Vol. 1132: Operator Algebras and their Connections with Topology and Ergodic Theory. Proceedings, 1983. Edited by H. Araki, C.C Moore, Ş. Strătilă and C. Voiculescu. VI, 594 pages. 1985.

Vol. 1133: K.C. Kiwiel, Methods of Descent for Nondifferentiable Optimization. VI, 362 pages. 1985.

Vol. 1134: G.P. Galdi, S. Rionero, Weighted Energy Methods in Fluid Dynamics and Elasticity. VII, 126 pages. 1985.

Vol. 1135: Number Theory, New York 1983–84. Seminar. Edited by D.V. Chudnovsky, G.V. Chudnovsky, H. Cohn and M.B. Nathanson. V, 283 pages. 1985.

Vol. 1136: Quantum Probability and Applications II. Proceedings, 1984. Edited by L. Accardi and W. von Waldenfels. VI, 534 pages. 1985.

Vol. 1137: Xiao G., Surfaces fibrées en courbes de genre deux. IX, 103 pages. 1985.

Vol. 1138: A. Ocneanu, Actions of Discrete Amenable Groups on von Neumann Algebras. V, 115 pages. 1985.

Vol. 1139: Differential Geometric Methods in Mathematical Physics. Proceedings, 1983. Edited by H.D. Doebner and J.D. Hennig. VI, 337 pages. 1985.

Vol. 1140: S. Donkin, Rational Representations of Algebraic Groups. VII, 254 pages. 1985.

Vol. 1141: Recursion Theory Week. Proceedings, 1984. Edited by H.-D. Ebbinghaus, G.H. Müller and G.E. Sacks. IX, 418 pages. 1985.

Vol. 1142: Orders and their Applications. Proceedings, 1984. Editec by I. Reiner and K.W. Roggenkamp. X, 306 pages. 1985.

Vol. 1143: A. Krieg, Modular Forms on Half-Spaces of Quaternions. XIII, 203 pages. 1985.

Vol. 1144: Knot Theory and Manifolds. Proceedings, 1983. Edited by D. Rolfsen. V, 163 pages. 1985.

Vol. 1034: J. Musielak, Orlicz Spaces and Modular Spaces. V, 222 pages. 1983.

Vol. 1035: The Mathematics and Physics of Disordered Media. Proceedings, 1983. Edited by B.D. Hughes and B.W. Ninham. VII, 432 pages. 1983.

Vol. 1036: Combinatorial Mathematics X. Proceedings, 1982. Edited by L.R.A. Casse. XI, 419 pages. 1983.

Vol. 1037: Non-linear Partial Differential Operators and Quantization Procedures. Proceedings, 1981. Edited by S.I. Andersson and H.-D. Doebner. VII, 334 pages. 1983.

Vol. 1038: F. Borceux, G. Van den Bossche, Algebra in a Localic Topos with Applications to Ring Theory. IX, 240 pages. 1983.

Vol. 1039: Analytic Functions, Błażejewko 1982. Proceedings. Edited by J. Ławrynowicz. X, 494 pages. 1983

Vol. 1040: A. Good, Local Analysis of Selberg's Trace Formula. III, 128 pages. 1983.

Vol. 1041: Lie Group Representations II. Proceedings 1982–1983. Edited by R. Herb, S. Kudla, R. Lipsman and J. Rosenberg. IX, 340 pages. 1984.

Vol. 1042: A. Gut, K. D. Schmidt, Amarts and Set Function Processes. III, 258 pages. 1983.

Vol. 1043: Linear and Complex Analysis Problem Book. Edited by V.P. Havin, S.V. Hruščëv and N.K. Nikol'skii. XVIII, 721 pages. 1984.

Vol. 1044: E. Gekeler, Discretization Methods for Stable Initial Value Problems. VIII, 201 pages. 1984.

Vol. 1045: Differential Geometry. Proceedings, 1982. Edited by A.M. Naveira. VIII, 194 pages. 1984.

Vol. 1046: Algebraic K-Theory, Number Theory, Geometry and Analysis. Proceedings, 1982. Edited by A. Bak. IX, 464 pages. 1984.

Vol. 1047: Fluid Dynamics. Seminar, 1982. Edited by H. Beirão da Veiga. VII, 193 pages. 1984.

Vol. 1048: Kinetic Theories and the Boltzmann Equation. Seminar, 1981. Edited by C. Cercignani. VII, 248 pages. 1984.

Vol. 1049: B. Iochum, Cônes autopolaires et algèbres de Jordan. VI, 247 pages. 1984.

Vol. 1050: A. Prestel, P. Roquette, Formally p-adic Fields. V, 167 pages. 1984.

Vol. 1051: Algebraic Topology, Aarhus 1982. Proceedings. Edited by I. Madsen and B. Oliver. X, 665 pages. 1984.

Vol. 1052: Number Theory, New York 1982. Seminar. Edited by D.V. Chudnovsky, G.V. Chudnovsky, H. Cohn and M.B. Nathanson. V, 309 pages. 1984.

Vol. 1053: P. Hilton, Nilpotente Gruppen und nilpotente Räume. V, 221 pages. 1984.

Vol. 1054: V. Thomée, Galerkin Finite Element Methods for Parabolic Problems. VII, 237 pages. 1984.

Vol. 1055: Quantum Probability and Applications to the Quantum Theory of Irreversible Processes. Proceedings, 1982. Edited by L. Accardi, A. Frigerio and V. Gorini. VI, 411 pages. 1984.

Vol. 1056: Algebraic Geometry. Bucharest 1982. Proceedings, 1982. Edited by L. Bădescu and D. Popescu. VII, 380 pages. 1984.

Vol. 1057: Bifurcation Theory and Applications. Seminar, 1983. Edited by L. Salvadori. VII, 233 pages. 1984.

Vol. 1058: B. Aulbach, Continuous and Discrete Dynamics near Manifolds of Equilibria. IX, 142 pages. 1984.

Vol. 1059: Séminaire de Probabilités XVIII, 1982/83. Proceedings. Edité par J. Azéma et M. Yor. IV, 518 pages. 1984.

Vol. 1060: Topology. Proceedings, 1982. Edited by L. D. Faddeev and A.A. Mal'cev. VI, 389 pages. 1984.

Vol. 1061: Séminaire de Théorie du Potentiel. Paris, No. 7. Proceedings. Directeurs: M. Brelot, G. Choquet et J. Deny. Rédacteurs: F. Hirsch et G. Mokobodzki. IV, 281 pages. 1984.

Vol. 1062: J. Jost, Harmonic Maps Between Surfaces. X, 133 pages. 1984.

Vol. 1063: Orienting Polymers. Proceedings, 1983. Edited by J.L. Ericksen. VII, 166 pages. 1984.

Vol. 1064: Probability Measures on Groups VII. Proceedings, 1983. Edited by H. Heyer. X, 588 pages. 1984.

Vol. 1065: A. Cuyt, Padé Approximants for Operators: Theory and Applications. IX, 138 pages. 1984.

Vol. 1066: Numerical Analysis. Proceedings, 1983. Edited by D.F. Griffiths. XI, 275 pages. 1984.

Vol. 1067: Yasuo Okuyama, Absolute Summability of Fourier Series and Orthogonal Series. VI, 118 pages. 1984.

Vol. 1068: Number Theory, Noordwijkerhout 1983. Proceedings. Edited by H. Jager. V, 296 pages. 1984.

Vol. 1069: M. Kreck, Bordism of Diffeomorphisms and Related Topics. III, 144 pages. 1984.

Vol. 1070: Interpolation Spaces and Allied Topics in Analysis. Proceedings, 1983. Edited by M. Cwikel and J. Peetre. III, 239 pages. 1984.

Vol. 1071: Padé Approximation and its Applications, Bad Honnef 1983. Prodeedings. Edited by H. Werner and H.J. Bünger. VI, 264 pages. 1984.

Vol. 1072: F. Rothe, Global Solutions of Reaction-Diffusion Systems. V, 216 pages. 1984.

Vol. 1073: Graph Theory, Singapore 1983. Proceedings. Edited by K.M. Koh and H.P. Yap. XIII, 335 pages. 1984.

Vol. 1074: E.W. Stredulinsky, Weighted Inequalities and Degenerate Elliptic Partial Differential Equations. III, 143 pages. 1984.

Vol. 1075: H. Majima, Asymptotic Analysis for Integrable Connections with Irregular Singular Points. IX, 159 pages. 1984.

Vol. 1076: Infinite-Dimensional Systems. Proceedings, 1983. Edited by F. Kappel and W. Schappacher. VII, 278 pages. 1984.

Vol. 1077: Lie Group Representations III. Proceedings, 1982–1983. Edited by R. Herb, R. Johnson, R. Lipsman, J. Rosenberg. XI, 454 pages. 1984.

Vol. 1078: A.J.E.M. Janssen, P. van der Steen, Integration Theory. V, 224 pages. 1984.

Vol. 1079: W. Ruppert. Compact Semitopological Semigroups: An Intrinsic Theory. V, 260 pages. 1984

Vol. 1080: Probability Theory on Vector Spaces III. Proceedings, 1983. Edited by D. Szynal and A. Weron. V, 373 pages. 1984.

Vol. 1081: D. Benson, Modular Representation Theory: New Trends and Methods. XI, 231 pages. 1984.

Vol. 1082: C.-G. Schmidt, Arithmetik Abelscher Varietäten mit komplexer Multiplikation. X, 96 Seiten. 1984.

Vol. 1083: D. Bump, Automorphic Forms on GL (3,IR). XI, 184 pages. 1984.

Vol. 1084: D. Kletzing, Structure and Representations of Q-Groups. VI, 290 pages. 1984.

Vol. 1085: G.K. Immink, Asymptotics of Analytic Difference Equations. V, 134 pages. 1984.

Vol. 1086: Sensitivity of Functionals with Applications to Engineering Sciences. Proceedings, 1983. Edited by V. Komkov. V, 130 pages. 1984

Vol. 1087: W. Narkiewicz, Uniform Distribution of Sequences of Integers in Residue Classes. VIII, 125 pages. 1984.

Vol. 1088: A.V. Kakosyan, L.B. Klebanov, J.A. Melamed, Characterization of Distributions by the Method of Intensively Monotone Operators. X, 175 pages. 1984.

Vol. 1089: Measure Theory, Oberwolfach 1983. Proceedings. Edited by D. Kölzow and D. Maharam-Stone. XIII, 327 pages. 1984.